GEOGRAPHIC MEASUREMENT AND QUANTITATIVE ANALYSIS

GEOGRAPHIC MEASUREMENT AND QUANTITATIVE ANALYSIS

ROBERT EARICKSON
University of Maryland—Baltimore County

JOHN HARLIN
Northern Illinois University

Macmillan College Publishing Company
New York

Maxwell Macmillan Canada
Toronto

Maxwell Macmillan International
New York   Oxford   Singapore   Sydney

Editor: Paul F. Corey
Production Editor: Rex Davidson
Text Designer: Jill E. Bonar
Cover Designer: Robert Vega
Production Buyer: Patricia A. Tonneman

This book was set in Times Roman by The Clarinda Company. The book and cover were printed and bound by Book Press, Inc., a Quebecor America Book Group Company.

Copyright © 1994 by Macmillan College Publishing Company, Inc.

Printed in the United States of America

All rights reserved. No part of this book may be reproduced or transmitted in any form or by any means, electronic or mechanical, including photocopy, recording, or any information storage and retrieval system, without permission in writing from the Publisher.

Macmillan College Publishing Company
866 Third Avenue
New York, New York 10022

Macmillan College Publishing Company is part of the
Maxwell Communication Group of Companies.

Maxwell Macmillan Canada, Inc.
1200 Eglinton Avenue East, Suite 200
Don Mills, Ontario M3C 3N1

Library of Congress Cataloging-in-Publication Data
Earickson, Robert.
    Geographic measurement and quantitative analysis / Robert Earickson and John Harlin.
       p.   cm.
    Includes bibliographical references.
    ISBN 0-675-21338-X
    1. Geography—Statistical methods.   I. Harlin, John M. II. Title.
G70.3.E22   1994
910'.21—dc20   93-18216
                                            CIP

Printing: 1 2 3 4 5 6 7 8 9   Year: 4 5 6 7 8

Preface

Since the 1960s, undergraduate majors in geography have been increasingly expected to have a working knowledge of elementary statistics and elementary mathematical models as applied to spatial processes. These expectations have come about in part from pressure of a discipline making rapid advances in quantitative research techniques and in part from increased expectations in the job market. Geography majors find themselves competing with students in mathematics, computer science, and information systems for positions in computer cartography, photogrammetry, geographic information systems, and planning.

There are more than a dozen textbooks on applied statistics in geographic research. They range from cookbooks, which teach how to do it but not why, to texts that present enough material to consume three or four academic terms at a mathematical level that can be fully understood only by someone with a working knowledge of calculus. Few of these texts provide any instruction on how to employ universally available computer programs to help solve numerical problems. Most also concentrate on either socioeconomic examples or physical examples to the near exclusion of the other.

Highly theoretical presentations have their role; however, we feel that this book fills a niche for use in a typical one-semester or two-quarter course. It is designed to give students a working knowledge of the procedures that geographers follow from the inception of a problem to the interpretation of results from statistical procedures. It has numerous examples of quantitative geographic problems with solutions based on data sets that are also supplied. The book also offers an abundant mix of both social and physical examples and exercises, so that instructors with expertise in either major facet of geography will find material that is familiar to them. To those instructors and their students who wish to delve deeper into the foundations of statistical procedures, such as underlying proofs, we recommend that a supplementary text be made available through a library reading reserve or equivalent.

For those students and their instructors who have computers and standard statistical software packages available and wish to use them, we provide examples of the control statements used in the Statistical Analysis System (SAS) and Statistical Procedures for the Social Sciences (SPSS-X) to solve the statistical problems. Students who can use computers will be better prepared to pursue graduate work and compete in the job market.

The text is divided into two sections. Part I contains four chapters beginning with (1) the connection of science to explanation of geographic phenomena, (2) a description of the measurement systems used in geography, (3) common ways in which numerical data may be presented on graphs and maps, and (4) descriptive statistics of spatial data. Part II is the heart of the text. Chapter 5 covers the fundamentals of probability, which is the basis for statistical inference, and its application to probability distributions. Chapter 6 describes the use of the normal and t distributions to test the significance of difference of means. Chapter 7 describes application of one-way analysis of variance to geographic research. Chapter 8 introduces the most common statistical procedure used to explain two-variable relationships, and Chapter 9 introduces a few of the many inferential tests that apply to data that do not conform to the rules for parametric statistics, including a discussion of the techniques for testing hypotheses about distributions (quadrat analysis). There may be more techniques here than the typical instructor can include in a single term, but our aim is to provide enough variety to cover most needs. Finally, each chapter provides a few references to further readings relevant to the techniques included.

A standard set of statistical tables is provided following the last chapter. One appendix contains a few sets of both physical and sociodemographic data, from which some text examples and exercises have been taken, and from which instructors may construct their own exercises. Another appendix that some readers will find useful contains the SAS and SPSS-X programs needed to solve all but a few of the statistical examples described in the text. The last two appendices contain answers to selected exercises and a list of the equations used throughout the text.

We wished to acknowledge the many people who helped us develop this text. Students at the universities of Maryland–Baltimore County, North Texas, and Alabama, where we have taught introductory quantitative techniques courses, have helped us refine virtually every chapter. We give special thanks to Paul Corey of Macmillan Publishing Company and to the following reviewers, who have commented on various drafts of the material:

- Robert R. Churchill
  Middlebury College
- Darrick R. Danta
  California State University, Northridge

- Timothy J. Fik
  University of Florida
- David C. Hodge
  University of Washington
- Sara L. McLafferty
  Hunter College
- Ron Mitchelson
  Morehead State University
- Morton O'Kelly
  The Ohio State University
- Bruce W. Pigozzi
  Michigan State University
- Dorothy Sack
  University of Wisconsin, Madison
- Dr. Anthony V. Williams
  The Pennsylvania State University

Contents

1 SCIENCES, STATISTICS, AND GEOGRAPHIC RESEARCH — 1
Statistical Applications to Global Systems 3
Systems and Theories 5
The Use of Statistics 6
Misuse of Statistics 8

2 MEASUREMENT AND SAMPLING — 13
Primary and Secondary Data 14
Variables 15
The Rudiments of Measurement 16
Measurement Scales 17
Geographic Primitives 24
Collecting Sample Data 41

3 GRAPHING AND MAPPING DATA — 55
Graphic Presentation of Data 56
The Frequency Distribution 60
Mapping Numerical Data 66

4 DESCRIPTIVE STATISTICS — 83
Measures of Central Tendency 83
Measures of Dispersion 89
Distribution Statistics for Spatial Distributions 104

5 PROBABILITY AND PROBABILITY DISTRIBUTIONS — 119
Probability of an Event 119
Probability Distributions 125

6 INFERENTIAL STATISTICS AND HYPOTHESES INVOLVING MEANS — 149
Basic Concepts for Inferential Statistics 150
Computing Test Statistics 158

7 ONE-WAY ANALYSIS OF VARIANCE — 183
Components of Variance 184
The F Distribution and the F Test 190

8 BIVARIATE CORRELATION AND LINEAR REGRESSION — 197
Correlation Analysis 198
Regression Analysis 207
Nonlinear Models 218

9 NONPARAMETRIC STATISTICS — 225
The Sign Test for the Median 227
Paired-Sample Test with Ordinal Data 229
Two Independent Samples Test 232
More Than Two (Independent) Samples with Ordinal Data 234
Nonparametric Goodness-of-Fit Tests 237
The Number-of-Runs Test for Randomness 253
The Kolmogorov-Smirnov One-Sample Goodness-of-Fit Test 254
The Kolmogorov-Smirnov Two-Sample Goodness-of-Fit Test 258
A Nonparametric Method of Correlation 260

APPENDIX TABLES

1	**Table of Random Numbers**	271
2	**Cumulative Normal Frequency Distribution**	272

3	Confidence Limits for Fractile Diagrams	273
4	t Distribution	274
5	Poisson Probabilities	275
6	Chi-Square (X^2) Distribution	276
7	5% and 1% Points F Distribution	277
8	Critical Values of U in the Mann-Whitney Test	280
9	Critical Values of H in the Kruskal-Wallis Test	281
10	Critical Values of D in the Kolmogorov-Smirnov One-Sample Test	282
11	Critical Values of D in the Kolmogorov-Smirnov Two-Sample Test	283
12	Critical Values of the Standard Normal Deviate z	284
13	Sample Sizes Required for 50% of the Population, for 95% Certainty, and for Specified Percent Accuracy	285
14	Run Statistic Critical Values for Sample Sizes n_1 and n_2 (Two-Tail Test) $\alpha = .05$	286

APPENDIXES

A	Instructions for the Use of SAS and SPSS-X in Statistical Analysis	287
B-1	Des Moines, Iowa, Mean Monthly Precipitation, 1900–1989	322
B-2	Percent Female Employment, Maryland Minor Civil Divisions, 1980	324
B-3	Selected Statistics from the 1980 Census Report "Provisional Estimates of Social, Economic, and Housing Characteristics," PHC 80-S1-1	326
B-4	AIDS Cases and Annual Rates per 100,000 Population, by State, Reported in 1989 and 1990; and Cumulative Totals, by State and Age Group, Through December 1990	327
B-5	Selected Population Characteristics, United States, 1990	328
C	Answers to Selected Exercises	331
D	List of Equations	337
	Index	343

Science, Statistics, and Geographic Research

Geography is the study of the places and processes associated with people, organizations, and the physical environment. A place, in all its richness and complexity, can only be understood by a thorough analysis of its complex structure. During the 1950s, the use of simple linear statistical techniques that purported to "explain" geographical phenomena began to appear more frequently in the literature. This methodological shift from qualitative description to quantitative analysis and modeling, in which associations were forged between numerical data and the places in which they were located, was hailed by many as a "revolution" in spatial analysis. Simple regression, correlation, and analysis of variance were the first commonly used techniques, but computation was slow and the only tool for solving statistical problems was the cumbersome desk calculator. In the 1960s census and other survey data became available on magnetic computer tape, and with the advent of second-generation electronic computers, multivariate and nonlinear statistical methods were adopted. Standard computer packages of statistical techniques were introduced. Social scientists embraced these statistical programs and applied them, sometimes indiscriminately and inappropriately, to the data. Today, the applications of statistics in geography go beyond explaining simple regularities and associations.

Problems of interest to geographers may be divided almost unambiguously into those of human (economic, cultural, and political) geography and those of physical (climatic, geomorphic, biological, and soils) geography. Clearly, the systems that are involved in global environment and local development overlap both human and physical geography. How do geographers quantify, map, and analyze these systems? Economic, cultural, and political geography are concerned with the spatial aspects of economic, cultural, and political systems. Climatology is concerned with the analysis of changes in temperature and rainfall over time. Geomorphology assesses landforms and the associated processes of weathering, erosion, transportation, and deposition. Biogeography attempts to explain variations in regimes of living organisms. Soils geographers may analyze soil formation, geographic variations in soil types, and problems of soil management. Using cartography, geographers map these phenomena in ways that enhance the verbal and quantitative analyses of distribution and change. Geographers also work with scientists from other disciplines to attempt to explain sources of conflict between lifestyle and economy based on the exploitation of natural and human resources. Two examples are considered below: (1) the complex problem of global climate change, and (2) the smaller-scale, but equally vexing problems of a local, metropolitan system.

STATISTICAL APPLICATIONS TO GLOBAL SYSTEMS

Average worldwide temperatures in the late 1980s were the highest recorded since the 1930s. A large number—though not all—scientists believe that unabated increases in greenhouse gas emissions might increase global mean temperature, raise sea level by 30 to 100 cm, and significantly alter weather patterns over the next century. The potential effects of global warming are:[1]

- Change in the climate patterns on earth.
- Decline in biological diversity and an increase in the extinction of plant and animal species.
- Potential increase in tropospheric ozone levels.
- Loss of most coastal wetlands due to sea level rise and the considerable expense of protecting developed coastlines.
- Reduction in forest ranges and changes in their composition.
- Shifts in agriculture to new locations.
- Increased water demand, posing problems for some water resource systems.
- Increased demand for electricity, particularly in southern regions, because of increased use of air conditioning.

Carbon dioxide (CO_2) is considered to be the chief greenhouse gas contributing to global warming. Concentrations of this gas in the atmosphere have been increasing by more than one part per million annually during the past decade. The National Academy of Sciences has predicted a potential increase in worldwide temperatures of between 1.5° and 4.5°C if the concentration of atmospheric carbon dioxide doubles by the year 2060. The burning of fossil fuels produces 5 billion tons of carbon emissions per year. In addition, a marked increase in CO_2 has accompanied the burning and clearing of forests. This has reduced both the filtering capacity of forest vegetation and the amount of CO_2 formerly stored by plants and soils.

---

[1] Adapted from "Focus on—Global Warming," Environmental Protection Agency, 1988.

Global climate change is one of the more vexing problems facing the world today. Climate change obviously impacts the entire world; however, a suggested cause is geographically more specific. For example, the industrialized nations produce most greenhouse gasses. Even the burning of the tropical rainforests in the otherwise undeveloped Amazon basin is designed in part to assist Brazil in becoming an "industrial" nation. Some social and economic activities in industrial nations have environmental consequences that are too dangerous to ignore.

The preceding vignette raises many questions that may be addressed with statistical methodology. To mention a few, is there a significant relationship between changes in temperature and fossil fuel use? Between temperature change and agricultural productivity? Between temperature and the increase in atmospheric CO_2? Is temperature actually changing at all? Has forest clearance changed significantly in the past fifty to one hundred years? Have patterns of agricultural activity changed significantly due to climatic change during the past fifty to one hundred years? Has there been a significant increase in demand for electricity, aside from that due to population increase alone, since the end of World War II? Statistical answers to all these questions depend on having valid data available, use of proper techniques, and precautions to recognize confounding multiple cause-and-effect relationships.

STATISTICAL APPLICATION TO LOCAL SYSTEMS

Global warming is a large-scale systemic problem, and the complexity of its interrelationships is overwhelming. By comparison, the problems of private sector and local government managers are smaller scale, although they may be no less troubling. Consider, for example, the interrelated systems of a growing metropolis such as Los Angeles (Figure 1.1). Migrants have been, and continue to be, drawn to this area of southern California. Until after World War II, Los Angeles exploited three natural resources—citrus, petroleum, and sunshine. Today, citrus production has mainly shifted elsewhere, the oil reserves have been depleted, and the sunshine is shrouded by smog. The bases for continued growth, besides tourism, have converted from natural resources to the aerospace, electronics, and communications industries. The city has suffered the synergistic effects of rapid immigration and overdevelopment in a relatively fragile physiographic and climatic environment. Consequently, political and economic decisions must routinely be made at the state and local level, decisions such as whether to increase the number of police officers, how to deal with the ever-increasing traffic congestion and concomitant air pollution, how to prepare for earthquakes, or how best to respond to urgent community demands for basic services.

Natural environmental hazards are sometimes as problematic as human modification of the landscape. A sequence of disastrous results can be triggered by a single hazardous event. For example, an intense thunderstorm can cause flash flooding that may lead to power outages, disruption of water supply and telephone service, fire, and environmental contamination (Figure 1.2).

Questions posed for local and global systemic problems often require decision makers to employ statistics to assess the social, economic, and spatial relationships be-

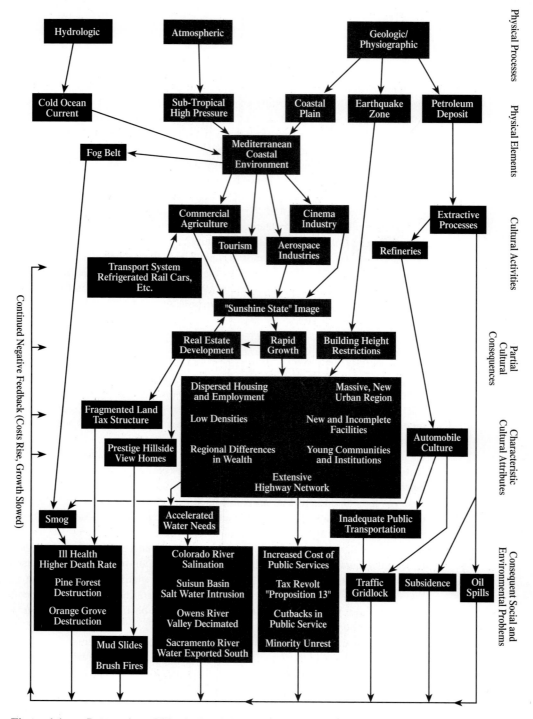

Figure 1.1 Interaction of Physical and Cultural Systems in the Los Angeles Basin

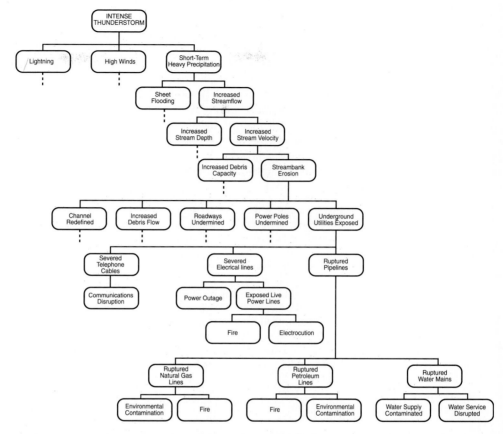

Figure 1.2 Natural Hazards Sequence: The Interrelated Effects of a Physical Phenomenon on Cultural Systems

Source: Fred May, Utah Comprehensive Emergency Management.

tween phenomena. The figures that are generated by the statisticians will enable decision makers to weigh outcomes based on different scenarios and, even if the ultimate decisions are dictated by politics, the statistics will usually be consulted, if not always used properly.

SYSTEMS AND THEORIES

You no doubt have noticed the frequent use of the word *system* in the foregoing text. The term is used here in the sense of a real-world system of interest (Wilson, 1981). Geographically, this could be a forest, a drainage basin, an economic development region, a city, or any other system. Patterns and/or processes found in and between systems are of interest and may be analyzed. When consistent interrelationships are found among characteristics in a system, an attempt is made to formalize a set of propositions that purport to explain the *structure* of that system.[2] This is a *theory*. A given level of smog in Los Angeles, for instance, is consistently associated with an illness threshold, in which predictable numbers of people will suffer acute respiratory problems. Scientists talk about *significant* levels of hydrocarbons, nitrogen oxides, and ozone, or *significant* increases in the number of automobiles in the region, while suggesting that ill health is associated with fossil fuel air pollution in this environment.

The level of mathematical and statistical analysis that must be employed to address many of these systemic problems is far beyond anything found within the pages of this book. However, the spatial structures to which elaborate systematic models pertain can often be broken down into simpler component parts. Consider the following situation described by Gould (1985):

> Almost every day of the week, in almost any city of reasonable size around the world, great tides of people wash into the centre in the early morning to their places of work, and wash back again in the evening to their homes. . . . Moving millions of people in two short bursts of activity every day causes enormous planning headaches involving many forms of transportation. In a city like New York, people pour into Manhattan by car, bus, train, ferry boat, hydrofoil, helicopter and underground railway, and perhaps we should not forget the few brave souls on bicycles, skateboards and roller skates! In Copenhagen and Amsterdam the proportions change, as great rivers of cyclists flow along their own well-marked paths forbidden to the bus and automobile, but the problems of congestion (and pollution) are still severe. Simply to describe such enormous systems of people moving twice each day almost boggles the imagination, for where do you start? No wonder so much urban planning is piecemeal, *ad hoc,* and uncoordinated. (pp. 66–67)

One must begin somewhere. Would the replacement of individual vehicles with mass transportation solve the problems of urban congestion and pollution? Mass transit is but one component of the urban system that is inextricably tied to the economic and technological state of society. Clearly, solutions to urban problems require a multifaceted approach.

Before attempting to walk, one must first learn how to crawl. Future theoreticians and practitioners must begin with basic skills and knowledge. This book is focused at that level. Upon mastering the techniques here, you may progress to multivariate statistical analyses, computer applications, geographic information systems applications, and eventually into complex model building.

THE USE OF STATISTICS

The uses of statistics include *description,* which is simply a summary of data that provides a useful snapshot of its structure; and *inference,* which involves making generalizations about some characteristic of a population based on a sample drawn from it. If repeated sampling of a population yields similar generalizations, we may become

---

[2]Structures tell us how things are connected, or ordered.

increasingly convinced of our findings, and we say that they become general *laws*. As Anselin (1988) stated:

> In philosophical discussions about the nature of progress in science it is often maintained that the development of theory cannot proceed without some form of reality-based validation. This leads to a continual process of interaction between theory formation and *empirical* assessment, motivated in part by the observation of phenomena which are not explained by existing theories, and by the failure of theoretical constructs to be reflected in experimental situations. Essential to this process is a set of clear standards or criteria of validity, and a uniform methodology to apply these to the theoretical constructs under scrutiny. Typically, this is based on a formal probabilistic framework. (p. 223)

Laws do not have to be quantitative or mathematical. A proposition may be expressed in any way that may be generally understood. As geographers, however, we should at least know how to answer simple mathematical questions about the patterns and processes found in any system.

The challenge we have been hearing since the 1960s is to apply this scientific and mathematical approach to *relevant* issues of the day, such as global climate change, poverty, and inequality. Many colleges and universities offer an upper-level undergraduate course or graduate seminar called geography and public policy. The motives behind offering such courses range from the notion that the relationships between spatial patterns and political economy need exploring, to the popular proposition that students will be better prepared for the job market if they have had at least some academic experience dealing with applications of geographic theory. However, it is generally accepted that the state of geographic theory is still far from adequate to deal with the full scope of real problems that nations and cities face. It has also been suggested that policy formation based on analyses of these systems has been neglected or ignored in the interest of economic gains for the private sector and political gains in the public sector.

Whether at the global or the local level, the challenge remains for modelers to build functional linkages between geographical systems. This may well begin with simple statistical tests of relevant hypotheses involving the relationships we have just described. The effects of decisions, whether they be individual household choices or multinational corporate actions, may be tested. All these choices and actions take place within a particular set of circumstances or events. Long-term, radical changes in society often occur through continual public exposure to systemic processes in the short term. The application of statistical expertise to local and global problems has the potential to achieve an immediate goal of answering questions, some critical, some relatively trivial. These positive incremental changes in applied quantitative analysis in geographic research have already made an impact beyond the discipline. As a result of what has been considered increased rigor in geographic research, the ideological status of geography is much more widely realized and current research is better informed (Johnston, 1981).

Science and statistics do not provide the only avenue to acceptable solutions to problems. The identification of solutions to problems may arise from qualitative methods. However, the proper application of science is absolutely necessary to the preservation of the social systems it has created. As Unwin (1981) puts it,

> techniques of quantitative analysis, I suppose . . . may be labelled "positivist." To the phenomenologist or whatever it may prove old-fashioned and, perhaps, also uninteresting and irrelevant. I can only say that in research over the past fifteen years, I have not found such an approach to be either. As a profession, geography has never lacked bold, imaginative ideas. Quite the contrary: what usually is lacking is the technical skill necessary to sharpen up and test these ideas. (p. 2)

Although the dollar figures do not approach those associated with national defense spending, science is supported by government and private enterprise alike with a level of financial investment that has paid a handsome dividend. This is true because scientists have usually been efficient at solving problems.

> Because scientists do work on problems of common concern and keep close track of each other's work, all aspects of scientific investigation and explanation are more objective and accurate than those encountered in other ordering systems. This does not mean to imply that scientists or science are objective according to any absolute criterion. Like other systems, science is order imposed upon the world, and even though we may feel that the order science produces is closer to whatever order exists in the world itself than that produced by other systems, this evaluation is itself subjective and teleological (based on the goals we have in mind). Even so, science has procedural safeguards which assure that the results of scientific analysis will be as accurate as is reasonably possible and more accurate than alternative systems. Here, again, the fact that the results of one scientist are replicable by another is extremely important. Without replication, it would indeed be difficult to attain the degree of objectivity which science has achieved. (Abler, Adams, and Gould, 1971, pp. 30–31)

Applied statistics involves explanation and prediction of phenomena based on probabilities which we obtain from accurate and properly collected data. Explanation and prediction are arranged around the past and present, with the assumption that the processes which produced known phenomena remain constant. Explanations are produced by applying methods, such as scientific investigation, to the past, and predictions are produced by applying the same methods to the future. In the following chapters, the fundamental concepts and techniques of applied statistics as used by geographers, are presented with the hope that they will help launch many more individuals on successful careers serving the societies that support them.

MISUSE OF STATISTICS

Statistics is not a method by which one can prove virtually anything one wants to prove. Statistics has carefully stated rules ensuring that interpretations do not exceed the data. Statistics may be used whether the data sample is large or small, but sample size aside, there is no substitute for good data. Statistics will not be able to overcome such weaknesses as measurement errors or a poorly constructed survey instrument. There is also no substitute for abstract theoretical reasoning, or for careful examination of exceptional cases. Statistics must be considered a *complementary* tool.

One all too common situation arises where a sample is drawn, not at random, but from a selected population—one that is likely to support or reject one's presupposition. Perhaps it is a group of "volunteers" who respond to a survey, such as persons who call in response to a television solicitation with a 900 number or people who vote for all-star baseball teams. Even if the sample is selected at random, bias may be present because many sample subjects will decline to respond, or be overly stimulated to respond, again producing a self-selected sample (Box 1.1). Survey documents routinely report sample responses well under 50 percent. This is partly because people tend to resent the intrusion of pollsters or because they are suspicious of how their responses may be used.

BOX 1.1

An Art, Not a Science

Political scientists and students of public opinion were a bit shaken by the performance of election pollsters in Virginia and New York City last month.

In each case, a black Democrat facing a white Republican had shown a commanding lead in the polls, but barely squeaked through on Election Day. Of course, this is not the first time that polls have failed to predict an outcome, but what made these two elections unusual is that the exit polls, based on samples of actual voters leaving the polling places, also erred in the same direction.

In discussing these two elections, Daniel S. Greenberg, the science columnist, recently referred to polling as "The Science of Getting It Wrong." But polling is not a science and I don't know anyone in the field who thinks it is. Polling is an art that applies elements of science, like weather forecasting or the practice of medicine.

The initial selection of our respondents is based on mathematical probabilities. But the final sample, because of the unavailability of some respondents or their unwillingness to be interviewed, is a very different population. The selection of probable voters is even more skewed and often explains why reputable polls taken at the same time on the same contest may not agree.

Further misunderstanding arises from the margins of error usually presented with newspaper polls. These apply only to the probable range attributable to sample *size*—that is, the mathematical probability of obtaining the same result with another sample of the same size—but this says nothing about validity of the samples or questioning techniques.

Errors of prediction are unacceptable if you believe the purpose of polling is to obtain in advance what everyone will find out anyway as soon as the votes are counted, but in fact polls do both more and less than that. While they may or may not predict the outcome correctly, what pre-election polls can do is to follow the dynamics of a contest, illuminating the election process by charting fluctuations over time and by comparisons among particular segments of the electorate. Polls are also ideal for showing which issues are important to which voters, as was made clear in some of last month's gubernatorial elections.

What we are discussing here are the familiar polls sponsored by mass media. The majority of political polls, however, are privately commissioned by candidates and are made public only if the sponsor wants them to be. Under the code of research ethics, a pollster is obligated to blow the whistle if such findings, even if privately commissioned, are disseminated in a misleading fashion (this writer has twice had to undertake the disagreeable task of publicly chastising hands that fed him). But those who conduct and report the major newspaper and television polls have no incentive to distort or mislead and are limited chiefly by time, budget and the state of the art.

In broader perspective, political polling is a small but conspicuous corner of a large industry of survey research or opinion polling, one of the few genuine innovations of the social sciences in this century.

Such surveys are used by managers and scholars alike in government, business, education, public health, finance, industrial relations, traffic management, criminal justice, public relations and so on and on. Election polling came early on because of its immediate value to candidates and media alike, and also because it provides a rare proving ground for survey technology—an opportunity to evaluate methodology at close range.

This is what makes it so instructive when polls are "wrong." In a Connecticut election several years ago, one newspaper poll showed a controversial senatorial candidate six percentage points ahead, while another newspaper found him six points behind. The two papers reviewed the results and eventually found the explanation: one newspaper asked for its respondents' preference in a concurrent gubernatorial contest *before* asking about the Senate race; the other asked about the Senate first. This surely says a great deal about "halo" effects in politics.

In a similar vein, the Gallup Organization has taken pains after some elections to examine voter rolls to ascertain whether those interviewed actually cast their ballots. In other studies, some survey respondents are asked to make their choices by marking ballots and dropping them into a box, while others follow the customary interview procedure. Comparison of results lets us see whether respondents are influenced by having to state their preferences aloud to an interviewer. Some believe this factor accounted for the recent discrepancies in Virginia and New York. From the standpoint of prediction, those polls were a fiasco; but, like other such aberrations over the years, they serve to illuminate our understanding of voter behavior.

The writer is president of Hollander, Cohen Associates, Inc., a marketing and opinion-research firm.

Source: Sidney Hollander, Jr., "An Art, Not A Science," *Baltimore Sun,* 22 Dec. 1989.

Even with adherence to proper survey techniques, the data may be flawed if respondents do not report truthfully on such facts as age, income, opinions, or behavior. Opinion surveys involving highly emotionally charged topics provide many examples of faulty methodology. Subtle differences in wording of questions can produce remarkable differences in answers given by respondents. There are cases in which a small sample yields a measure that is suspect to the researcher. More data are collected until the results fit the analyst's notion of an acceptable outcome, at which time the survey is concluded and the results reported. Television is full of advertising and "documentaries" that purport to demonstrate, through surveys, how popular a product, service, or political position is relative to its rivals.

Another villain in geographical analysis is the *ecological fallacy*. This problem arises as a result of comparing statistics across scales or translating results between disparate environments. It is, for instance, fallacious to use events observed at the metropolitan area level to predict what will happen to individuals, and you cannot interview friends and neighbors to predict opinion in the county population. Even if there is a bona fide cause-and-effect relationship at one scale, it might not hold at another scale, although the change in pattern of association may suggest cause and effect. Generally, however, patterns of association vary at different scales of analysis and it is an error to infer that an association found at one scale, or in a unique environment will also be found at any other scale, or in any other environment.

Statistics is not simply a collection of facts, nor is it a recipe book. It is possible for an individual researcher, either through carelessness or intellectual dishonesty, to draw erroneous conclusions in spite of the data. There are numerous statistical techniques that may be applied to a data set, each of which will describe some aspect of the data. Unfortunately, if the wrong technique is employed, either out of ignorance or self-service, then the research is invalid.

QUESTIONS

1. To what technological advance can we credit the increased use of multivariate statistics in geographic research?

2. Based on the list of potential effects of global warming, what kinds of numerical data would we require to answer questions raised about these potential effects?

3. What kinds of numerical data would we require to monitor adverse effects of environmental change in a metropolitan region such as Los Angeles?

4. List four or five questions involving numerical measurement and analysis that would involve systems of interest to both human and physical geographers.

5. How do social scientists arrive at a theory?

6. Describe several things that can go wrong with a sampling experiment.

7. Write out definitions of the following terms:
   ecological fallacy   significant
   inference   structure
   scientific law   system

8. What can a geographic perspective contribute to our understanding of major global, national, and regional problems that capture daily headlines in the newspapers of industrialized nations?

REFERENCES

Abler, R., Adams, J. S., and Gould, P. (1971). *Spatial Organization.* Englewood Cliffs, NJ: Prentice-Hall. Chapter 2, "Science and Scientific Explanation."

Anselin, L. (1988). *Spatial Econometrics: Methods and Models.* Dordrecht: Kluwer Academic Publishers.

Environmental Protection Agency (1988). "Focus on—Global Warming." *The Information Broker.* November/December.

Gould, P. (1985). *The Geographer at Work.* London: Routledge & Kegan Paul.

Johnston, R. J. (1981). "Ideology and Quantitative Human Geography in the English-speaking World," in *European Progress in Spatial Analysis,* ed. R. J. Bennett. London: Pion Limited, pp. 35–50.

Unwin, D. (1981). *Introductory Spatial Analysis.* London: Methuen.

Wilson, A. G. (1981). *Geography and the Environment: Systems Analytical Approaches.* Chichester: John Wiley.

Measurement and Sampling 2

Social and physical scientists face the fundamental problem of *measurement:* assigning words, symbols, or numbers to persons, objects, or places according to some set of conventions. We do not always agree about the way in which such conventions should be stated, the way in which different measurement rules should be applied, or the procedure for judging whether or not a given set of conventions has been correctly applied. Despite these problems, however, there is sufficient consensus so that measurements can be obtained and scientific inquiry can proceed.

Should geographers be able to quantify everything? Certainly not. The usefulness of applying a number to something is that it fulfills a particular descriptive need. For most people, a description of a place as "humid" or "dry" may be all they want to know about their weather. But to a farmer or manufacturer, the relatively imprecise description "dry" is less useful than a more exact description such as that relative humidity is 45 percent or average rainfall is 10 inches per year. To the farmer, the amount of rainfall determines whether the land is suitable for commercial grain farming or is only useful for grazing. To a manufacturer of paper or wood products, knowledge of humidity is absolutely crucial to decisions about how much to invest in air conditioning or humidification equipment.

To a variety of people, from farmers and grocers to stock market analysts who trade in agricultural futures, the statement "You can't depend on rainfall" is no help. For the agricultural industry, an index of precipitation variation is a far more meaningful statistic on which to assess risk, whether it is for planting wine grapes or investing in corn futures.

In Chapter 4, the derivation of useful statistics, such as *average* and *variation,* will be shown. These statistics allow us to compare phenomena at places in a more precise way. Thus, if rainfall in Arkansas varies less than rainfall in Kansas, all other things being equal, Arkansas farmers should have more confidence in **reliability** of their rainfall from year to year than do Kansas farmers.

Sample data involving human characteristics are obtained by questionnaire, from other existing records, or simply by observation. If an observer is involved, the *reliability* of the sample may be questioned. The United States census of 1950, for instance, reported "deteriorated" and "dilapidated" residential structures in census enumeration districts throughout the nation. Structural status was determined by enumerators who simply visited assigned residential districts and recorded qualitative values to residences based on appearance alone. Since enumerators had to make subjective judgments and different persons might have reported different frequencies of houses in poor condition, critics of the 1950 census considered these counts to be unreliable statistics. This information was dropped from subsequent censuses.

Scientists have devised systems of measurement in order to lend **precision** and **accuracy** to our sensory observations because our desire for comparison and contrast requires these qualities. When an appropriate instrument is available, a *precise* numerical description of a geographical phenomenon is possible. An expensive compass might read direction within a degree, whereas a cheap compass might err as much as five degrees. The first compass is said to be more precise than the second. *Accuracy* is exactness. If we use a sensitive thermometer that has not been properly calibrated, we will obtain precise but inaccurate temperatures. In the case of a cheap or faulty instrument or a poorly designed questionnaire, the measurements will not reflect what they are intended to measure and the statistics we report from them will be inaccurate.

As we attempt to assign values to phenomena, we must exercise caution because the usefulness and power of statistical techniques are limited by the level of precision possible. Our interpretation of the measurement is also limited by the precision. For instance, 0°C does not indicate that there is no temperature. Zero degrees is merely the value assigned to a thermometer by Anders Celsius, an eighteenth-century Swedish astronomer, to designate the temperature at which ice melts.

One qualitative and obviously ambiguous "scale" used by urban and economic geographers and others consists of a set of terms including "village," "town," "city," and "metropolitan area." There is no uniform agreement on exactly how many people or what population density is necessary to qualify for any of these descriptive categories. Furthermore, population counts are undependable because the accuracy of such data depends on sometimes careless enumerators and a citizenry that is not uniformly conscientious about filling out census forms.

PRIMARY AND SECONDARY DATA

Sometimes, data must be gathered directly from the field, because the values that we are looking for have not been previously recorded. Field observations have the advantage of being specifically tailored for a research problem. Examples of field observations include operating a campus weather station, surveying slopes in a national park, collecting vegetation samples, or asking people how long it took them to get from home to their workplaces or which shopping centers they frequent. These are all called **primary** data.

At least as often, however, the required data have been previously collected and are found in tabular form. Census and other government documents are frequently used data sources. We refer to these as **secondary** data. Occasionally the distinction becomes fuzzy. For example, suppose we measure the distances between trees of a certain species on a map. We then employ these unique distance values as data in an experiment. Although the map itself is a secondary source, these distances, because they have not been previously collected and recorded for any reason, are arguably primary data.

VARIABLES

Chapter 1 discussed concepts as abstractions representing empirical phenomena. In order to shift from the conceptual to the empirical level, we must think in terms of variables. A **variable** is a quantity capable of assuming any of a set of measured values. A tree is not a variable, but species and physical measurements of trees are variables, and are measurable quantities. Distance, altitude, precipitation, shape index, area, and travel time are but a few examples of variables used in geographical research. The measurement scale we use is often dictated by the variable and will be a determining factor as to what statistical technique, or type of analysis, will be employed to answer a question or solve a problem.

A **constant** is a single derived value which is applied within an equation, such as the mathematical constants e and π. The base of the natural system of logarithms is e having a numerical value of approximately 2.718. Pi (π), with a value of approximately 3.14159, represents the ratio of the circumference to the diameter of a circle and appears as a constant in a wide range of mathematical problems.

Variables are of two major types—**discrete** and **continuous (metric).** Discrete variables, also known as *discontinuous variables,* have only certain fixed whole number values with no intermediate values possible. *Frequencies* are considered discrete measures. Frequencies are simply the numbers of objects in a set. If X represents a population of trees in a study area with N observations, and X is partitioned into tree species A, B, and C, then n_A, n_B, and n_C are the number, or frequency, of trees of species A, B, and C. The number of spruce trees in a study area may be 12, 35, or 50, but never 12.2 or 37½. Other examples of discontinuous variables are the frequency of meteorites sighted monthly during a year or the number of automobiles passing through an intersection each hour within a twenty-four-hour period. A variable, such as gender, that is assigned binary values, say 0 and 1, is a discrete variable. We may also refer to *discrete locations.* Two houses occupy unique locations on a spatial continuum. Between these houses, assuming there are no other houses, every other location on that continuum has zero houses.

Conversely, continuous (metric) variables, at least theoretically, can assume an infinite number of values between any two fixed points. Temperature or humidity can be measured *everywhere along the spatial continuum.* Between the two temperature measurements 19° and 20°C there are an infinite number of values that could be measured. The precision of the measurement is restricted only by the quality of the instrument.

THE RUDIMENTS OF MEASUREMENT

Measurement is a procedure by which we assign numbers or other symbols to objects, events, or variables according to certain conventions. Say we wish to distinguish between two places O and P with respect to some characteristic x at a given time or over a time period. One or more of the outcomes in Table 2.1 will apply. With these succinct guidelines, we may compare places or objects with qualitative or quantitative measures. Under conventions 1, 2, and 3, we can simply state that the properties of x at place O are approximately the same as, the same as, or different from place P. Under 4 and 5, we can designate greater or lesser quantities of x in place O than in place P. Convention 1 takes into account our inability to precisely measure some phenomena.

Measuring a **concept** is much more difficult. The concepts of accessibility or quality of life cannot be measured directly and simply. Instead, we invent **surrogate** indicators: something that might be easier to measure that could "stand in" for what would otherwise have to be found in nature or by survey. Examples of surrogates include driving time as a measure of the difficulty of getting to and from the workplace; occupation as a substitute for income data (which are often confidential); and infant mortality or low birth weight as a substitute for community health. But, how *valid* are surrogate measures? In its simplest form, **validity** is the degree to which a variable measures what it is supposed to measure. For surrogates, much depends on the operational definition given to them. Many concepts, such as "prestige" or "deteriorated," are ambiguous and open to argument. If a surrogate variable is to be used in an analysis, several questions should be asked about its validity. Is the variable unambiguously defined? Is it the best measure that could have been used? Is it sufficiently precise? Is it sufficiently accurate? Could it be re-

Table 2.1 Possible Outcomes of Comparisons on Characteristic x at Places O and P

Outcome	Characteristic (x)	Place/Time, Quantity/Quality
1. $O(x) \approx P(x)$	Location U.S. violent crime rate (per 100,000 population)	St. Charles, Canada $\approx$ (70°57′W) Connecticut (455) $\approx$ Ohio (452) 1988
2. $O(x) = P(x)$	Mean annual rainfall (in.) Number of measles cases	Brazzaville, Congo (50) = Madras, India (50) Maryland, USA (8) = Tennessee, USA (8) 1990
3. $O(x) \neq P(x)$	Rock types Population change (%)	Sandstone $\neq$ Granite Birmingham, Alabama, USA (3.7) $\neq$ Atlanta, Georgia, USA (24.3) 1980–1987
4. $O(x) > P(x)$	Elevation above sea level (ft.) Fertility rate	Simla, India (7,200) > Madras, India (40) Kenya (6.7) > United Kingdom (1.8) 1990
5. $O(x) < P(x)$	Streamflow (in.) Personal income per capita ($)	Flow in 1960 (.0024) < flow in 1957 (.4938) Walnut Gulch, Arizona, USA Mississippi, USA (11,116) < Connecticut, USA (23,059) 1988

sponsible for any misleading inferences? Could unmeasured variables be *confounded* with the measures we have available to us (see Nachmias and Nachmias, 1987, Chapter 7, and Blalock, 1982, Chapter 6, for more on this subject)? If a business is profitable in a particular location, we might infer that this is an indicator of the location's accessibility to its market. If affluent families reside in a neighborhood where there is little if any social disorganization, such as poverty, unemployment, or crime, we may infer that their quality of life is relatively high. Thus, some identifiable and measurable value can be used as an *indicator* of the underlying concept.

MEASUREMENT SCALES

The scientist strives to describe objects and events as precisely and unambiguously as possible. Out of this have reemerged many rulers, both qualitative and quantitative. All measurement conforms to one of four scales—nominal, ordinal, interval or ratio.

The Nominal Scale

Often, the only way to measure something is to classify it with a name or symbol. The simplest form of this scale is the binary (dichotomous) classification, on which an individual or event is assigned to one of two discrete classes. Using (typically) the values 0 and 1, we can quantify such things as rural–urban, male–female, yes–no, and wet–dry. When more than two classes are possible, we simply use more numbers or names. All that is required is that the classes are inclusive and mutually exclusive. By *inclusive*, we mean that we should be able to assign all observations to some category or other. For categories to be *mutually exclusive* requires that if people, places, or objects belong to one category, they cannot belong to another. Each and every category is represented by a different symbol or name. One example of a nominal variable is "settlement size," which was mentioned earlier as an example of an ambiguous set of terms. Another nominal example is shopping center type ("neighborhood," "community," and "regional"), which is based on the shopping space, number of stores, and parking area of these centers. Rocks might be categorized as sandstone, shale, and granite, or simply as rock types (1), (2), and (3). Assignment of numerical values (1, 2, 3, etc.) to nominal classes serves merely as a shorthand device. These numbers should not be manipulated mathematically. It would be meaningless, for instance, to add, subtract, multiply, or divide numerical values assigned to rock types, ethnic groups, or any other nominal categories.

The states and census divisions in Tables 2.2 and 2.3 are a nominal classification of places. Nominal data represent the lowest and weakest level of measurement in the sense that they are imprecise and usually mask information that would be available at higher levels, like the metric scales.

The Ordinal Scale

In contrast to the nominal scale, where items are simply allotted to categories, the ordinal scale involves distinguishing among people, places, or objects on the basis of

Table 2.2 Ranking of Top Ten States on Per Capita State and Local Government Direct General Expenditures, 1988

State	Per Capita Dollars	Rank
Alaska	9,546	1
District of Columbia	6,221	2
Wyoming	4,279	3
New York	4,200	4
Minnesota	3,470	5
New Jersey	3,297	6
Delaware	3,294	7
Massachusetts	3,286	8
Connecticut	3,264	9
California	3,240	10

Source: U.S. Bureau of the Census (1991), pp. 286–287.

some relative position, grade, or standing, such as age, proximity, porosity, or quality, and then assigning rank values to each observation.[1] Although ties are possible for any pair of observations, in most cases one is ranked higher than the other. Rivers, streams, gullies, and rills may be given *stream order* values such as "4," "3," "2," and "1." Rankings are a weak form of measurement, and they also often mask attributes that could produce more valid statistics.

In Table 2.2, for instance, ten states are ranked on the basis of their per capita state and local government direct general expenditures in 1988. In the table we may readily see that Alaska's state and local governments spend substantially more per capita than do other state governments. If all we had were the rankings, we would have no way of knowing *how much* more Alaska's governments spent than other state governments.

Such ambiguous concepts as "quality of life" have been quantified with ordinal scales. Liu (1975) used advanced statistical techniques to produce quality-of-life rankings of U.S. metropolitan areas based on 1970 census and other data. Figure 2.1 shows Liu's rankings of sixty-five metropolitan areas on the social component of the analysis. The scores that produced this ranking are composite indicators that include health, education, race, and recreational variables. The statistical term "standardized score" will be explained in Chapter 6.

---

[1]The term *hierarchy* refers to a set of people, places, or things organized or classified according to rank, capacity, importance, or authority. A common use of this concept is found in central place theory in economic geography. Specialized activities of every kind—business, governmental, social, educational, etc.—concentrate in places like cities, special districts, or shopping centers. Depending on their capacity, importance, authority, or any other criterion, these places may be ranked from highest to lowest. A hierarchy of American cities, for instance, commonly is headed by New York City, followed by cities such as Chicago, Los Angeles, and Philadelphia. Major retail centers in cities are at the top of the retail business hierarchy while a single shop in a predominantly residential neighborhood is at the bottom of that hierarchy.

Table 2.3   **Rankings of Nine United States Regions on Seven Indices of Psychological Well-being**

	Indices of Well-Being*						
Region	A	B	C	D	E	F	G
West South Central	1	2	1	3	7	3	2
West North Central	2	7	2	4	1	2	3
New England	3	1	7	7	2	5	1
Mountain	4	6	6	1	4	1	6
Pacific	5	5	3	2	8	4	8
South Atlantic	6	4	4	5	5	7	5
East South Central	7	3	5	9	3	9	4
East North Central	8	9	8	6	9	6	7
Middle Atlantic	9	8	9	8	6	8	9

*The indices are titled as follows: A = Overall rank; B = Outlook on life; C = Stress; D = Positive feelings; E = Negative feelings; F = Personal competence; G = Life satisfaction. A rank of 1 is highest, 9 lowest.

Source: Rubenstein (1982), p. 25.

Figure 2.1 contains ordinal information that is of interest to businesses and chambers of commerce. All cities, states, and regions stress their ranking relative to those of other places on any characteristic that might affect their attractiveness to business and tourism. Often this is the only information about these places that is reported by the popular media. Using data from a survey conducted by the Institute for Social Research at the University of Michigan, Rubenstein (1982) identified regional variations in psychological well-being (Table 2.3). The survey contained responses to thirty-nine questions, which were aggregated into six indices ranging from people's outlook on life to their perceptions of personal competence. For example, respondents of New England were found to have the greatest life satisfaction, with the Middle Atlantic region providing the least satisfaction. The Middle Atlantic's highest ranking (sixth) was awarded to "negative feelings." Overall, the West South Central region was found to be highest in psychological well-being.

Perceived differences among regions or metropolitan areas may explain migration patterns, but such a notion is highly speculative. Like all such indices, those produced by Liu and Rubenstein are open to multiple interpretations, but they receive considerable media attention and are often accepted at face value by business organizations and the public.

People sometimes become uncomfortable upon being confronted with ranks. This feeling of unease occurs because the requirement of simple order does not support our intuitive ideas of what "real" measurement should be. Everyday examples include newspaper ranking of sports teams or someone's ranking of desirable places to live.

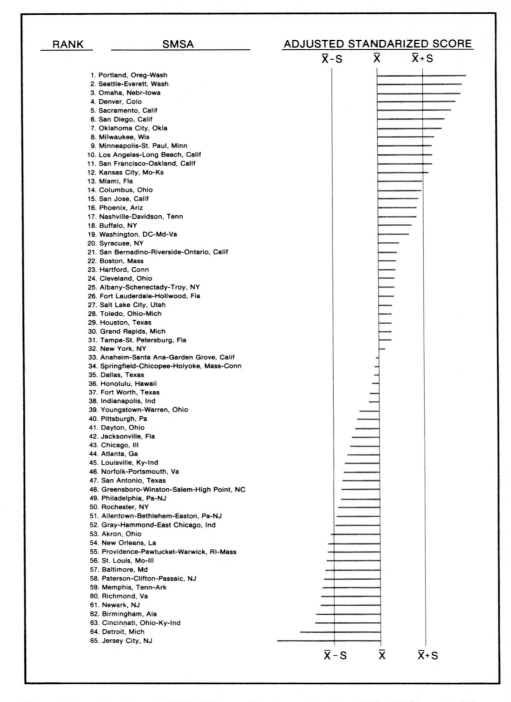

Figure 2.1   Rankings of 65 U.S. Metropolitan Areas, Based on 22 Social Characteristics, 1970

Source: Liu (1975), p. 21.

The Metric Scales

The most familiar forms of measurement are the **interval** and **ratio** scales. We refer to these as **metric** scales because, unlike ordinal data, values on these scales may be based on an equal interval of measurement and a continuous distribution of values. The metric scale of measurement should not be confused with the *metric system* of measurement (kilometers, hectares, kilograms, etc.). Some of the most common uses for the metric scale in geography are location upon the earth's surface using latitude and longitude, elevation, and distance. The metric scale represents a higher level of measurement than either nominal or ordinal data. Measures such as altitude, humidity, and temperature can be measured to any desired number of decimal places, depending on the precision of the instrument.

We often use a Cartesian coordinate system to locate places on the earth's surface. The term *Cartesian* is derived from the name René Descartes, who devised a method to locate positions arbitrarily on a plane surface using a point of origin at the intersection of two perpendicular axes. This plane may then be divided into a grid by an infinite number of equally spaced lines parallel to each axis. The position of any point on the plane with reference to the point of origin may then be stated by indicating the distance from each axis to the point, measured in each case parallel to the other axis. A vertical scale, designated Y, and a horizontal scale, designated X, must be constructed. Figure 2.2a illustrates the use of this system where latitude is substituted for Y, and longitude is substituted for X. Figure 2.2b shows two more examples of the metric scale: sea-surge elevations in centimeters above sea level (Y), and a transect, divided by metric distance, through tropical storm Carla in the Gulf of Mexico (X).

Measurement on an **interval** scale consists of allocating a number to an observation to indicate its precise position on a metric scale. The starting point, however, is arbitrarily set. We noted at the beginning of this chapter that 0°C referred to the freezing point of water. On a Fahrenheit thermometer, this point is established at 32°. Consider the following demonstration of Fahrenheit to Celsius conversion. Since the temperature interval from water's freezing to boiling is represented by 0° − 100° (an interval of 100 degrees) on the Celsius scale, and by 32° − 212° (an interval of 180 degrees) on the Fahrenheit scale, a conversion technique is:

$$\frac{F - 32}{180} = \frac{C}{100} \tag{2.1}$$

and

$$C = \frac{5}{9}(F - 32) \tag{2.2}$$

It is immediately evident from these equations that when F = 32°, then C = 0°. The lower limit of an interval metric variable may be below zero on whatever ruler is employed. For example, temperatures below 0°C are below the freezing point of water and are designated with negative signs, for example, −10°C. The point of this example is that the zero point for temperature has been set arbitrarily; 0°C is not the same as 0°F, and 0° does not mean there is no temperature.

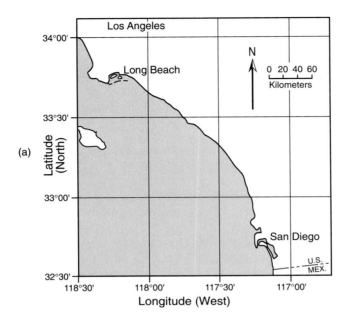

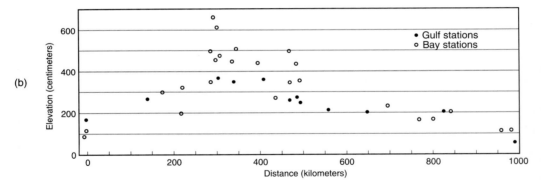

Figure 2.2 Graphic Examples of Metric Data: (a) Cartesian Coordinates and Distance, and (b) Elevation and Distance

Ratio variables, such as areas of counties, densities of population, population, and rainfall are distinguished by our ability to designate an absolute, nonarbitrary zero point for them. As a result, we can compare values by taking their ratios. For example, we can say that 10 miles is exactly twice as far as 5 miles, 2 liters of water is exactly twice as much water as 1 liter, and 0 liters of water means exactly no water. Had the zero point been arbitrary, such as in the case of the temperature scales, this would not have been a legitimate comparison. Thus, we do not say that 100°F is twice as hot as 50°F, although we can say that the *difference* between these temperatures is the same as that between 20°F and 70°F. Therefore, it is inappropriate to manipulate the base of these numbers (see *data transformations* in Chapter 8).

One ratio variable that is frequently encountered in economic and population geography is the **rate.** A rate is computed by dividing the number of cases or events in a given category by the total number of observations. Symbolically, a rate is expressed by Equation 2.3:

$$\frac{f}{P} \tag{2.3}$$

where f = the frequency of a phenomenon observed during a specified time period, and
P = the total number of persons or objects that could be observed exhibiting this phenomenon during the same time period.

Some of the rates used in comparative studies by geographers include rates of speed, fertility rates, wage rates, disease rates, and sediment transport rates. For instance, the *frequency* of measles cases in one year in the United States versus that same value for China is a meaningless comparison, as the population at risk in China is a much higher number than in the United States (the population of China is about 4.5 times as large as that of the United States). Therefore, we standardize the comparison by dividing the measles frequencies of the two nations by their respective populations.

A *rate of change* is useful in comparing the distribution of a variable through time. For instance, if we wish to compute the gain in population between 1980 and 1990 of California, we subtract the population of the *earlier* time from the *current* population and divide by the value of the earlier time. The result can be multiplied by 100 to reflect percentage change. This is symbolically expressed as

$$\frac{f_{T+1} - f_T}{f_T} \times 100 \tag{2.4}$$

California's (April 1) 1980 population was 23,669,000, and its (April 1) 1990 population was 29,760,000. Therefore, the rate of increase for the decade is computed as,

$$\frac{29,760,000 - 23,669,000}{23,669,000} \times 100 = 25.7$$

If a population doubles exactly, the percentage change is 100 percent. Population gains and losses of greater than 100 percent are sometimes observed when there is massive migration to or from a region.

Table 2.4 summarizes what we know about measurement scales. Note that we can always transform data from a higher scale (interval or ratio) to a lower scale (ordinal or nominal), but it is impossible to transform in the other direction. For instance, in the ordinal or nominal scales, we have no idea of how much difference exists between units,

Table 2.4 Summary of Measurement Scales

Scale	Empirical Test	Geographic Example
Nominal	Places, people, or objects are in one of several qualitative categories; no fixed interval; zero point arbitrary	Soil classification; land uses; ethnic group classification
Ordinal	Something is greater or less than something else; no fixed interval; zero point arbitrary	Ranking of place preference; ranking of cost-of-living indices for cities
Interval	Intervals are the same between people, places, or things; fixed interval; zero point arbitrary	Contour lines on maps; calendar time; Celsius and Fahrenheit temperature scales
Ratio	Intervals are the same between people, places, or things; fixed interval; true zero	Distances between places; bushels per acre; death rates; per capita incomes

say, ranks 1 and 2, or 4 and 5; or, how much larger is a given regional shopping center than a community shopping center. Consider the table of pH, the relative acid/alkaline scale, for example (Table 2.5). A pH value of 5.7 is not only more precise than the nominal counterpart "moderately acid," but it can be used in more rigorous statistical tests, as we shall see in later chapters. Underlying all forms of measurement is the ability to assign qualities or numerical values to things according to the conventions we specified in Table 2.1.

GEOGRAPHIC PRIMITIVES

As we contemplate the properties of the things we measure in geographic research, we are forced to consider the basic primitives of the field. That is, what are the most essential, most primitive things we have to consider, the basic blocks from which we build all our other constructs?

Points, Lines, Areas, and Surfaces

Measurement in geography involves assignment of values to *points, lines, areas,* and *surfaces*. How such values are assigned depends on the scale at which the research is being conducted. **Points** can take on a variety of descriptions (metric or otherwise) depending on the frame of reference. For example, in geometry a point has no property but location, that is, no length or width. In geographic research points can refer to cities on a map, individual plants, or industrial parks. Obviously, the phenomena just described do

Table 2.5 Interval and Nominal Scales of pH

pH	4.0	4.5	5.0	5.5	6.0	6.5	6.7	7.0	8.0	9.0	10.0	11.0
Acidity	Very strongly acid	Strongly acid	Moderately acid	Slightly acid				Neutral	Weakly alkaline	Alkaline	Strongly alkaline	Excessively alkaline

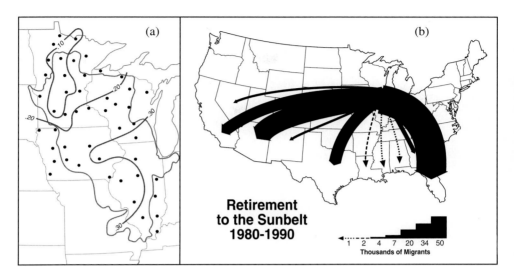

Figure 2.3 Examples of Point and Line Representation of Geographic Facts: (a) Rainfall, and (b) Migration

Source: Part (a) adapted from Keables (1988), pp. 78, 88.

not occupy points as points are described in Euclidean geometry. However, we may define the location of a tree by assuming the geometric center of the tree is defined by an X–Y coordinate on a map. It follows then that "attributes" of points can be measured in either qualitative or quantitative terms. For example, the city represented by a point on a map has 3,200,573 residents, or the city represented by this point is an industrial city. Figure 2.3a shows a set of points that represent places in the American Midwest at which rainfall was measured under certain conditions.

You should also be aware that some characteristics of points may form *discrete variables* with fixed whole number values or a place's rank on some quality, such as a retirement locale or vacation spot. Most of the characteristics that follow refer to *continuous* (interval or ratio) measurements.

A **line** has the single dimension of length l but no width. In geographic research, we use lines to measure length (to link places), but we also allow lines to carry measurable quantities of people, information, disease, and other things. The lines in Figure 2.3b show the relative number of retirees that migrated to the American Sunbelt between 1980 and 1990.

Areas are bounded by lines, just as lines are bounded by points. A study area might be the state of Maryland. To study the distribution of ethnicity in Maryland, one might compute the proportion of African-American population within each county. Maryland counties then become *unit areas* within the study area (state). Displays of these phenomena are called *spatial distributions.* Depending on the scale of the research, areas may range from a few square inches of space to a nation, or beyond.

Figure 2.4 is an example of a spatial distribution of motor vehicle accident rates in the United States in 1989. The range of the data (10–35.2 percent) has been subdivided

26 *Geographic Measurement and Quantitative Analysis*

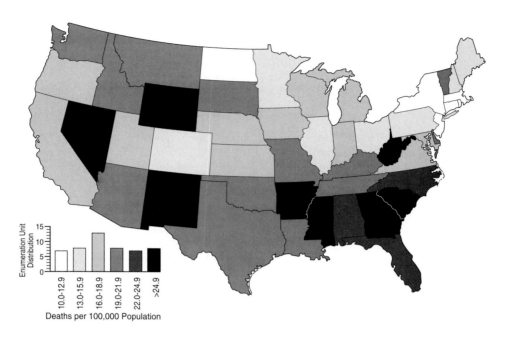

Figure 2.4   Motor Vehicle Accident Deaths per 100,000 Population

into six *class intervals* to illustrate the variation in accident rates throughout the nation. Map patterns are then assigned to each interval. The range of data within each class interval is critical to the message portrayed by the map. In this case, the interval range is 3 percent. As we will see in Chapter 4, there should be a rationale for assigning specific data ranges to each interval. Here the object is to standardize the interval in order to emphasize the relative distribution of high and low accident rates. As you might have surmised, varying the range of data within categories can produce a radically different visual pattern of a situation, and intentional misuse of this technique could paint a distorted picture of reality.

Surfaces are the display of volume or altitude. They are maps with a third spatial dimension. For example, consider the simple *topographic map* in Figure 2.5. This map shows the altitude of the ground *surface,* but in so doing also defines the three-dimensional volume of landforms ranging from some 1,100 feet to over 1,200 feet above sea level (Figure 2.5b). Surface mapping involves the use of the mathematical concept called **scalar.** A scalar is any quantity characterized only by its magnitude. An example of a scalar is altitude; one number represents its magnitude and this remains the same irrespective of how we transform its geographic position with, say, different map projections. A scalar field is any graph of the scalar as a function of its X–Y position, as shown in Figure 2.5b. Scalar fields can be represented mathematically by the general equation $Z = f(X, Y)$ where Z = scalar magnitude; X, Y = spatial coordinates; and f denotes "function of."

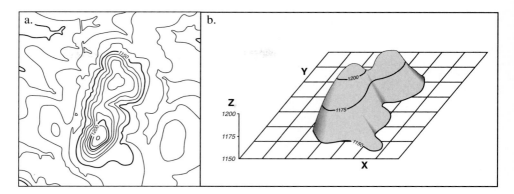

Figure 2.5 (a) A Contour Map of Altitude and (b) Its Scalar Field

For the hill in Figure 2.5b, the altitude (Z) is related to coordinate position (X, Y) with two critical assumptions having already been made. First, a continuity assumption, is that there is a Z value everywhere on the surface, and second, that there are no sharp discontinuities. This means that on the hill there are no bottomless pits or vertical cliffs. Second, we assume that there is only one value of Z for each X–Y location. Thus, on our hill there are no cuts or overhangs. As we will see in Chapter 3, contour maps of other phenomena, such as rainfall, travel time, and land values, can have more complex fields.

The most basic geographic concepts are *distance* and *direction*. Other concepts such as area and shape are derived from them. Distance and direction may be illustrated with points, lines, areas, or surfaces.

Distance

The concept of **distance** is an important geographic primitive because it characterizes the discipline that has at its core a concern for spatial pattern, arrangement, and juxtaposition. Distance seems at first to be easy to understand and measure, but on closer inspection, it turns out to be much more complex. Basically, a scale for measuring distance involves specific conventions for putting numbers on something, but it is not necessary to assume that we must adhere to that scale. We might say that two towns are 10 miles apart. The number 10 is a value of distance on a particular scale agreed upon in the United States. The same distance relationship may be represented on a metric scale, whose unit name is the kilometer. Then that same 10-mile distance becomes 16.09 kilometers on the metric scale. Both are ratio scales having a true zero. Zero has the same meaning, whether it is expressed in miles or in kilometers.

We begin by illustrating Euclidean distance, which is based on a linear metric scale. Before we can proceed to compute linear distance, however, we need a system for the measurement of length between places on the earth's surface. There are several systems for measurement of distance along north–south and east–west directions, but the most commonly used is our familiar terrestrial form of the Cartesian coordinate system—

latitude and longitude (Figure 2.2). For small-scale geographic studies it is not unusual to bound the study area and establish a coordinate system with a base point (X = 0, Y = 0) at the lower left intersection of the study area boundary.

Figure 2.6 illustrates the flight paths between Boston and New York as a straight-line distance. To calculate this distance, we turn to the well-known Pythagorean theorem:

$$c^2 = a^2 + b^2 \qquad (2.5)$$

where c is the length of the hypotenuse of a right triangle, and a and b are the lengths of the legs. Lengths or distances may be derived from the coordinates of X and Y that define the locations of the cities in question. Note that the length of side a is $X_1 - X_2$ and side b is $Y_1 - Y_2$.

$$C = \sqrt{(X_1 - X_2)^2 + (Y_1 - Y_2)^2} \qquad (2.6)$$

Equation 2.6 provides the Euclidean distance between the points represented by X_1, Y_1 and X_2, Y_2. Box 2.1 illustrates how straight-line distances are computed between two pairs of hypothetical towns with the following [X, Y] coordinates: Northtown [4, 12], Southtown [3, 0], Easton [8, 6], and Weston [1, 7].

Note that the accuracy of distance is subject to two important assumptions. First, we must assume that a Euclidean geometry is appropriate. At distances no longer than we would encounter within a city or average county, Equation 2.6 would present no problem. For larger areas we must consider distortion introduced by the convergence of lines of longitude as we move toward either pole. For experiments involving human interaction over space, cost or time, which are not Euclidean, might be more appropriate than kilometers or miles. Second, we have only found the straight-line distance; again this might not be appropriate to the research problem at hand. We might be more interested in highway, rail, or river distance. Measuring distance along irregular curves like a coastline is a classic problem in mathematics, the difficulty being that the more accurately we measure, the longer the line becomes.

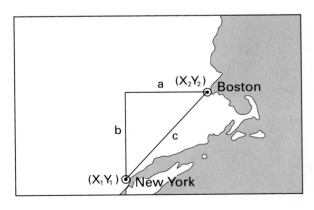

Figure 2.6   Simple Euclidean Distance: Flight Line between Boston and New York

Measurement and Sampling

> **BOX 2.1** **Computation of Euclidean Distances between Northtown and Easton, and Southtown and Weston**
>
>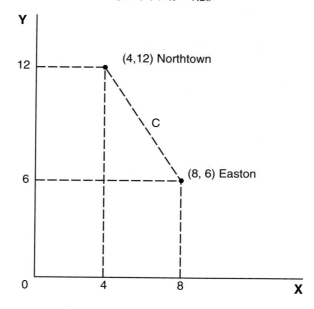
>
> Northtown–Easton
>
> $$C = \sqrt{(X_1 - X_2)^2 + (Y_1 - Y_2)^2}$$
> $$C = \sqrt{(4 - 8)^2 + (12 - 6)^2}$$
> $$C = \sqrt{16 + 36} = 7.21$$
>
> Southtown–Weston
>
> $$C = \sqrt{(3 - 1)^2 + (0 - 7)^2}$$
> $$C = \sqrt{4 + 49} = 7.28$$

A map of distances between points may be constructed on nonlinear scales. In transforming distance from the standard to the nonlinear scale, the spatial order of things is retained, but magnitudes are not additive as they would be on a linear scale. One widely used transformation in geographic research is the logarithmic scale. Figure 2.7 is one example of how we may represent relative spaces in useful and informative ways. This map was constructed by Edgar Kant for use by Torsten Hägerstrand in his research on migration to and from Asby, a district in Sweden. On a logarithmic map, distance away from the center of the district decreases in proportion to the logarithm of physical dis-

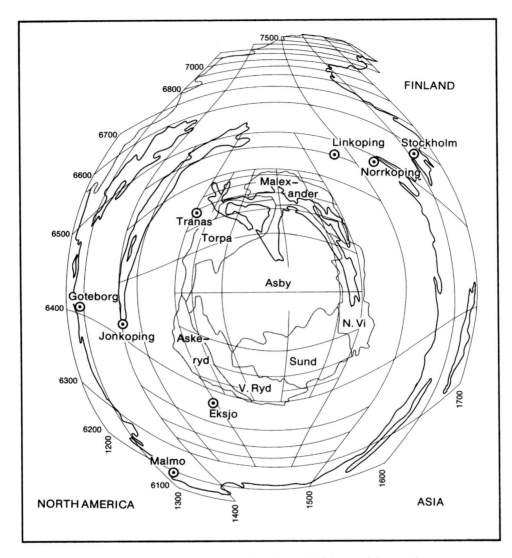

Figure 2.7 A Map of the Asby District in Sweden, with Distance Measured on a Logarithmic Scale

Source: Hägerstrand (1957).

tance.[2] A map that reports migration moves, most of which are short as in this example from Sweden, magnifies the space in which human interaction is most intense. Peripheral areas of limited importance are small and do not detract attention from the intensity of

---

[2]The logarithm of a number is the *power* to which ten must be raised to make it equal to the raw number. For example $10 = 10^1$, so the log of 10 is 1; $156 = 10^{2.19312}$, so the log of 156 is 2.19312.

Table 2.6 Manhattan Distances (miles) and Approximate Driving Times (minutes*) between Four Cities in Southern California

	Inglewood	Redondo Beach	Long Beach	South Gate
Inglewood	0	10.4 (32)	21.6 (40)	8.9 (45)
Redondo Beach		0	16.6 (45)	17.9 (67)
Long Beach			0	12.7 (33)
South Gate				0

*Driving times are in parentheses.

movements near the center. Further consideration of relative space and the problems of measurement is available in books authored by Lewis, and Abler, Adams, and Gould, which are referenced at the end of the chapter.

For another variation on physical distance, consider the data in Table 2.6. The nonbracketed numbers show the *Manhattan distance,* in miles between four points in the Los Angeles metropolitan area in southern California. Manhattan distance is the physical distance accumulated by traveling between points on a grid of streets that intersect at right angles. Manhattan distance between two points is computed with the expression,

$$d_M = |X_1 - X_2| + |Y_1 - Y_2| \tag{2.7}$$

which is simply an addition of the **absolute value** (where the sign + or − is ignored) distances in both directions. The numbers in parentheses represent minutes of driving time between the four cities.

Further expanding the definition of distance, geographers routinely refer to terms such as *economic distance,* which address distance in terms of difficulty or cost. For example, the U. S. Bureau of the Census published a table in 1976 showing median time and distance statistics for trips to work by Americans in five cities (Table 2.7). It is apparent from this table why many Americans in these cities consider automobile travel to be superior to public transit, at least in terms of *time* spent in travel to work. In fact, the advantage of the automobile is probably understated, since these figures are restricted to work trips. The use of time in lieu of physical distance in many instances provides a more cogent explanation of why people locate their homes or businesses where they do, or why they travel when and where they do for various purposes. Inglewood and Long Beach are closer together than Manhattan distance would indicate because one can travel most of the space between Long Beach and Inglewood on a high-speed freeway, whereas space between the other cities can be traversed only by ordinary arterial streets with much stopping and slowing for traffic signals and congestion.

Table 2.7   Work Trip Times and Distances, by Mode of Transport for Selected U.S. Metropolitan Areas, 1976

City	Private Auto or Truck	Public Transit	Walk	Other Means
Baltimore				
Median time (min.)	23.1	39.7	9.1	15.8
Median distance (miles)	10.0	6.5	0.6	3.3
Denver				
Median time (min.)	19.5	28.4	9.4	16.1
Median distance (miles)	18.2	7.8	0.6	4.0
New York				
Median time (min.)	21.9	42.0	10.4	17.2
Median distance (miles)	8.6	9.4	0.6	3.8
Houston				
Median time (min.)	21.7	35.8	7.5	14.1
Median distance (miles)	9.6	9.5	0.6	3.5
Sacramento				
Median time (min.)	18.0	30.3	8.1	13.8
Median distance (miles)	7.8	10.3	0.5	2.7

Source: U. S. Bureau of the Census (1976).

Area, Shape, and Density

In some geographic research, the size and shape of areas are of interest in their own right, irrespective of any phenomena that we might find within them. In political geography, the Gerrymander problem is a case in point; the size and shapes of drainage basins are of interest to geomorphologists; and so on.

Most students learn by eighth grade that the area of a rectangle is computed simply by multiplying the values of its length and width ($l_L \cdot l_W$). Using the symbolism introduced in the discussion of distance, we may simply restate the area of a rectangle as the absolute value of the product of the length of the two perpendicular sides, or $|(X_1 - X_2) \cdot (Y_1 - Y_2)|$. Another elementary area computation is that of a circle, which involves the use of the constant π: $A = \pi r^2$.

Complications set in, however, when the shape of the polygon varies from the rectangular or circular. The simplest and most accurate method for area measurement is with a device, such as a *digitizing tablet,* that automates the operation.[3] Unwin (1981, Chapter 5) suggests several less precise methods. If you have a ruler and hand calculator you may compute the area of a relatively simple polygon with the following technique.

To illustrate, we arbitrarily analyze Howard County, Maryland, which appears to be about halfway between a circular shape and a straight line (Figure 2.8a). We begin by redrawing the county outline, replacing the curvilinear outline with an irregular polygon.

---

[3]Most modern cartography laboratories have a digitizing tablet. The user traces the features of interest on a map with either a pen-like stylus or a flat cursor. The electronics in the tablet system convert the position of the stylus or cursor to a computer-compatible digital signal. After the shape has been digitized, the system will provide, among other things, the area of that shape. For further explication on this technology, see Star and Estes (1990).

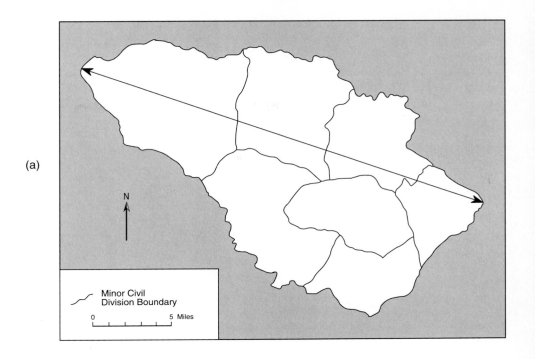
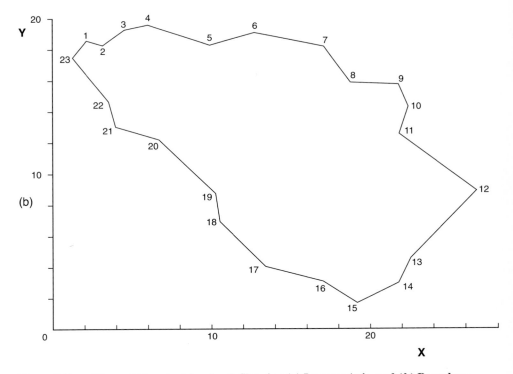

Figure 2.8 Howard County, Maryland, Showing (a) Longest Axis, and (b) Boundary Reduced to 23 Straight-line Segments

33

The number of segments in the polygon should be a function of the required accuracy of the area and labor involved in computing that area. The greater the number of sides, the more accurate the result. Figure 2.8b illustrates how this complex shape is reduced to a polygon with twenty-three sides. The axes of this diagram are scaled in miles. The origin of the (X, Y) coordinates is located in the lower lefthand corner. Working clockwise from any point on the polygon, X and Y coordinates are then recorded for each point (Table 2.8).

To find the "contribution" of each point to the area the X coordinate of the vertex prior to the point of reference is subtracted from the X coordinate that follows in sequence, and multiplied by the value of the Y coordinate of the referent point. For the first point the X coordinate of the second vertex is 3.2 and the X coordinate of the prior vertex (point 23) is 1.3. The Y coordinate for point 12 is 19.0. The contribution for Point 1 is thus

$$19.0(3.1 - 1.3) = 36.1$$

For point 13, the contribution is

$$4.5(22.0 - 27.0) = -22.5$$

and so on. The contributions are then summed. The area of the polygon is computed by Equation 2.8:

Table 2.8 X and Y Coordinates of the 23 Vertices of the Howard County Outline (Figure 2.8b)

Point	X	Y
1	2.2	19.0
2	3.2	18.2
3	4.6	19.4
4	6.0	19.7
5	10.0	18.4
6	12.7	19.3
7	17.3	18.4
8	19.0	16.0
9	22.0	16.0
10	22.4	14.7
11	22.1	12.7
12	27.0	9.1
13	22.7	4.5
14	22.0	2.7
15	19.5	1.4
16	17.5	2.9
17	13.7	3.8
18	11.0	6.7
19	10.5	8.6
20	7.0	12.0
21	4.0	12.8
22	3.7	14.5
23	1.3	17.3

$$A = 0.5 \sum_{i=1}^{n} Y_i(X_{i+1} - X_{i-1}) \qquad (2.8)$$

In Box 2.2, the sum of the individual vertex contributions is 495.38, which is then multiplied by 0.5 to yield an area for Howard County of 247.69 square miles. A census publication states that Howard County's area is 251 square miles. Allowing for cartographic error in the original map of Figure 2.8, plus the error introduced in computation of the contributions of the vertices, the discrepancy is negligible. Of course, there may be a substantial monetary or political price associated with just over a three-square-mile discrepancy!

Areal units all have two-dimensional shapes with constant relationships of position and distance of points on their perimeters. **Shape** is a fundamental property of many objects and spaces of interest in geography, such as drumlins, drainage basins, voting districts, and others. Some shapes, notably the hexagons of central place theory, have theoretical implications or relationships within the economy of a nation or region. Traditionally, shapes were described verbally, using analogies such as spherical (basin morphometry), prorupt (voting districts), linear (settlements), and so on.

An obvious quantitative approach is to devise indices which relate the real-world shape to some regular geometric figure, such as a circle, hexagon, or square. The most commonly used shape index is that of **compactness**, which is based on deviations from the most compact possible shape, a circle. For simple applications a shape index will permit rudimentary comparisons between shapes. Taylor (1977) provided a shape index that demonstrated boundary changes for Germany, Poland, and Czechoslovakia between 1920 and 1945. He noted fewer elongated or prorupted shapes and more compact, circular shapes. Unwin (1981) and Ebdon (1985) also provided shape-index formulas that measure the extent of a shape's deviation from a circle. The index assumes a value of 1.0 for a perfect circle (maximum compactness), and a value less than 1.0 for any less compact shape. At the opposite end of the range of the shape index is the straight line, which has no area, hence $S = 0$.

Measurement of shape is indirectly obtained from length. A simple measure of the shape index is computed with Equation 2.9:

$$S = \frac{d}{l} \qquad (2.9)$$

where l is the length of the longest diagonal of the shape. This may be determined by enclosing the referent shape in a circle. The distance between the points of the shape touching the circle that are farthest apart yield l; d is defined as $2(A \div \pi)^{1/2}$, where A is the area of the shape.[4]

---

[4]An expression taken to the power of 1/2 simply means the square root of that expression. An alternative computation for shape is

$$S = \left[\frac{A}{A_C} \right]^{1/2}$$

where A_C is the area of a circle having the same perimeter as the referent shape. This is seldom employed, since it is more difficult to compute circular area from the perimeter.

BOX 2.2 — Computation of the Area of a Polygon: Howard County, Maryland

Point	X	Y	Contribution
1	2.2	19.0	36.10
2	3.2	18.2	43.68
3	4.6	19.4	54.32
4	6.0	19.7	106.38
5	10.0	18.4	123.28
6	12.7	19.3	140.89
7	17.3	18.4	115.92
8	19.0	16.0	75.20
9	22.0	16.0	54.40
10	22.4	14.7	1.47
11	22.1	12.7	58.42
12	27.0	9.1	5.46
13	22.7	4.5	−22.50
14	22.0	2.7	−8.64
15	19.5	1.4	−6.30
16	17.5	2.9	−16.82
17	13.7	3.8	−24.70
18	11.0	6.7	−21.44
19	10.5	8.6	−34.40
20	7.0	12.0	−78.00
21	4.0	12.8	−42.24
22	3.7	14.5	−39.15
23	1.3	17.3	−25.95

Sum the contributions to area:

$$\sum_{i=1}^{n} Y_i [X_{i+1} - X_{i-1}] = 495.38$$

Apply Equation 2.8:

$$A = 0.5 \sum_{i=1}^{n} Y_i [X_{i+1} - X_{i-1}] = 0.5(495.38) = 247.69$$

Figure 2.9 illustrates the approximate shape of an electoral district in New York City. This district is an example of *gerrymandering*. Gerrymandering is a method of drawing voting district boundaries in such a way that one political party can contain or disperse voters of the opposing party. This practice gives the party in power an advantage at the polls. One measure of "fairness" in the geography of electoral districts is that they be of compact shape. If a district is redrawn into an elongated or meandering shape, the opposing party has legal grounds for protest.

Figure 2.9 **A Gerrymandered Voting District in New York City**

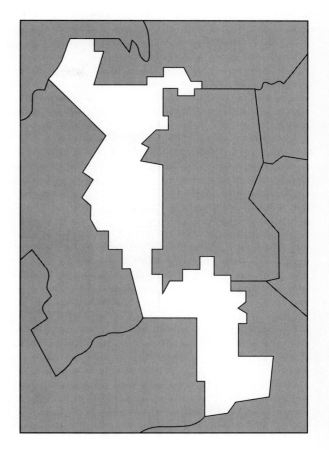

The area of the electoral district in Figure 2.9 is given as 6.77 square miles and the length of the long axis is 6.80 miles. From Equation 2.9, we compute the shape index to be $d \div l = 2.93 \div 6.80 = .43$. The electoral district is indeed suspiciously long. As a basis for comparison consider the shape index for Howard County. The length of its longest axis is 22.5 miles. The value of $d = 2(247.69 \div 3.14)^{1/2} = 17.76$. Thus, $S = d \div l = 17.76 \div 22.5 = .79$.

The value of the shape index is fraught with problems of accuracy and precision and can vary considerably. The use of the perimeter in a shape index yields values that are oversensitive to the sinuosity of the perimeter (Unwin, 1981, Chapter 5). It is easy to introduce bias and error into the measurement of line segment lengths, either because of the scale at which we are working, or because our instruments are too crude (either in terms of size or calibration) for the level of precision to which we aspire. Lack of precision may be problematic, but it may be reduced by simply repeating the measurement operations several times and taking the average. On the other hand, inaccuracy is a far greater problem because we might not know that our instruments are biased. For example, Taylor (1977) points out that over their history rain gauges have been redesigned to counteract splash, making comparability impossible over a long time period. Error may also be compounded if measurements are derived from multiplication, division, or transformation. Taylor (1977) mentions one interesting source of error involving the nominal scale:

> rural land use mapping from a hilltop is likely to produce an accurate but imprecise classification, whereas classification by a conscientious student who confuses wheat with barley will give systematic biases and result in a precise but inaccurate classification. (p. 45)

For simple applications, a shape index will permit rudimentary comparisons between shapes.

Density plays a broad role in geographic research in both physical and human geography. In soil structure, *bulk density* increases with clay content and is a measure of soil compactness. Geomorphologists pay attention to *drainage density* when modeling hillslope and channel flows. Density is also an important variable in economic development and urban and regional planning. The U.S. Department of Agriculture produces *dot density maps,* which show the distribution of crops and farms by state or county (Figure 2.10). Each dot represents a fixed number of farms, bushels, or other units. The resulting map gives a good visual impression of farm or crop density variation across the study

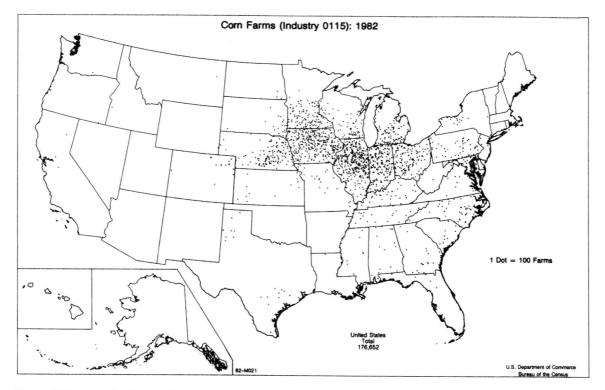

Figure 2.10   A Dot Density Map
Source: U.S. Bureau of the Census (1985), p. 18.

area. Transportation planners pay attention to the effects of residential density on the number of automobile and transit trips, or the number of dwelling places per residential acre. Meteorologists create dot maps of severe weather events to show areas of heightened risk.

An alternate use of density, which is useful in physical as well as human geography, involves counting the number of objects found in a grid of identically shaped and equal-size subareas, or *quadrats*, overlain on a map. In effect quadrat analysis allows for a study of density variation across the study area, but its greatest usefulness is in the statistical summary of the pattern of density of dots in the quadrats. We will demonstrate quadrat analysis in chapters 5 and 9.

Direction

Direction is another fundamental measurement that can be extracted from maps and measured on the interval scale. **Direction** from any origin may be specified as an angle, or *azimuth*, usually measured clockwise from north, which is designated zero degrees (0°) (Figure 2.11). Of course, direction could just as easily be measured clockwise from a west bearing. Thus, direction is a *relative* measure that depends on the frame of reference used to take the bearing. There are 360 degrees in the polar coordinate system. An angle

Figure 2.11 A Polar Coordinate System, Showing Directions and Distances of a Set of Points from an Origin

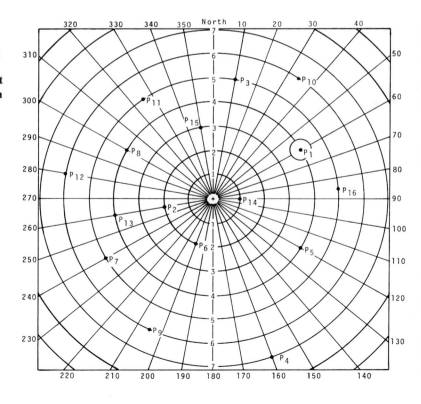

of 45° is clearly recognizable as half a right angle, 90°. Of course, a distance scale must be specified in order to know the exact distance of each point from the origin. In Figure 2.12, City B is located at azimuth $\Theta_1°$ and 60 miles from the origin. Without an instrument (protractor), however, we would require a trigonometric equation to compute the exact direction of any angle in this diagram.

Although north is the *usual* frame of reference, it is possible and may even be desirable to have a different frame of reference. For instance, in the ridge and valley region of Appalachia, streamflows and human spatial interaction are either parallel to the ridges, which run roughly northeast to southwest, or nearly perpendicular to them. In this case it might make more sense to measure direction with reference to a northeasterly or northwesterly direction. As another example, an area where measurements of glacial striations are taken might occur within a broad topographic depression aligned northwest–southeast. A statistical question would be: Does the mean direction of ice movement, as indicated by the striations, coincide with the axial direction of this depression?

One problem unique to direction is its circular pattern and its nonlinear number sequence. Angles increase from 0° to 360°. As north is the usual frame of reference, 180° is the opposite direction, or south. Larger angles, 181° to 359°, bring us through the westerly directions back to north. All these bearings, although seemingly real numbers, cannot be translated into the natural number system for mathematical manipulation: 1° and 359° vary equally from 180°. This problem has been overcome with vector arithmetic, which

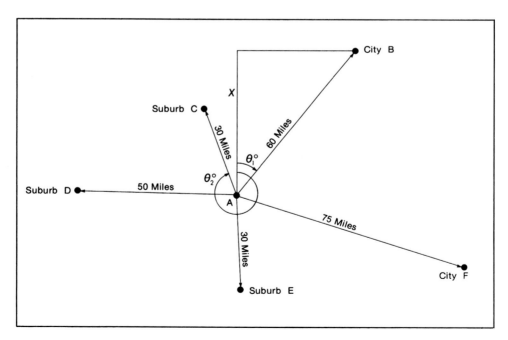

Figure 2.12 **Measures of Distance and Direction from an Origin: Hypothetical Cities and Suburbs from an Origin**

is beyond the scope of this book. The interested reader is referred to Taylor (1977, Chapter 2) and Unwin (1981, Chapter 4).

COLLECTING SAMPLE DATA

The collection of data is a crucial operation in the execution of a good research plan. The quality of your research results rests upon the quality of the data. Most of the methodological detail on collecting data from the library, field, and laboratory is beyond the scope of this book.[5] Basic geographic sampling strategies follow.

From a single researcher working in a forest in China to a consulting group collecting data on global air pollution, the scheme for data collection must be carefully considered. A geomorphologist interested in fluvial processes may wish to investigate differences in the size, shape, and directional orientation of pebbles along a stream channel. The number of pebbles in the channel, though not infinite, is too large to allow observation of all of them. For an urban geographer to describe the statistics (mean, standard deviation, etc.) of a population of characteristics, the number of observations required is beyond their reach unless they are fortunate enough to live in a wealthy nation with a large, well-financed census agency. Researchers must sometimes rely on samples from which we may *infer* population statistics. It is important that samples be as unbiased as possible.

The key to accurately projecting the characteristics of a sample onto the population is to draw the sample with care to ensure that it is truly representative of its population. It would be unrealistic to expect an accurate estimate of rural attitudes from a sample of households on the fringe of an urban region, or to expect a reasonable estimate of grassland plant characteristics from a sample of plants taken from a study area covered with trees.

Not all sampling methods are equally suitable for choosing samples that are representative of their population. Certain methods are more appropriate to physical experiments in open areas or along rivers, while other methods are preferable for surveying residents in urban neighborhoods. The choice of a sampling scheme should be made so as to avoid bias. Three methods of sampling are commonly used: random sampling, systematic sampling, and stratified sampling. Any and all of these methods may involve sampling either points (places), lines (traverses), or areas (spaces).

Point and Area Samples

A distinction is made in geographic experiments between individual sampling units and areal sampling units, termed **point** and *area* **samples.** A *sampling unit* is the smallest subdivision of the referent population, whether it be points or areas. When the sample is drawn from fixed and discrete locations, we consider these as units in a point-sampling

---

[5]For more comprehensive coverage, see books by Goddard, Miller, and Sheskin, found in the list of references.

scheme. Some variables effectively measured at points include soil acidity (pH), moisture content, or grain size.

For certain experiments it becomes necessary to divide a study area into smaller sized areas. These small areas may be called *subareas*, and they may or may not be uniform in size. In biogeography, vegetation cover may be measured over an area. In urban research, subareas may be called blocks, enumeration districts, or census tracts. Census tracts are predetermined by the census bureau. By using these already bounded subareas, researchers are relieved of the trouble of defining their own subdivision boundaries. In addition, census tract boundaries remain fixed, which allows comparison of population characteristics from one census to the next. In other places and for certain kinds of research it is necessary to superimpose a cross-ruled grid over the area to be sampled. In these cases, it is first necessary to decide on the most effective size and shape of the sampling areas.

Random Sampling

A **random sample** is one that is drawn in such a way that every member of the population has an equal chance of being included. A random sample also implies that every possible sample of a given size, n, has an equal chance of being our *selected* sample drawn from the population. Picking place-names printed on small slips of paper from a box or using numbers that come up on the spin of a roulette wheel are random sampling methods.

The selection of one individual sampling unit must in no way influence the selection of the next. This quality is known as *independence*. This restriction is not always possible to meet. Meteorological observations, for instance, are usually not independent of preceding conditions; a high or low pressure system needs time to develop or disappear. Also, the daily amount of rainfall may not have any relationship to the amount in a preceding year, but it frequently is influenced by the conditions the day before.

Formally stated, the probabilities of inclusion in the sample must be *equal* and *independent* of each other. Thus, if the study area contains 300 households and the goal is to pick a random sample of 50, every household should have one chance in six (a probability of $50 \div 300 = .17$) of being selected. The choice of a particular sampling unit (individual, household, area, etc.) must in no way be influenced by the researcher at any stage in the sampling process.[6]

If every sampling unit in the study population can be assigned a unique number, then a table of random numbers like that of Appendix Table 1 may be used to select sampling units. The numbers in the table have been selected using a computer that is programmed to draw from a large pool (population) of numbers in such a way that each of the numbers has an equal probability of occurrence. Thus we should expect each of the values, say 00 to 99, to occur approximately the same number of times in the table as a

---

[6] Examples of sampling bias are rife. They apply mainly to surveys, in which the respondents are less likely to represent random samples of the populations for which generalizations are to be made.

whole. Furthermore, the numbers occur in random order, whether they are read down a column, across a row, or on a diagonal.

The digits may be used singly to pick a sample from a population of 10 items (numbered 0 to 9), in pairs to pick a sample from a population of 100 (numbered 0 to 99), and so on. One complication is that sample sizes are seldom of such convenient numbers as 10 or 100; more likely is a sample size of 78 or 562. If the sample size is 78, it would be necessary to use the random digits in pairs, rejecting any numbers drawn from the table that exceeded 77 (00 is a valid number which may be used as the first or last sampling unit).

During the process of selecting a sample using the table of random numbers the same sampling unit may be drawn more than once, especially if one- or two-digit numbers are being drawn. This is explained by what is termed the *replacement rule*. Recall that for a number to be truly random, it must have had the same probability for selection from a pool as any other number. Therefore, when the computer selects a number from a pool of all numbers, that number is then returned to the pool. Assume that the pool contains values from 0 to 999. As the first number is about to be drawn, each number in the pool has a probability of 1 ÷ 1000 (.001) of being selected. If the first number drawn is not replaced, each of the remaining numbers has a probability of 1 ÷ 999; after the fifth number is drawn (and not replaced), the probability increases to 1 ÷ 995, and so on. Such an alteration of the probabilities clearly violates the equal probability criterion of a random sample. When a sampling unit is selected more than once, should the data for that unit be included in the sample? In practice this is seldom done, as the bias introduced into the results by not replacing is unlikely to be of significance for large samples.

There are several ways to draw a random point sample. For example, suppose you are collecting data related to some agricultural practice and your sampling units are farms. It is necessary to compile a list of all farms in the area and then sample at random from that list. Alternatively, a base map which contains the locations of all farm residences might be available. In any case, each farm must first be assigned its own unique number. Suppose we wish to draw a sample of 30 farm households at random from a study area population of 200 farms. The 200 farms are all assigned a unique three-digit number (say 000–199). We then consult our table of random numbers. As each number is taken from the table, the farm with the corresponding number is selected for the study. If the random number is greater than 199, it is rejected. In the unlikely event that the same number is drawn more than once, the duplicate is rejected. This procedure is followed until 30 separate farms have been identified.

If the distribution over the study area is fairly even, another technique, which will eliminate the need to number every sampling unit, may be used. Obtain a base map of the area and superimpose a grid thereon. Number each grid line from zero to the maximum number of lines in both directions (Figure 2.13a). Suppose that a grid of twenty-five to thirty lines in both directions is found to yield a sufficient number of vertices. The table of random numbers is consulted twice for each sample, the first time to select a number that would fall between 00 and 30 for the Y coordinate, and again to select a random number between 00 and 25 for the X coordinate. Find the intersection of these two lines and select the sampling unit (farm, home, street corner, or whatever) nearest that intersection on the grid. This latter technique may be used where samples of any phenomenon

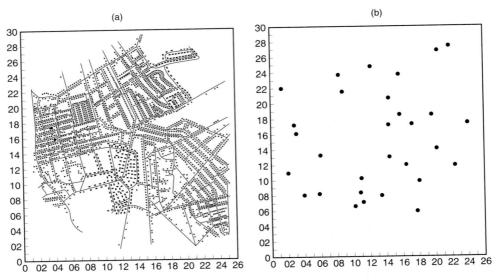

Figure 2.13 Using (a) a Cartesian Grid to Select (b) a Random Point Sample

must be collected at random locations throughout the study area. The resulting sampling pattern might then appear something like that of Figure 2.13b.

Suppose a biogeographer determines that vegetation must be sampled at random within a study area. A decision is made to select plant samples within 20-square-meter quadrats. One could first superimpose a 20-square-meter grid over a base map of the area. As was true for the point sampling procedure, each quadrat must have an equal chance of being selected. Each quadrat is assigned a unique number. The random number table is then consulted, and the sample quadrats are selected in the same way as described for points. How vegetation is then selected from within the sample quadrats depends on the objectives of the research. Applications of point and quadrat sample analysis are found in Chapter 9.

Systematic Sampling

A **systematic sample** is one that is selected with numerical or spatial regularity in the sampling technique. Depending on the density of the population to be sampled, every tenth, fiftieth, or one hundredth unit in the study area may be selected. If the research is based in a city, using a base map like that of Figure 2.13a, a decision would first be made on how to cover every street in the study area systematically. Having made that decision, the researcher would proceed to traverse the study area and select every tenth, fiftieth, (or other interval) house, noting the address or location of each, until the required number of sampling units is obtained. This procedure would appear to produce an unbiased sample in that it would involve even coverage of the study area and avoid the clustering of sample observations that can occur with the random-point sample technique. It is true that no house would be picked more than once with this procedure, but every household in the study area would not have the same chance of being selected. Every tenth, fiftieth (or other interval) house would have a probability of 1.0 of selection, while all other houses would have a zero probability of selection. In other words, if thirty houses are to be selected, once the first house has been selected, all but twenty-nine of the houses on the map have lost all chance of representation in the sample. This bias might not be serious if we know that the population within the study area is homogeneous in terms of those aspects that addressed the objectives of the research. It is usually quicker and easier to obtain a systematic sample than a random one.

Sampling on a Transect

When there is no reference system apparent on the surface (an open field, for example) as there would be in a city (with streets and addresses), the sampling strategy must be revised. If soil samples are to be collected in the field, it may be more convenient to extend a *transect* across the study area. A *transect*, or *traverse*, is a line established in the study area along which samples will be collected. Following a transect and stopping every so many meters to collect samples could be easier than preselecting points on a base map, then discovering that it is very difficult to be sure that you have indeed found the exact spot on the ground that corresponds to the point on the map. The transect is generally a straight line, established in such a way as to insure that it will yield a representative sample for the population of whatever attribute is being examined within the study area. The line may be curvilinear as long as this does not introduce bias into the sample. If water samples were being collected from a curving stream, sampling every 10 meters along the stream might insure that the data are unbiased. Another use for the traverse could be applied to Figure 2.13a. By selecting every tenth household along certain streets (transects) that crossed the study area, one could obtain a sample that is representative of the population attribute of interest. The directional orientation of traverses in a systematic sampling scheme may be selected at random if the study area is homogeneous in terms of the attribute of interest.

Stratified Samples

Random sampling methods assume that the composition of the population, either human or physical, is not known and that a representative sample will be best approximated by a strictly random selection. In some cases the more or less exact composition of the population with respect to some characteristic is known before a sampling scheme is designed.

A **stratified sample** is usually a *proportionate* sample. That is, it is one in which every sampling unit is selected according to the proportions of a particular known characteristic. For example, one may know the ratio of minority population to total population within a study area. In such cases the chances of selecting a representative sample may be increased by drawing samples of both the minority and nonminority groups proportionate in size to their numbers in the population. In another example, suppose it is obvious from a topographic map of our study area that 40 percent of the area contains north-facing slopes, while 30 percent of the area has slopes that face some other direction, and the remaining 30 percent contains flat areas. In order to assure equal slope representation from the entire study area, it will be necessary to sample south-facing slopes according to their proportionate share within the study area. Thus, the sample must be designed such that 40 percent of the samples taken are from north-facing slopes. In other words, the sample has been *stratified* according to the slopes within the study area. Samples in each slope zone are collected in the same proportion as they are represented throughout the study area.

Cluster Sampling

A variant on areal random sampling is what is called a **cluster sample.** This method can be applied when it is more convenient and cost effective to select neighborhoods or quadrats at random from a very large study area; examples that come to mind are grasslands, forests, and metropolitan suburbs. It is not necessary that each cluster be internally homogeneous. The disadvantage to cluster samples is their potential for bias. If the sampled clusters misrepresent the study area population, the sampling error is likely to result in misleading statistics. To assure a random selection of quadrats or neighborhoods, we might first subdivide the study area into smaller units. Unique numbers are assigned to each subarea, then a random number table is used to select a sample of these subareas. Samples may then be drawn at random from within each cluster, or all of the available cases in the cluster may be included in the sample. This approach assumes that the study area is homogeneous with respect to whatever characteristic is being investigated. In Chapter 6, we will demonstrate techniques for testing homogeneity.

Sampling can become a creative, complicated exercise. Imagine, for instance, a *stratified cluster sample.* Such a scheme might be useful when surveying some aspect of housing in urban neighborhoods. Since these neighborhoods vary in most respects (income, race, age of housing, etc.), the sampling scheme must be stratified along the lines of income, race, age of housing, and so forth, depending on the characteristics that are crucial to the research question. Clusters of houses may be picked in random fashion from each stratum. There is usually more than one sampling scheme that may be used for any geographic research experiment and still insure a minimizing of bias.

It is not always possible or practical to draw a strictly random sample. Often the population will be very large, and it will be virtually impossible to enumerate every member. There are alternatives to this procedure, such as overlaying the study area with a grid of points and selecting from those points at random. While practical and cost considerations often obviate the drawing of a completely random sample, a researcher should always strive for a sampling scheme that is in the *spirit* of random sampling; that is, drawing a sample that is believed to be representative of the parent population to which one wants to generalize.

How large a sample would you require to make a statement about the population within your study area, and what form does that statement take? These are separate but related questions that require us to first establish some groundwork and understand some further terminology. The procedures for making formal statements and specifying sample sizes are found in Chapter 6.

According to Kerlinger (1970), we do not know *why* the principle of randomization works. Somehow random samples do give dependable, representative samples. If we were omniscient beings, knowing all the contributing causes of events, then there would be no randomness. Therefore, randomness is a way of coping with ignorance. Researchers use this ignorance and turn it to knowledge, as you will discover in the chapters ahead.

KEY TERMS

absolute value 31	distance 27	ratio 21, 22
accuracy 14	interval 21	reliability 14
area 25	line 25	scalar 26
area sample 41	metric 15, 21	secondary 15
cluster sample 46	nominal 17	shape 35
compactness 35	ordinal 17	stratified sample 45
concept 16	point 24	surface 26
constant 15	point sample 41	surrogate 16
continuous 15	precision 14	systematic sample 44
density 38	primary 15	validity 16
direction 39	random sample 42	variable 15
discrete 15	rate 23	

REFERENCES

Abler, R., Adams, J. S., and Gould, P. (1971). *Spatial Organization.* Englewood Cliffs, NJ: Prentice-Hall.

Ahrens, C. D. (1982). *Meteorology Today.* St. Paul, MN: West Publishing Co.

Blalock, H. M. (1982). *Conceptualization and Measurement in the Social Sciences.* Beverly Hills, CA: Sage Publications.

Centers for Disease Control. (1991). State Tobacco Prevention and Control Activities. *Morbidity and Mortality Weekly Report.* Vol. 40, Number RR-11. Atlanta, GA: Centers for Disease Control.

Ebdon, D. (1985). *Statistics in Geography,* 2d ed. New York: Basil Blackwell.

Goddard, S. (1983). *A Guide to Information Sources in the Geographical Sciences.* New York: Barnes and Noble Books.

Goudie, A. (1990). *The Human Impact on the Natural Environment,* 3rd ed. Cambridge, MA: The MIT Press.

Hägerstrand, T. (1957). Migration and Area, in *Migration in Sweden.* Lund Studies in Geography, Series C, No. 1. Lund: Gleerup.

Haggett, P. (1983). *Geography: A Modern Synthesis,* 3rd ed. New York: Harper and Row.

Keables, Michael J. (1988). Spatial Associations of Midtropospheric Circulation and Upper Mississippi River Basin Hydrology, *Annals of the Association of American Geographers* 78: 74–92.

Kerlinger, Fred N. (1970). *Foundations of Behavioral Research.* New York: Holt, Rinehart and Winston.

Kuhn, G. G., and Shepard, F. P. (1983). Beach Processes and Sea Cliff Erosion in San Diego County, California. In *CRC Handbook of Coastal Processes and Erosion,* ed. Paul D. Komar, 267–287. Boca Raton, FL: CRC Press.

Lewis, Peter. (1977). *Maps and Statistics.* New York: Halsted Press.

Liu, Ben-Chieh. (1975). *Quality of Life Indicators in the U.S. Metropolitan Areas, 1970 (Summary).* Kansas City, MO: Midwest Research Institute.

Miller, D. C. (1983). *Handbook of Research Design and Social Measurement,* 4th ed. New York: Longman.

Nachmias, D., and Nachmias, C. (1987). *Research Methods in the Social Sciences.* New York: St. Martin's Press.

Nummedal, D. (1983). Barrier Islands. In *CRC Handbook of Coastal Processes and Erosion,* edited by Paul D. Komar, 77–122. Boca Raton, FL: CRC Press.

Population Reference Bureau, Inc. (1991). *The United States Population Data Sheet,* 9th ed. Washington, DC: Population Reference Bureau.

Rubenstein, C. (1982). Regional States of Mind, *Psychology Today,* February, pp. 22–30.

Sheskin, I. M. (1985). *Survey Research for Geographers.* Washington, DC: Association of American Geographers, Resource Publication.

Star, J., and Estes, J. (1990). *Geographic Information Systems.* Englewood Cliffs, NJ: Prentice Hall.

Taylor, Peter J. (1977). *Quantitative Methods in Geography.* Boston: Houghton Mifflin.

Unwin, David. (1981). *Introductory Spatial Analysis.* London: Methuen.

U. S. Bureau of the Census. (1976). *Current Population Reports: Selected Characteristics of Travel to Work in 20 Metropolitan Areas.* Washington, DC: U. S. Government Printing Office.

U. S. Bureau of the Census. (1985). *1982 Census of Agriculture,* Volume 2, *Graphic Summary.* Washington, DC: U. S. Government Printing Office.

U. S. Bureau of the Census. (1990). *Statistical Abstract of the United States, 1990.* Washington, DC: U. S. Government Printing Office.

U. S. Bureau of the Census. (1991). *Statistical Abstract of the United States, 1991.* Washington, DC: U. S. Government Printing Office.

World Bank. (1988). *World Development Report, 1988.* New York: Oxford University Press.

World Resources Institute. (1992). *World Resources 1992–93.* New York: Oxford University Press.

EXERCISES

1. Identify the measurement scales used to produce the numerical data in the tables and map that follow. Specify: **N** = nominal, **O** = ordinal, **I** = interval, and **R** = ratio for each of the examples (a) through (g) on the next two pages..

(a) Change in Per Capita Commercial Energy Consumption 1950–1989 (percent change per year)

Country	1950–60	1960–70	1970–80	1980–90
Brazil	5.4	2.6	5.8	0.7
Chile	1.6	5.1	−1.2	2.0
Mexico	4.1	3.8	4.8	0.4
Indonesia	9.3	−0.5	6.8	2.6
Malaysia	×	×	6.2	3.6
Thailand	11.3	11.9	7.2	6.5
World Total	3.2	3.0	0.9	0.4

Source: World Resources Institute (1992), p. 51.

(b) Gas and Electric Energy Consumption in Central Europe, 1989

Country	Gas	Electricity
Albania	16	11
Bulgaria	238	36
Czechoslovakia	403	114
Germany (Dem Rep)	303	53
Hungary	427	90
Poland	397	20
Romania	1,346	74
Yugoslavia	239	113
U.S.S.R.	22,970	1,429

Source: World Resources Institute (1992), p. 61.

(c) Average Monthly Temperatures for Selected North American Cities

J	F	M	A	M	J	J	A	S	O	N	D
				Albuquerque, NM							
35	40	46	56	65	75	78	76	70	58	44	37
				Boston, MA							
30	30	38	48	59	68	74	72	65	55	45	33
				Calgary, CANADA							
13	17	26	40	50	56	62	60	51	42	28	19
				Denver, CO							
28	32	36	46	56	66	73	72	63	51	38	32
				Miami, FL							
67	68	70	74	78	81	82	82	81	78	72	68

Source: Ahrens (1982), pp. 482–483.

(d) Land Area (square miles) of Selected U. S. Cities

City	Land Area
Phoenix, AZ	419.9
Seattle, WA	83.9
Virginia Beach, VA	248.3
Washington, DC	61.4

Source: U. S. Bureau of the Census (1991), p. 36.

(e) The Ten Most Populous U. S. Metropolitan Areas, 1980 and 1990

Metropolitan Area	1980 Rank	1990 Rank
New York	1	1
Los Angeles	2	2
Chicago	3	3
San Francisco	5	4
Philadelphia	4	5
Detroit	6	6
Boston	7	7
Washington	8	8
Dallas	10	9
Houston	9	10

Source: U. S. Bureau of the Census (1991), pp. 29–31.

(f) Per Capita Gross Domestic Product of Selected Countries (Current International Dollars), 1987

Country	Dollars
All less developed countries	2,019
Low income countries	650
Middle income countries	2,242
Highly indebted countries	3,082
Industrial market economies	14,264

Source: World Bank (1988), p. 188.

(g) Total Precipitation (inches), October 1992

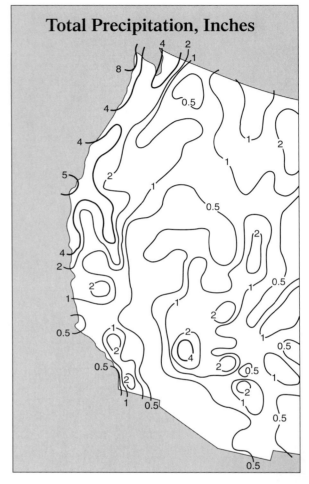

Source: U. S. Geological Survey (1992), National Water Conditions. Washington, D.C. p. 25.

2. In Appendix B-5 are found selected variables for the fifty states of the United States (and the District of Columbia) for 1990. Compute the (primary and secondary) student rate per 100,000 population for any ten states and write your results in descending rank order.

3. Using the Euclidean distance equation, compute the distances (in miles) between the following subarea control points on the hypothetical study area map on page 51. Estimate coordinates on the X and Y scales to the nearest tenth of a mile. Note that one unit on the X or Y axis equals one mile.

   $1 \rightarrow 4$ _____ $8 \rightarrow 18$ _____ $4 \rightarrow 11$ _____ $9 \rightarrow 3$ _____

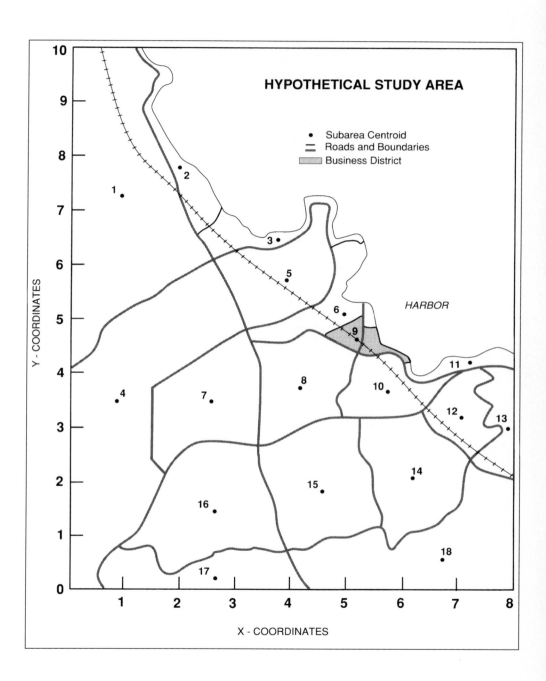

4. Using the method described in the text for computing the area of a polygon and on this map of Washington, D.C., compute the area (in square miles) of the polygon bounded by Massachusetts Avenue, 14th Street, Constitution Avenue, 23rd Street, and New Hampshire Avenue. Assume that 1 inch on the map equals 1 mile.

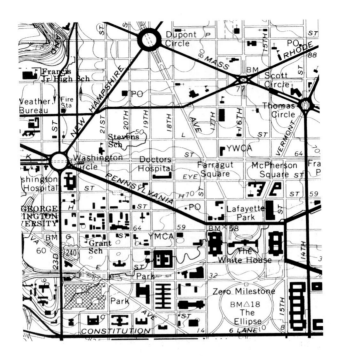

5. Using the method described in the text for computing shape indices, compute the shape index for any area on the hypothetical study area map.

6. What geographic primitives are demonstrated in Figures 2.3a and 2.5b? What distinguishes these two maps?

7. What is the purpose of a map, such as Figure 2.7, constructed on a logarithmic scale?

Measurement and Sampling 53

8. Using the contour map below, use one of the following sampling methods to show the locations of either thirty sampling points or ten sampling subareas.
   (a) random sample (b) systematic sample (c) cluster sample

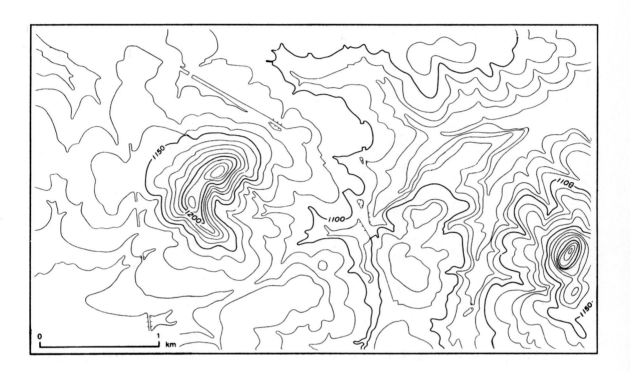

9. Referring to the map on page 54 of the Long Green–Manor View area, use a random stratified point sample scheme to select twenty homes for a survey. The two strata are: (1) the area west of the main north-south road, and (2) the area east of the main north-south road. Sixty percent of your sample homes should be selected from the western stratum. Each home in the study area must have an equal probability of being selected.

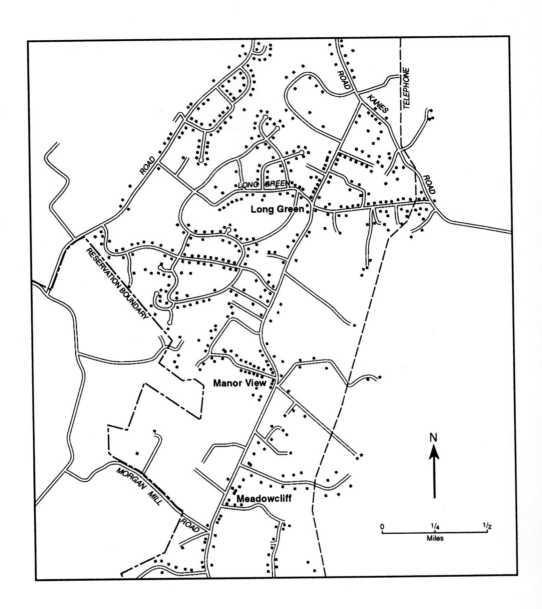

Graphing and Mapping Data   3

To study landscape systems, whether humans are seen to be involved or not, it is helpful to reduce the reality of the globe to a geometry of points, lines, areas, and volumes. Points and lines are employed to project maps, and maps in turn can be used to portray population centers, links connecting centers, streams, or stream networks. Phenomena such as volcanoes, lakes, human settlements, coal mines, smokestacks spewing forth fumes, or whatever form patterns of points on the earth's surface. Rivers, ocean currents, oil pipelines, roads, and satellite beams form networks of movements and interaction between points. Air masses, pressure cells, a tropical savanna, and a city neighborhood are examples of areas, often distinguished by homogeneity or functional cohesion. Maps of these spatial distributions are called *thematic maps*.

Like the road map, a thematic map orients the reader, but it also displays phenomena that illustrate a concept or demonstrate a process. An accurate description of where things are is a prerequisite to explaining their presence. We may describe qualitatively, that is without numbers, or quantitatively. You may have surmised from Chapter 2 that the more precise our descriptions are, the better we may understand the processes that yield the points, lines, and areas that we observe.

Chapter 2 presented techniques for measuring phenomena and suggested procedures for collecting sample data. Once a study has been proposed and data collected, it is often useful to put these data into graphic form, or to construct maps that display them in their spatial context.

GRAPHIC PRESENTATION OF DATA

An unorganized list of numbers, termed **raw data,** is about as meaningful as an anagram puzzle on the Sunday comic page. It is very difficult to detect any patterns among the numbers. A simple tabular organization or graph of the data brings some order to the numbers and may even suggest some relationships. However, tables, maps, and graphs can be misleading, intentionally or otherwise. Something that appears obvious in a table, graph, or map may be an artifact of the construction of the graphic or its legend.

Suitably constructed tables, like some of those introduced in Chapter 2, can be used as a basis for further investigation because many statistical tests may be based on them directly. For instance, the regional rankings of 1986 per capita government expenditures (Table 2.2) may be compared with regional government expenditures several years before or after 1986 to verify whether or not those expenditures have increased or decreased significantly. Graphs and maps may be even more useful and effective for quickly presenting quantitative information and suggesting statistical analysis. Graphs and maps, however, require a level of skill above and beyond that of table construction. Not only do you need to conceptualize the graphic presentation of data, but you also need some rudimentary training in the use of drafting equipment, such as T-squares, pens, and scribers, or training and experience with computer mapping software.[1]

Computer graphics programs, such as those embodied in some word processing programs and to a lesser extent in computer-based statistical programs like SAS and SPSS, speed the process of producing tables, graphs, and even maps. Assuming that one accurately keys into the computer all data and commands, neat and accurate results will be forthcoming, although they may not always be of publishable quality. For those readers who are acquainted or wish to become conversant with SAS and SPSS, we have included program code statements and printouts of results for most of the procedures that are described in this text. This material is found in Appendix C.

Distance Decay

Distance decay, in a cultural sense, is based on the simple notion that people tend to minimize travel distance for most necessary shopping trips and some routine recreation journeys, such as to local parks or health clubs. For example, if one needs to make a trip to purchase a common food item, logic dictates that it is unnecessary to travel beyond the nearest shop that sells that item. Repeatedly, empirical research has verified that, in *aggregate,* the proportion of household trips made to purchase common food items and other necessities declines exponentially with increasing distance to the available sales outlets. A graphic device that neatly summarizes this economic behavior is the distance decay diagram.

Figure 3.1 illustrates a distance decay curve for trips from hypothetical neighborhoods to a hypothetical shopping mall. As this figure shows, graphs of this type are con-

---

[1]Readers may wish to know more about the design and mechanics of construction of tables, graphs, and maps than we are able to present here. For that level of instruction, we refer those readers to one of the standard cartography textbooks.

Graphing and Mapping Data 57

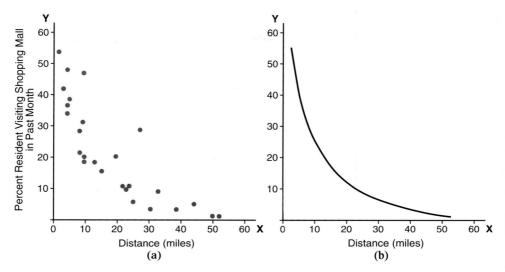

Figure 3.1 **(a) Proportion of Neighborhoods' Residents Visiting Shopping Mall at Given Travel Distances, and (b) Generalized Distance Decay Relationship**

structed by plotting points that represent the intersection of a subarea's proportion of persons that were attracted to the destination and the travel time or distance from the center of the subarea to the destination. The same curve might well describe those neighborhoods' aggregate trip behavior to a state park, or the drop in frequency of a certain species of plant as one moves away from a favorable habitat for that plant.[2] Box 3.1 illustrates the procedure for plotting points on a distance-decay graph.

In Chapter 2, vertical and horizontal axes representing two-dimensional space were introduced, and locations were mapped as points (recall Box 2.1). When we wish to graph the functional relationship between two phenomena, in this case attraction and distance, the vertical axis is designated as the variable that we are *attempting to explain*. The variable represented by the vertical axis is conventionally called the *dependent variable*, and is usually assigned the symbol Y. The horizontal axis represents the *explanatory* data, formally termed the *independent variable*, and conventionally labeled X.

An example of a relational graph with dependent and independent variables is shown in Figure 3.2. The dependent variable is altitude, and the independent variable is temperature. Altitude is a ratio variable, while temperature is measured on the interval scale. A simple graphic demonstrates that, in general, as altitude increases, the temperature of air decreases at a standard rate. This method of analysis will be further demonstrated in Chapter 8.

---

[2]This simple concept is the basis for the kind of complex spatial interaction models that are used to explain many of the processes in large-scale systems, such as the one discussed in Chapter 1. See Abler, Adams, and Gould (1971) and Haynes and Fotheringham (1984).

> **BOX 3.1**
>
> **Procedure for Plotting Points on a Distance-Decay Graph**
>
> Suppose data on use of state parks are collected by a state planning agency by means of a sample survey. At parks throughout the state, nearly 1,300 visitors are asked where they live and which parks they have visited over the past month. Home locations are generalized to minor (sub-county) census districts. A "control point" is established in the center of population of each census district. The distances from census-district control points to the parks are measured on a map and recorded in miles. The planners extrapolate their sample to the populations of each census district in the state. The proportion of all visitors to one park over the past month from each civil district is then calculated and the graph shown below is constructed. Each dot represents the proportion of visitors in a census district to the referrent park and the distance from the district control point to the park. For instance, one park-to-district distance is 8 miles; the proportion of the district's population that visited this park was 22 percent. Another distance-proportion pair was 25 miles and 6 percent. After plotting dots for all census districts, a smooth line is "fitted" visually to the scatter of points that appears to represent the average expected differences in visitor proportions at various distances from the park.
>
>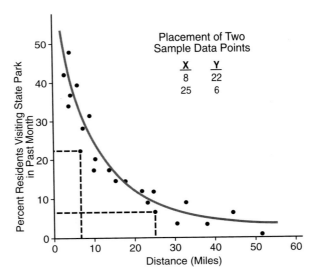

Classification

The first and most elementary step in any measurement procedure is to define the characteristic of interest in our research. Having done that, we would like to know as much about that characteristic as possible. Suppose we are interested in the landward migration rates of barrier islands. Are migration rates all about the same, or is there a wide variation in those rates? Are most of the migration rates close to an average rate, or do these rates cluster around one or two extreme rates? A sample of one or two migration

Figure 3.2 The Environmental Lapse Rate

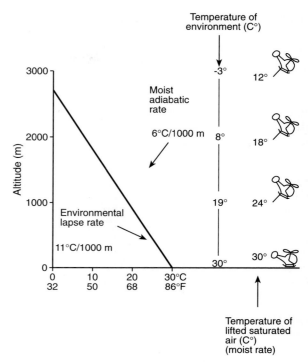

rates is not only unlikely to reveal anything about the population of all barrier islands, it is also quite probable that we will obtain a distorted image from so small a sample. Therefore, samples that are of adequate size are collected to demonstrate a process, and, as we will soon see, it is then helpful to classify the objects of interest to us. We may put objects into pigeonholes. This procedure usually gives us a general picture of the distribution of the characteristic in which we are interested. Classification aids in the formation of hypotheses and guides further investigations.

There are several graphic tools available to simplify the task of summarizing a set of data. Consider U. S. government economic data on personal income per capita by state for 1987 (Table 3.1). In their original published form, these income data were in alphabetical order by state, not in the order that they appear in Table 3.1. There are several reasons to reorder the data, all preparatory to further statistical analysis. In a data set numbering more than about twenty items, the simple task of locating the lowest or highest value can become tedious. After sorting the income data, the maximum, minimum, and middle values are readily identified. If the data are to be grouped into classes, it is a much easier task to count frequencies within each class.

It is conventional in statistics to indicate the total frequency of items in a *population* with the letter N. It is also standard practice to use the symbol n to specify the total of items in a *sample*. In the per capita income example of Table 3.1, the entire population of the nation has been aggregated into fifty states and the District of Columbia. There are ($N =$) 51 *observations*. Most of the values cluster in the range from $11,000 to about

Table 3.1   Per Capita Personal Income (in Current Dollars), United States, 1987

State	Income	State	Income	State	Income
Mississippi	10,301	Texas	13,840	Washington	15,634
West Virginia	11,013	Oregon	13,906	Colorado	15,680
Arkansas	11,421	Indiana	13,987	Rhode Island	15,683
Louisiana	11,506	Maine	13,996	Minnesota	15,789
Utah	11,530	Iowa	14,028	Delaware	16,305
Idaho	11,797	Nebraska	14,100	Nevada	16,359
New Mexico	11,889	Vermont	14,267	Illinois	16,394
Kentucky	11,996	Arizona	14,322	Virginia	16,539
Alabama	12,039	Georgia	14,387	California	17,770
South Carolina	12,078	Ohio	14,575	New York	17,943
Montana	12,304	Missouri	14,630	New Hampshire	18,083
South Dakota	12,414	Wisconsin	14,674	Maryland	18,217
Oklahoma	12,607	Kansas	15,089	Alaska	18,461
North Dakota	12,825	Pennsylvania	15,198	Massachusetts	19,131
Wyoming	12,836	Michigan	15,558	District of Columbia	19,543
Tennessee	12,977	Hawaii	15,569	New Jersey	20,277
North Carolina	13,353	Florida	15,594	Connecticut	21,258

Source: U. S. Bureau of the Census (1990), p. 437.

$15,000. To visualize the structure of the data set, it helps to group the data into several *classes*. But, where should the class boundaries be placed? This is a decision that must be made every time data are summarized for tabular or graphic presentation.

THE FREQUENCY DISTRIBUTION

Figure 3.3 shows one form of a **frequency distribution**, called a **histogram**. A histogram is a vertical or horizontal bar chart on which each bar represents a certain value range of the distribution, and the length of the bar shows the frequency of values that fall within that value range. The purpose of a histogram is to show how observations are *distributed* between the minimum and maximum values in the data set. When data are divided into several classes, or categories, a notational convention in statistics is to designate the number of classes with the letter k. In order to construct a histogram, we must determine the width of each class interval.

Most histograms feature a **class interval** (minimum to maximum value of the characteristic being represented *within* the class) that is uniform from one interval to the next. It can be misleading to have a different numerical range in each interval, unless it happens to be the first or last interval, which often contain only one or two extreme values. With highly variable data that have small frequencies over a considerable part of the range, using intervals of unequal widths in different parts of the distribution is acceptable.

Suppose the amount of atmospheric carbon monoxide is measured each day at some urban location. Over a year, one might expect that there would be days (weekends

Figure 3.3 **Histogram of Per Capita Personal Income in the United States, 1987**

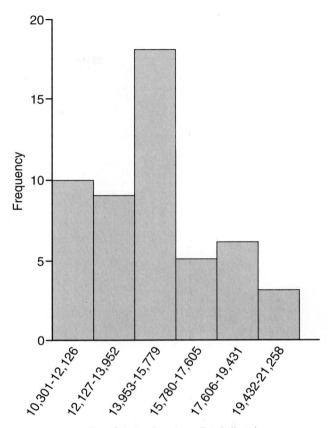

and holidays) on which we would find low levels of the pollutant and days that, due perhaps to unusual atmospheric conditions, we would find high levels, but more often we would expect the levels to hover about the average value for all of the days in the year. In the situation just described, a frequency histogram in which each class contained an equal magnitude of carbon monoxide (0.5 to 1.5 micrograms, 1.51 to 2.5 micrograms, 2.51 to 3.5 micrograms, etc.) would show that frequencies within class intervals below the average are about the same as frequencies within class intervals above the average; in other words, it would be a nearly symmetrical distribution (Figure 3.4a). However, if the width of the intervals of the bars were structured so that each bar contained about the same frequency as all other bars, the result would be a *rectangular distribution* (Figure 3.4b). Reading this graph uncritically could lead to the conclusion that it was just as likely for any daily carbon monoxide measurement to be high as for that measurement to be low!

Suppose we chose an interval of $1,000 for the per capita income data shown in Table 3.1. There would be two observations in the first class ($10,301 to $11,300), eight observations in the second ($11,301 to $12,300), and so forth. The $1,000 class interval produces twelve classes. The number of classes into which a frequency distribution

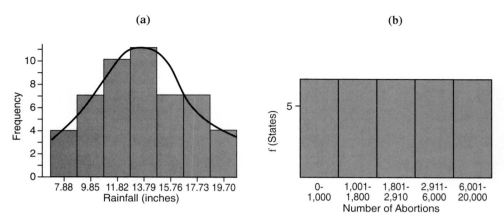

Figure 3.4 Histograms of (a) Frequencies of Years with Given Annual Rainfall Classes, Denver, 1921–1970, and (b) Frequencies of States with Legal Abortions among Married Women, Selected United States, 1989

should be divided is a function of the number of observations and the size of the difference between the minimum and maximum values of the characteristic being measured. Generally, as the number of observations increases, more intervals, each with smaller ranges, may be used. Box 3.2 demonstrates a procedure to create six class intervals of the per capita income data, which determines the frequency for each class.

The vertical bar in Figure 3.3 shows the frequency of observations within six class intervals. A glance at this distribution readily shows the most frequent income classes. Clearly, there are more states in the lower income levels. Figure 3.4a is a histogram of annual precipitation data at Denver, Colorado. As in Figure 3.3, the horizontal axis in Figure 3.4a represents the phenomenon (in this case precipitation) with class intervals equal to just under 2 inches.

If the variable we are describing happens to be measured on a qualitative, or nominal scale, for example, soil type, then a classification already exists in the form of established guidelines for soil identification and naming. But, when the characteristic is measured on the metric scale, class intervals must be determined arbitrarily as was done with the income and precipitation examples.

Another way to display class interval data is in what is known as a **frequency polygon**. Figure 3.5 demonstrates the polygon technique with frequencies of infant death rates in twenty-five metropolitan areas (with a uniform class interval of 1). The horizontal axis is broken between 0 and 6 simply to show that no observations occur between those values. The graph is interpreted as follows: there is one observation with 6, none with 7, four with 8, and so on.

From the histogram or the frequency polygon, a frequency distribution can be created with a smooth curve that displays the general shape of the data. The curve follows a path from the top center of one bar of a histogram to the next. The bars may then be eliminated and the intervals on the horizontal axis may be replaced with the middle value of

Graphing and Mapping Data

BOX 3.2

Procedure for Establishing Six Uniform Class Intervals for a Histogram of Personal Income Per Capita, United States, 1987

Minimum income value = $10,301

Maximum income value = $21,258

Data range = $10,957

$$\frac{\text{Range}}{6} = \$1{,}826$$

Interval number 1: $10,301 to (+ 1,826) $12,127

Interval number 2: $12,128 to $13,953

Interval number 3: $13,954 to $15,779

Interval number 4: $15,780 to $17,605

Interval number 5: $17,606 to $19,431

Interval number 6: $19,432 to $21,258

Frequency of states in each interval:

Interval	Frequency
10,301–12,127	10
12,128–13,953	9
13,954–15,779	18
15,780–17,605	5
17,606–19,431	6
19,432–21,258	3

---

its interval. In Figure 3.6, for instance, the frequency of cities in the interval 150–170 thousand population is about ninety. From this graph the population size around which cities tend to cluster may be readily identified, and it is apparent that the distribution is almost symmetrical with a few cities of over 200,000 causing a slight extension of the upper end of the curve.

Cumulative Frequencies, Proportions, and Ogive

A frequency distribution shows the number of observations in each class interval. The proportion of cases in each class interval may be obtained by dividing the frequency of cases in each interval by the total number of observations. The frequency histogram of per capita income in the United States was shown in Figure 3.3. Suppose it is desirable to show the proportion of observations with per capita income values less than or equal to

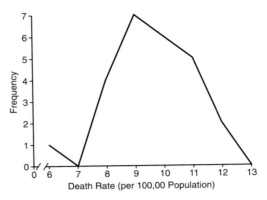

Figure 3.5 A Frequency Polygon of Infant Death Rates in a Sample of 25 U.S. Metropolitan Areas, 1987

Data Source: U. S. Bureau of the Census (1991), Table A.

$15,779. This is accomplished by portraying the data in *cumulative* form. If the frequency data from Box 3.1 were accumulated by class interval, the result would be Table 3.2. The cumulative frequency column is calculated by summing the frequencies of all intervals below and including the current interval. At the end of the first interval, for an income value of $12,127, there are ten observations. There are nineteen observations including values up to $13,953, ten in the first interval and nine in the second interval, and so forth. The last entry in the cumulative frequency column is always equal to the total number of observations, or the sample size, n.

The righthand column shows the cumulative proportion of observations by class interval. For the first interval, the cumulative percentage of observations less than or equal to $12,127 is calculated as the cumulative frequency divided by the total number of observations, multiplied by 100. The result is $10 \div 51 \times 100 = 19.6$ percent; for the second interval, $19 \div 51 \times 100 = 37.3$ percent; and so forth. The answer to our earlier question, what was the proportion of the observations with per capita income values less than or equal to $15,779, is 75.5 percent.

A graphic representation of Table 3.2 is shown as Figure 3.7. The cumulative histogram is constructed in the same way as a simple frequency histogram except that the height of each bar is proportional to the *cumulative* frequency at the *end* of that interval. A series of line segments is then drawn as follows: connect the lower bound of the first

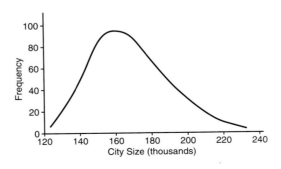

Figure 3.6 A Frequency Distribution of Cities Classed According to Population Size

Table 3.2 Cumulative Frequencies of Personal Income Per Capita, United States, 1987

Class	Interval	Frequency	Cumulative Frequency	Cumulative Relative Frequency
1	10,301–12,127	10	10	.196
2	12,128–13,953	9	19	.373
3	13,954–15,779	18	37	.725
4	15,780–17,605	5	42	.824
5	17,606–19,431	6	48	.941
6	19,432–21,258	3	51	1.000
	Total	51		

interval to the cumulative frequency at the end of the first interval, then to the cumulative frequency at the end of the second interval, and so on. This segmented line forms a curve called an *ogive*. The end of the ogive is always at the upper righthand corner of the histogram. In Figure 3.7, the left vertical axis is labeled with the cumulative frequency and the right vertical axis is labeled with the cumulative proportion.

The ogive may be used to supply a ready answer to a question such as: What level of per capita income is exceeded by 50 percent of the observations? By extending a horizontal line from the cumulative proportion axis at 50 percent to the ogive, then dropping a vertical line from this point to the income axis in Figure 3.7, we may estimate that 50 percent of the observations have a per capita income of less than about $14,500.

With some reflection, it becomes apparent that the shape of the ogive is related to the shape of the frequency distribution. Distributions that have the same number of observations in each class interval result in a linear ogive; bell-shaped distributions result in asymmetrical S-shaped ogives; distributions with two peaks result in ogives with two

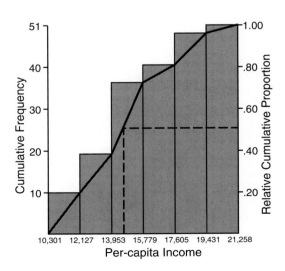

Figure 3.7 A Cumulative Frequency Histogram and Ogive for Per Capita Income, United States, 1987

S-shaped curves placed end-to-end. The usefulness of this association will become apparent in Chapter 5.

MAPPING NUMERICAL DATA

Tables and charts help to condense and give order to data. But the geographer often uses another tool to represent data—the map. Thematic maps, consisting of networks or spatial distributions, play an important part in geographical research. Such maps are often used to suggest spatial relationships between phenomena within the same space or from place to place, and to assist in the identification of the processes that produce spatial order and spatial differentiation. According to Unwin (1981), maps have four major advantages over other forms of representation:

1. Maps yield a synoptic or simultaneous expression of all the information represented.
2. Because of this synoptic view, a map permits the isolation of a number of important spatial properties that were not initially measured. These include the orientation of phenomena, their shape, and, most important of all, their relative location. By relative location, we mean where things are in relation to others, making up spatial patterns that cannot be seen unless a map is drawn. Maps, therefore, contain what has been called *spatial structure.* If spatial structure did not exist, it would be difficult to speak of "geography"; its detection and analysis is at the heart of the geographical sciences.
3. A map that contained all the real things it mapped would be indistinguishable from reality itself, so as we move from the world of people on the earth to lines and symbols on paper, there must be a scale change, a selection of material and some generalization. Maps are, then, representations, or *models,* of the real world made in order to facilitate our understanding. We can use them much as we use theories, to supply information, to predict, and to analyze relationships in and about the real world.
4. Finally, maps can be *communications devices* used to express and exchange ideas about the world. A poorly designed map is like a poorly written book, giving its reader a distorted view of what its author intended; a good map is like a good book, having a clear message that is almost impossible to misinterpret. (p. 6)

Dot Maps

Although there are many kinds of maps, examples will be shown of four of the most common thematic types: dot, network, contour, and choropleth. A **dot map** containing symbols that represented deaths from a communicable disease would be of more use to a public health official than would a list of names and addresses of disease victims. Figure 3.8 shows John Snow's famous nineteenth-century map of cholera deaths around the Broad Street water pump in London. The clustering of cholera in the vicinity of a well

Figure 3.8 A Dot Map Showing an Outbreak of Cholera Deaths, London, 1848

Source: Meade, Florin and Gesler (1988), p. 20.

supported Snow's hypothesis that cholera was a waterborne disease, with the pump the obvious source of infection.[3]

Dots may represent one or more observations of any phenomena on a nominal scale. On Snow's map, each household either has or has not had a death from cholera. The dot represents the "has" condition. Any symbol may be used, such as a square, circle, or triangle. Dot maps may also be used for ratio data. Each dot might represent, for example, a given number of occurrences per capita. In Figure 2.10, we demonstrated how agricultural product distributions may be shown using this technique, each dot representing a number of farms.

A dot map representing earthquake epicenters gives seismologists clues about fault zones and potential earthquakes. One such map, compiled by the United States Geological Survey, is shown in Figure 3.9. Such a map could have employed variable-sized dots or open circles to indicate intensity of tremor. Dot maps with large numbers of observa-

---

[3] Recall our earlier discussion of distance decay. Obviously, cholera incidence falls off with distance from the pump because the farther one is from that pump, the more likely one will obtain water from a different source.

Figure 3.9 Location of Seismic Tremors in Western North America, 1989

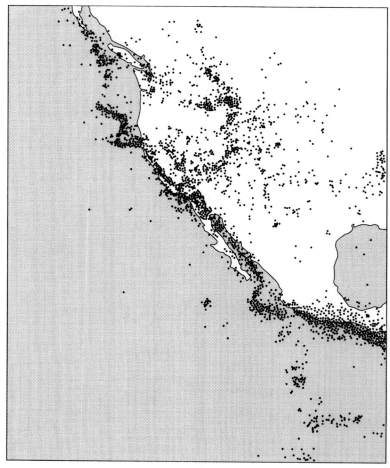

Source: Simkin et al. (1989).

tions are difficult and time-consuming to construct and are best produced with a computer and plotter if possible.

Network Maps

Two closely related subfields in geography are *communication* and *transportation*. These fields are concerned with *access* to activity sites—voice and visual contact among places, proximity to places of work, recreation, socializing, shopping, medical care—and *mobility*, the ability to communicate and move between these activity sites. These subfields of geography often require the types of graphs and maps shown in this chapter. A **network map** usually shows volume of movement such as product shipments, people transported, or telephone calls made. One technique in producing network maps is to make the lines proportional in thickness to the quantities involved; an example of this is the world migration flow map of Figure 3.10. Although the volume of movement cannot be judged with much accuracy from the scale of this map, the general pattern of movement is clearly shown.

It is also possible to map *economic distance* by changing the scale from a physical distance, such as miles or kilometers, to time or cost. For example, using data from Table 2.7, maps with scales based on both driving time and mileage between cities in Southern California can be constructed (Figure 3.11). The map in Figure 3.11a depicts the absolute mileage between the cities as they would appear on a typical road map. The map in Figure 3.11b shows travel time between the cities, where, for example, Inglewood and Long Beach appear closer because of the high-speed expressway linking the two cities.

To be informative, it is not necessary for a network map to show actual distances between places. Transportation authorities often use a **topological map** to display places in terms of their position on a network (Figure 3.12). The places shown on the map in Figure 3.12 may bear little relationship to their actual location in southern England. The intervals between towns, stations, and terminals do not reflect physical distances, only the fact that these places are connected in the network. Geographers have employed a useful and interesting set of related techniques emanating from graph theory. These techniques are beyond the scope of this text and are demonstrated elsewhere.[4]

Contour Maps

Geographers make extensive use of **contour maps.** These maps are used for a variety physical and cultural phenomena. The information of interest is contained in **isolines,** which are lines connecting points of equal value. By convention, any of the lines we have described may also be called contours. Interpretation of these maps can reveal slopes or gradients, or rates of change in slope. Lines connecting points of equal elevation on a topographic map are called *contours* or *isohypse;* lines connecting points of equal depth below sea level are called *isobaths;* lines connecting points of equal barometric pressure on a weather map are termed *isobars;* and lines that show equal temperature are called *isotherms* (Figure 3.13).

Figure 3.14 is an alternative way of presenting travel time on a map. Here *isochrones* are used to illustrate spatial relationships. This kind of map illustrates a traditional mapping technique which retains true physical distances while demonstrating the friction of travel time away from a point. The heavy lines are major arterial streets, while the lighter, numbered lines are close approximations of driving time from center A. The isochrones dip inward between arterials because travel times are slower on more congested streets.

Construction of contour maps, like dot maps, requires considerable cartographic skill as well as intelligent choices of numbers of control points and contour intervals, and

---

[4] Fundamental expositions and applications are found in Abler, Adams, and Gould (1971), Chapter 8, Tinkler (1977), Wallis (1978), and Unwin (1981).

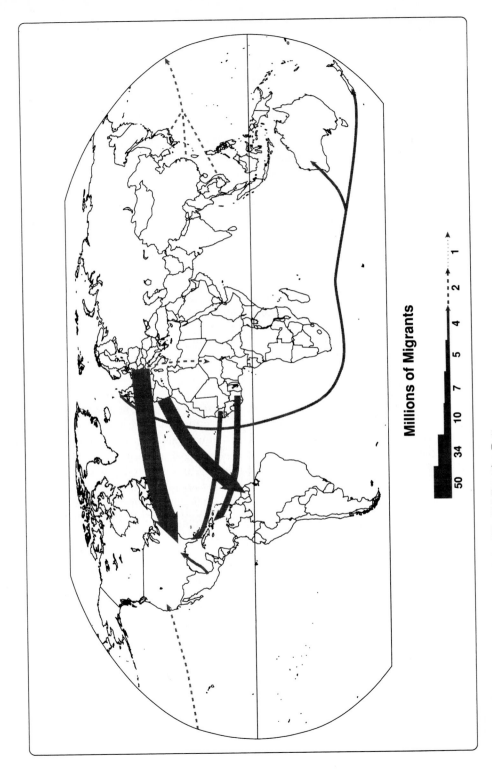

Figure 3.10 Pre–World War II Major World Migration Patterns

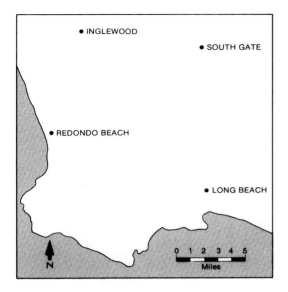

Figure 3.11 Illustration of Absolute and Relative Distances: Distance between Cities in Southern California

are more quickly and accurately produced with a computer, plotter, and appropriate software. Simple contour maps, however, may be constructed with drafting instruments. The most difficult task is fitting contour lines to data points accurately. The "fitting" process may involve time-consuming scaling with a linear dividing grid, or if accuracy is unimportant it may be done by visual reckoning. Figure 3.15 illustrates the latter method for positioning of a line with a value of 100 to a set of hypothetical data points. Note how the line is proportionately closer to 103 than to 95 and to 96 than to 105.

Choropleth Maps

The **choropleth map** uses colors, shades of gray, or line and dot patterns to show the variations in magnitude of phenomena within an area. In order to construct a choropleth map it is first necessary to divide a set of data into class intervals, then to choose patterns, colors, or shades of gray to portray each interval. In Figure 3.16, for example, patterns are used to indicate level of streamflow in the United States and southern Canada in September 1989. A glance at this map makes it easy to distinguish where streams were discharging more water than normal.

In choosing class intervals, the instructions offered earlier in the chapter are applicable here, except that the number of mapping intervals may be limited by the variety of shades, patterns, or colors that can be used effectively. A cartographic convention is to use a gradation in shading (or line/dot patterns that simulate gradation) to illustrate differences in the range of the variable being mapped. If percentages are being shown, for example, areas on the map having the highest percentage of the phenomenon might be

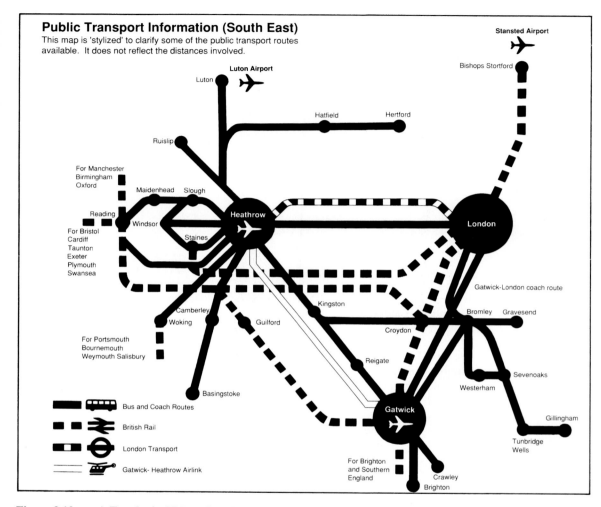

Figure 3.12 A Topological Map of an English Public Transit Network

shown in a solid color or black. The pattern should be progressively lighter as lower intervals are mapped so that the lightest shade shows the lowest percentage.

An example of shading patterns is shown in the map of per capita income, constructed from the data of Table 3.1 (Figure 3.17). A glance shows that the least affluent states cluster in the relatively rural Southeast and intermountain West; the most affluent states are those that contain the largest metropolitan populations. With the exception of Alaska, 90 percent of the state population resides in metropolitan areas in the eleven states that represent the two highest categories.

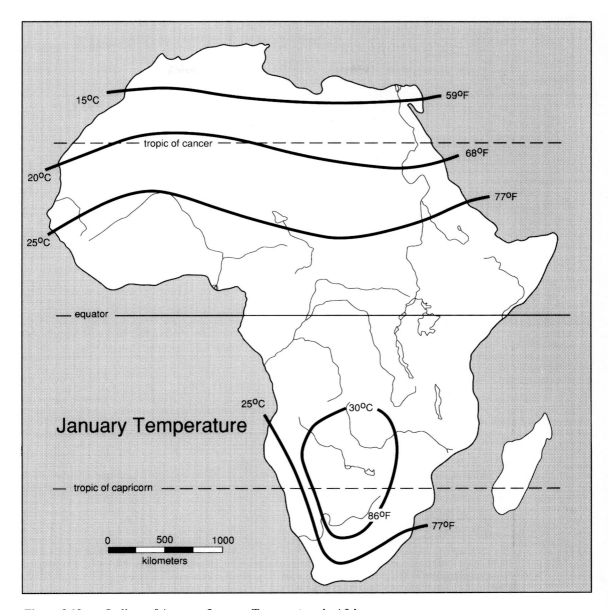

Figure 3.13 Isolines of Average January Temperature in Africa
Source: Adapted from Monk and Alexander (1977), p. 67.

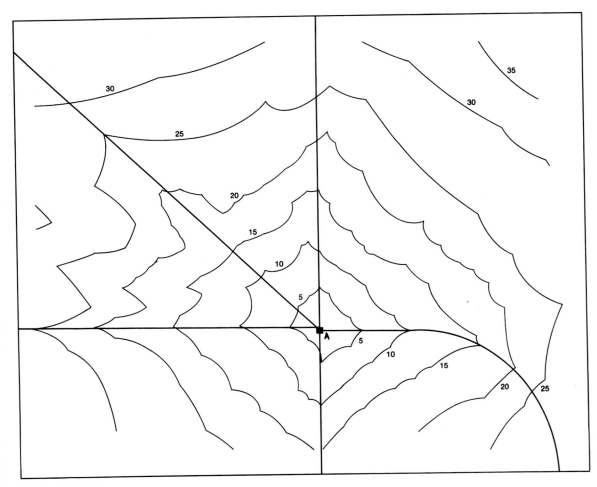

Figure 3.14 An Isochrone Map of Driving Time
Source: Western Management Consultants (1965), p. 262.

Finally, we present an example from a study conducted in Dallas, Texas, by Harries, Stadler, and Zdorkowski (1984), which attempted to describe and explain some causes of urban violence. The authors compiled a socioeconomic status index, which they used to classify twelve residential community areas. The community areas were evaluated on three ordinal measures—substandard housing, black population, and median household income (Table 3.3). By summing the ranks for each community, the authors arrived at an "urban pathology index" (UPI), which is also shown in the table. High scores indicate a high level of urban pathology, and low scores indicate a low level. The authors then noted that the UPI values tended to cluster around values of 6.0, 19.5, and 33.0, so

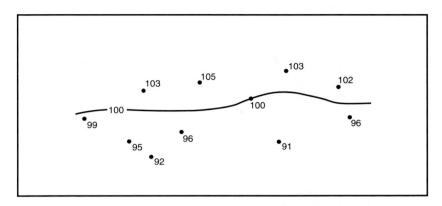

Figure 3.15 The Position of an Isoline among Control Points

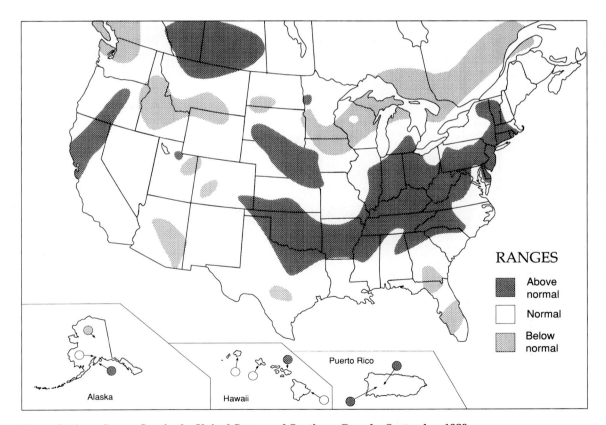

Figure 3.16 Streamflow in the United States and Southern Canada, September 1989
Source: U. S. Department of the Interior (September, 1989).

they established a three-way classification of UPI which they labeled high-, medium-, and low-status. Communities 1, 2, and 12 fell into the high-status group, and the remaining communities composed the medium-status group. The three UPI groups were then assigned choropleth patterns. The result is the map of Figure 3.18.

The use of different measurement scales can produce maps that portray radically different impressions to the viewer. For instance, consider the table of reported AIDS cases in the United States in 1990 (Appendix B-4). A few states, such as New York and California, reported thousands of AIDS cases, while some northern interior states, such as North and South Dakota, reported fewer than ten cases. The average number reported for the fifty states and the District of Columbia was about 815 cases. These data present a dilemma for graphing and mapping because most of the observations are in the range between 50–600 cases. Clearly, it would be impractical to employ class intervals in which each class had the same range.

Furthermore, it is not true that states contain AIDS patients in strict proportion to their populations. This is verified by Table 3.4, in which the top ten states by *frequency* of reported AIDS cases are juxtaposed with the top ten states in *rate per 100,000 population*. The District of Columbia has a rate more than twice that of New York, the leader in frequency. Nevada and Louisiana appear in the top ten in the rate column, but are ranked thirty-third and twelfth respectively in frequency. Meanwhile, Illinois and Massachusetts rank sixth and tenth in frequency, but are nineteenth and eleventh respectively in rate. Thus, a map of a frequency distribution will often convey a misleading impression be-

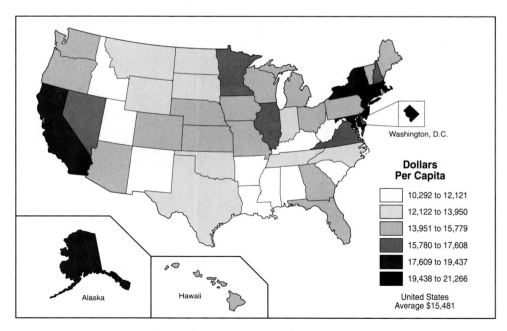

Figure 3.17 Per Capita Personal Income, United States, 1987
Data Source: U. S. Bureau of the Census (1990), p. 437.

Table 3.3 Ranking of Dallas Community Areas on Three Socioeconomic Measures and Urban Pathology Index

Community[b]	Ranks[a]			
	Substandard Housing	Black Population	Median Household Income[c]	Urban Pathology Index[d]
1	2.5	2.5	2	7.0
2	2.5	1	3	6.5
3	5.5	6.5	8	10.0
4	11	12	12	35.0
5	7.5	5	7	19.5
6	10	11	10	31.0
7	4	9	4	17.0
8	9	4	9	22.0
9	5.5	6.5	5	17.0
10	12	10	11	33.0
11	7.5	8	6	21.0
12	1	2.5	1	4.5

[a]High values mean high absolute values of original variables.
[b]Community locations are shown in Figure 3.18.
[c]Ranks reversed so that high value means low income to conform with other variables.
[d]Measure of socioeconomic status; higher values indicate lower status.
Source: Harries, Stadler, and Zdorkowski (1984), p. 597.

Table 3.4 The Top Ten States in Frequency of Reported AIDS Cases, and Rate per 100,000 Population, United States, 1990

State	Frequency	State	Rate
New York	8,399	Dist. of Columbia	121.1
California	7,346	New York	46.7
Florida	4,047	New Jersey	31.5
Texas	3,361	Florida	31.2
New Jersey	2,464	California	24.9
Illinois	1,278	Maryland	21.2
Georgia	1,223	Texas	19.2
Pennsylvania	1,197	Georgia	18.6
Maryland	1,002	Nevada	17.1
Massachusetts	844	Louisiana	15.8

Source: Centers for Disease Control [1991], p. 6.

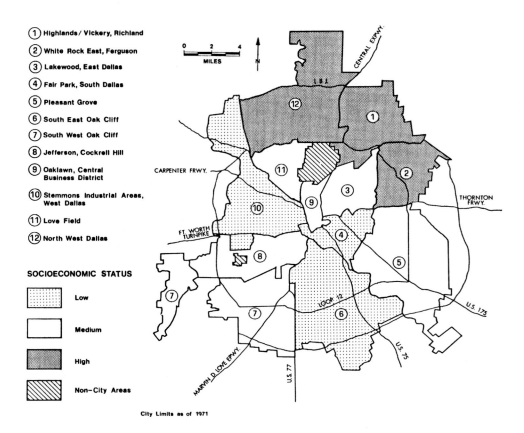

Figure 3.18 Socioeconomic Status of Neighborhoods in Dallas, Texas, 1980
Source: Harries, Stadler, and Zdorkowski (1984), p. 597.

cause the underlying at-risk population differs among observations. The *per capita* distribution may provide a more relevant picture for public policy purposes.

An interesting cartographic technique, called a cartogram, depicts the map in such a way that the area of each unit is proportionate to its contribution to the whole. Figure 3.19 illustrates this concept, using population data. If we were to use this technique with per capita income, the District of Columbia would appear to be about twice the area of Mississippi.

In conclusion, the application of geographic techniques in research, business, or government involves not only statistical analysis and modeling, but the effective use of graphs and maps. If courses on statistical and thematic cartography or geographic information systems will not be part of your curriculum, an excellent reference to the techniques and pitfalls of map construction, such as Unwin (1981), should be considered for your bookshelf.

Graphing and Mapping Data

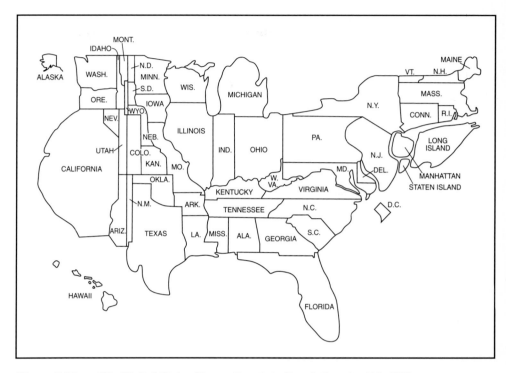

Figure 3.19 The United States, Proportionate to Population, April 1, 1990
Source: Population Reference Bureau, September 1991.

KEY TERMS

choropleth map 71
class interval 60
contour map 69
distance decay 56
dot map 66
frequency distribution 60
frequency polygon 62
histogram 60
isoline 69
network map 68
raw data 56
topological map 69

REFERENCES

Abler, R., Adams, J. S., and Gould, P. (1971). *Spatial Organization.* Englewood Cliffs, NJ: Prentice-Hall.

Ahrens, C. D. (1982). *Meteorology Today,* 4th ed. St. Paul, MN: West Publishing Co.

Centers for Disease Control. (1991). *HIV/AIDS Surveillance, Year-End Edition.* Washington, DC: U. S. Department of Health and Human Services.

Centers for Disease Control. (1992). Abortion Surveillance, United States, 1989. *Morbidity and Mortality Weekly Report, Surveillance Summaries,* September 4, 1992. Atlanta, GA: Centers for Disease Control.

Harries, K. D., Stadler, S. J., and Zdorkowski, R. T. (1984). Seasonality and Assault: Explorations in Inter-Neighborhood Variation, Dallas, 1980, *Annals of the Association of American Geographers* 74: 590–604.

Haynes, K. E., and Fotheringham, A. S. (1984). *Gravity and Spatial Interaction Models.* Scientific Geography Series, Vol. 2. Beverly Hills, CA: Sage Publications.

Héquette, A., and Ruz, M-H. (1991). Spit and Barrier Island Migration in the Southeastern Canadian Beaufort Sea, *Journal of Coastal Research* 7: 677–698.

Meade, M. S., Florin, J. W., and Gesler, W. M. (1988). *Medical Geography.* New York: Guilford Press.

Monk, J. J., and Alexander, C. S. (1977). *Physical Geography, Analytical and Applied.* North Scituate, MA: Duxbury Press.

Population Reference Bureau. (1991). Speaking Graphically. *Population Today* 19, No. 9.

Simkin, T., Tilling, R. I., Taggart, J. N., Jones, W. J., and Spall, H. (1989). *This Dynamic Planet: World Map of Volcanoes, Earthquakes, and Plate Tectonics.* Denver, CO: U.S. Geological Survey.

Tinkler, K. J. (1977). *An Introduction to Graph Theoretical Methods in Geography.* Norwich, England: Geo Abstracts, Ltd.

Unwin, D. (1981). *Introductory Spatial Analysis.* London: Methuen.

U.S. Bureau of the Census. (1990). *Statistical Abstract of the United States: 1989,* 110th ed. Washington, DC: U.S. Government Printing Office.

U.S. Department of the Interior. (1989). *National Water Conditions, September 1989.* Reston, VA: U.S. Geological Survey.

Wallis, W. D. (1978). Graph Theory and the Study of Activity Structure, in *Human Activity and Time Geography,* ed. T. Carlstein, D. Parkes, and N. Thrift. London: Edward Arnold.

Western Management Consultants. (1965). *The Economy of Maricopa County, 1965 to 1980.* Phoenix, AZ: Western Management Consultants, Inc.

EXERCISES

A metropolitan area has been divided into eighteen "shopping districts," as defined by a local newspaper marketing survey. We wish to construct a graph that will depict the "distance decay" concept. The table at the top of page 81 contains the estimated percentage of households living in each shopping district who have shopped at each of two shopping centers and in the central business district in the thirty days preceding an annual marketing survey.

1. Obtain a sheet of graph paper. At the top of the graph, print *"Graph of Relationship Between Shopping Participation and Driving Time"* in block letters. Place dots at points on the graph that coincide with percent shopping and driving time between the shopping districts and the 1.6 million-square-foot Regional Shopping Center. Sketch a smooth curve, as shown in Figure 3.1, that approximates the participation trend. On the same graph, repeat this procedure for the 140,000 square foot Neighborhood Shopping Center and for the Downtown Business District. (It is suggested you use different color pens/pencils, or employ different symbols for the points for each size center.) Which of the shopping destinations appear to have the widest geographic appeal in this metropolitan area? Why do you think it draws people from greater distances?

2. From Appendix B-4, which shows AIDS incidence rates per 100,000 population in the United States for 1990, the data have been rearranged and presented in the table at the right. Using graph paper, construct a frequency histogram like that of Figure 3.3. Use six equal class intervals for all states, but eliminate the District of Columbia. Construct a table like Table 3.2 to show the frequency of observations in each interval, the cumulative frequencies and cumulative

Shopping Participation and Driving Times between District Centroids and Shopping Places in a Hypothetical Metropolitan Area

	Regional Center		*Neighborhood Center*		*Downtown*	
Shopping District	Driving Time	Percent Shopped	Driving Time	Percent Shopped	Driving Time	Percent Shopped
1	44	2	55	0	40	0
2	34	15	37	0	35	9
3	9	62	17	1	33	6
4	5	80	12	8	28	12
5	22	27	20	1	36	6
6	16	34	14	7	30	11
7	8	49	1	40	13	23
8	10	47	9	12	27	10
9	37	12	19	0	26	12
10	31	16	9	15	5	50
11	9	57	4	28	8	37
12	19	28	20	2	15	18
13	18	30	13	6	1	66
14	42	8	24	0	14	19
15	32	17	22	0	8	35
16	40	9	24	1	13	22
17	47	4	33	0	32	11
18	54	2	36	1	37	5

relative frequencies. Construct a cumulative frequency histogram and ogive for these data, modeled on Figure 3.7. What would the corresponding approximate AIDS rate be for .50 relative cumulative proportion?

3. Refer to the data at the top of page 82. On a base map of the United States, produce a choropleth map of AIDS rates in 1990. Use five colors or line patterns that cover the following intervals: 0–4.9, 5–9.9, 10–14.9, 15–29.9, and 30+. Entitle the map, *AIDS Incidence Rates per 100,000 Population, United States, 1990.* Leave out the District of Columbia.

4. Appendix B-5 gives the live birth rate per 1,000 population in 1988 for the United States. Use a base map of the United States to produce an isoline map of birth rates. Begin by placing a pencil dot near the center of each state. Label each point with its birth rate, using small print. Use the technique described in Figure 3.15 to produce birthrate isolines for 9, 10, and 11 per 1,000. Label the isolines following the example of Figure 3.13. Entitle the map, *Live Birth Rates per 1,000 Population, United States, 1988.* Leave out the District of Columbia.

5. Appendix B-5 gives the estimated percent minority population for each of the United States in 1990. Using a base map of the United States, produce an isoline map that depicts the proportion of minority population. Begin by placing a dot near the center of each state; label each point with its minority percentage, using small print; use the technique described in Figure 3.15 to produce isolines for 10, 25, and 40 percent minority populations. Label the isolines following the example of Figure 3.13. Entitle the map, *Percent Minority Population, United States, 1990.* Leave out the District of Columbia.

AIDS Rates per 100,000 Population, United States, 1990

State	Freq	Rate	State	Freq	Rate
ND	2	0.3	AZ	315	8.6
WY	3	0.6	AR	208	8.6
SD	9	1.3	RI	88	8.8
MT	17	2.1	SC	342	9.6
IA	69	2.5	PA	1197	9.9
ID	28	2.8	MS	279	10.6
WV	62	3.3	CO	364	10.7
NE	58	3.6	IL	1278	11.0
VT	22	3.9	MO	583	11.2
WI	209	4.3	VA	738	11.9
AK	25	4.5	OR	335	12.0
MN	204	4.7	CN	425	13.0
KY	189	5.0	WA	6378	13.3
IN	282	5.1	HI	156	13.8
KN	137	5.4	DE	94	13.9
ME	67	5.5	MA	844	14.2
UT	98	5.6	LA	703	15.8
AL	239	5.8	NV	191	17.1
NH	66	5.9	GA	1223	18.6
OH	660	6.1	TX	3361	19.2
MI	577	6.2	MD	1002	21.2
OK	203	6.2	CA	7346	24.9
TN	342	6.9	FL	4047	31.2
NM	109	7.0	NJ	2464	31.5
NC	558	8.4	NY	8399	46.7

6. Make a photocopy of the base map of the hypothetical metropolitan area in Chapter 2 exercises. Then produce an isoline map of percent shopping downtown using data from the first problem. Use 20 percent as a contour interval. Begin by placing a dot near the center of each district. Use a pencil to label each point with its percentage, using small print. Use the technique described in Figure 3.15 to produce isolines. Label the isolines following the example of Figure 3.13.

Descriptive Statistics   4

A number set can be described visually by using a graph or map. Data may also be described numerically. This chapter will introduce the use of quantitative techniques to describe data and the distributions that these data sets form. The term *statistics* derives from the Latin root *status,* which implies that statistics are used to describe the condition of people, places, and things. Descriptive statistics provide the foundation upon which more complex inferential methods can be constructed. The most basic descriptive statistics are designed to provide values for *central tendency* in data sets and the *variability* within them.

MEASURES OF CENTRAL TENDENCY

Measures that describe a data set may be computed from sample data or from population data. To distinguish between them, we describe measures computed from sample data as *statistics,* and measures computed from the data of a population as *parameters.*[1] There are several types of descriptive measures that can be computed from a data set. In this chapter, we will consider measures of central tendency and measures of dispersion. In each of the measures of central tendency we have a single value that is considered to be typical of the set of data as a whole. The three most commonly used measures of central tendency are the *mean,* the *mode,* and the *median.*

---

[1]The types of statistical tests scientists employ are based on certain assumptions about populations, which will be explained in Chapter 6.

The Arithmetic Mean

The most common measure of central tendency is the simple average or **arithmetic mean.** Arithmetic means are commonly used in describing sports performances, climate, and stock market performance. The mean describes the point about which a distribution of numbers is centered.

> *Definition:* The **arithmetic mean** of a sequence (or set) of measurements X_1, X_2, X_3, . . ., X_n is equal to the sum of measurements divided by the number of measurements in the sequence, or n.

Summation Notation. Many algebraic expressions of descriptive statistics involve the addition of measurements (or elements) in a sequence. Statisticians use a lexicon of symbols, or **summation notation,** to specify mathematical operations. Summation is one of the more common symbols encountered in statistics. The Greek letter Σ (sigma) is used to initiate the summation of numbers *(elements)* over a defined *sequence.* Each element is ordered according to its position in the sequence, or data set, and not according to its magnitude. For instance, for the data set in Table 4.1, the term X_3 refers to the third item in a variable X. Suppose there are only ten observations for a variable X. If we wish to sum the ten elements of X, we would specify,

$$\sum_{i=1}^{10} X_i$$

which we read as, $X_1 + X_2 + X_3 + . . . + X_{10}$. Thus, we may identify a particular observation's position in the distribution by its subscript. The value of X_8 in Table 4.1 is 2 miles. X_n is the "nth," or last element in a distribution of n observations. We may then refer to any observation *(i)* in a data set as element X_i. Additional definitions and examples of summation notation are described in Box 4.1.

Table 4.1 **Hypothetical Travel Distances, X (in Miles), and Their Summation**

i	X_i
1	5
2	7
3	1
4	15
5	43
6	3
7	6
8	2
9	21
10	17
Sum	120

BOX 4.1 Additional Definitions of Summation Notation

The following definitions will help comprehend several descriptive techniques that follow in subsequent chapters. In practice, the term

$$\sum_{i=1}^{n}$$

is usually simplified to Σ, where it is implied that the summation includes *all* elements of the data set.

Let C be a constant and i the index of summation.

$$\sum_{i=1}^{n} C_i = nC$$

where C is summed n times, or where the sum involves n elements, with each element equal to C. For example

$$\sum_{i=1}^{n} C = nC$$

$$\sum_{i=1}^{5} 3 = 5(3) = 15$$

Also, in the following case, since 8 is a common factor in each term, then

$$\sum_{i=1}^{3} 8i = 8(1) + 8(2) + 8(3)$$

$$= 8(1 + 2 + 3) = 8 \sum_{i=1}^{3} i$$

Thus, the summation of a constant times a variable is equal to that constant times the summation of the variable.

Following from the last example, let C be a constant. Then

$$\sum_{i=1}^{n} CX_i = C \sum_{i=1}^{n} X_i$$

And,

$$\sum_{i=1}^{n} (X_i + Y_i + Z_i) = \sum_{i=1}^{n} X_i + \sum_{i=1}^{n} Y_i + \sum_{i=1}^{n} Z_i$$

Two subscripted variables appearing together, as in Equation 4.11, $w_i X_i$, indicates that the *i*th value of w is to be multiplied by the *i*th value of X. This will sometimes appear separated by a *center dot*:

$$\sum_{i=1}^{n} w_i \cdot X_i$$

to indicate a multiplication operation for each observation.

The terms $\sum_{i=1}^{n} X_i^2$ and $\left(\sum_{i=1}^{n} X_i \right)^2$ require different operations. The left-hand term requires that all values of X_i be squared before a summation is executed. The right-hand term requires that all values of X_i first be summed, and then that value is squared.

Brief descriptions of *population* and *sample* statistics were given in Chapter 2. A population of state average incomes (Table 3.1) could be found by summing all fifty-one values and dividing by 51. The symbol for the *population mean* is μ (mu). A sample mean for state income could be found by taking a sample of 20 of the 51 states and calculating that mean value. The sample mean is designated by the symbol $\overline{X}$ (x-bar). The equation for the sample mean is

$$\overline{X} = \frac{X_1 + X_2 + X_3 + \ldots + X_n}{n} \tag{4.1}$$

or, more briefly, using the summation symbol, Σ,

$$\overline{X} = \frac{\sum_{i=1}^{n} X_i}{n} \tag{4.2}$$

As an example, we compute the mean for the sequence,

$$4, 8, 3, 9, 5$$

Since $n = 5$,

$$\overline{X} = \frac{\sum_{i=1}^{5} X_i = (X_1 + X_2 + \cdots + X_5)}{5}$$
$$= \frac{4 + 8 + 3 + 9 + 5}{5} = 5.8$$

On the familiar linear arithmetic scale, the sum of 4, 8, 3, 9, 5 yields a mean value that is analogous to a fulcrum about which the distribution would balance if objects of equal weight were placed at each of the object's locations on the scale (Figure 4.1). The mean, therefore, does not represent the midpoint of an ordered sequence, but the point where the sum of distances (deviations) between each measurement and the mean on the left side of the fulcrum would equal (or balance) the sum of deviations on the right.

To demonstrate further, suppose we subtract the mean of the travel distances in Table 4.1 from each observed distance of the set and sum the differences. The result should be zero. The mean, or fulcrum, is located at $(120 \div 10 =) 12$.

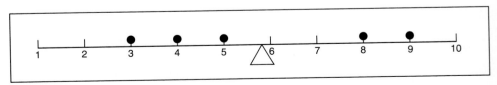

Figure 4.1 **The Mean Value as a Fulcrum**

$$\sum_{i=1}^{10}(X_i - \overline{X}) = 0$$

$(5-12) + (7-12) + (1-12) + (15-12) + (43-12)$
$+ (3-12) + (6-12) + (2-12) + (21-12) + (17-12)$
$= (-7) + (-5) + (-11) + 3 + 31 + (-9) + (-6) + (-10) + 9 + 5 = 0$

The Mode and the Median

Definition: | The **mode** is defined as the peak of the frequency distribution, or the most frequent class.

There are other measures of central tendency, none of which are very commonly used in geographical research.[2] One such measure is the **mode,** which is simply the most frequent score or scores. The mode is seldom used because it depends to some degree on the class interval chosen and the shape of the distribution. Bimodal distributions have two modes and say little about central tendency. The mode is at best only an approximation of the distribution's center. Consider the following series of Fahrenheit temperature readings:

1. 61, 65, 73, 65, 51, 58
2. 61, 65, 63, 64, 51, 58
3. 61, 65, 73, 65, 73, 58

The first series has a mode of 65, since there are two readings of this magnitude and no other temperature appears twice. There is no mode in the second set of readings, but there are two modes in the third (65 and 73). A distribution with two modes is *bimodal*. In bimodal distributions, it is not necessary that both peaks be of exactly the same height. The mode is more useful when there are more observations and when data have been grouped. Then, a modal class interval becomes apparent, and the mode is the midpoint of the interval.

Definition: | The **median** of a set of n observations, $X_1, X_2, \ldots, X_n$ is defined to be the central value when the observations are arranged in order of magnitude. If there is an even number of observations, the median value is the midpoint between the two center observations.

In order to identify a **median,** data must first be reordered from smallest to largest values. We will refer to this data set as an **array.** The median is defined as the observation in the middle of the ranked set of observations. There must be an equal number of observations above and below the median. There is no established symbol for the median of a distribution. Here, one of several conventional alternatives, Md, will be adopted.

---

[2] A particularly understandable treatment is found in Chapter 4 of Stahl and Hennes (1980).

Perhaps the most important and valuable property of the median is that it is influenced by the position of the items in the array but not by the magnitude of individual values in the data, as is the average. Referring again to the per capita personal income data of Table 3.1, in which the states are listed in ascending order, note that there is an uneven number ($N = 51$) of observations, so the median becomes the value of the 26th observation, or $15,302.[3] The median income is considered to be more representative of the "typical" family income in the nation than the mean. The mean per capita personal income in the United States in 1988 was just over $16,000, or about $1,000 higher than the median. Means are influenced by extreme values, whereas medians are not.

Before the median of the travel distance data of Table 4.1 can be computed, it is first necessary to reorder the ten observations from shortest to longest travel distance, or 1, 2, 3, 5, 6, 7, 15, 17, 21, and 43. The median must divide the distribution into two equal parts. Because there is an even number of cases, the value of the median is the average of the *two middle travel distances* (half of 6 + 7 = 6.5).

In any distribution where the values of the mean, median, and mode are identical, or nearly so, that distribution is symmetrical about its mean and approximately bell shaped. This is known as a **normal distribution.** On the graph of personal income in Figure 3.3, the median fell to the left of the mean indicating that there were some high income values that were not balanced by equally low income values below the median. When this occurs, an elongated "tail" is observed on the right-hand side of the distribution. The term used in this case is **positively skewed.** If the median appears to the right of the mean, this means that there are some low values in the distribution that are not balanced out by high values. In that event, the tail appears on the left-hand side, and distribution is described as **negatively skewed.** Figure 4.2 illustrates each of these situations.

In practice, it is desirable to determine which measure of central tendency will most accurately portray a particular distribution. When skewness is not severe but is due to one or two observations deviating substantially from the mode, the median will be only slightly affected. The arithmetic mean, however, is affected by the value of every item in the series, and the presence of one or a few extremely large or small items in a data set may result in a skewed distribution. In inferential statistics (Chapter 6) the importance of the shape of the distribution to operations involving sample means will become apparent. The median may also be used in statistical testing and its use will be explained in Chapter 9.

In many geographic applications, positively skewed distributions are more common than normal or negatively skewed ones. This is generally true because, with ratio data, zero is the lower limit on many distributions. For instance, the variable *percent population below the poverty level* in United States metropolitan areas, if measured at the census tract level, is skewed toward the higher percentages (positive skewness) because a minority of the tracts contain most of the poor population. Most tracts contain small propor-

---

[3]This does not signify that the median income for persons in the United States in 1988 was $15,302. Data at one scale, here the *state* scale, must not be confused with data at another scale—in this case, income for *persons.* This is also an example of the ecological fallacy, described in Chapter 1.

Figure 4.2 Frequency Distributions with (a) Identical Mean and Median, (b) Mean Larger than Median, and (c) Mean Smaller than Median

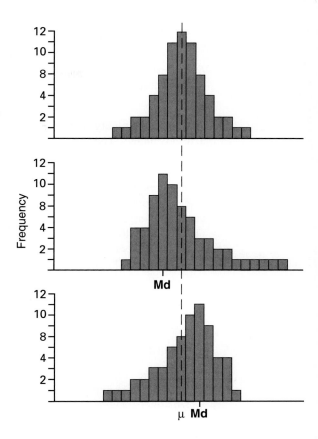

tions of poor population, including many tracts with nobody below the poverty line. In geomorphology, there are many more small drainage basins than large, and in climatology most daily observations of rainfall in a dry region will record little or no rainfall, which again yield positively skewed distributions. On the other hand, in the rainy tropics the rainfall distribution is likely to be skewed to the left.

MEASURES OF DISPERSION

Measures of dispersion are concerned with the distribution of values around the mean in a set of data. In some ways a measure of dispersion is a qualification of the mean. The mean of monthly temperatures in Omaha is not as representative of the temperature regime as the mean of monthly temperatures in Miami. In Omaha the monthly temperature mean from winter to summer ranges from 23°F to near 80°F, whereas the temperature range over the year in Miami is from 67°F to 82°F. Used alone, the mean may not be very informative in research problems. Several measures of dispersion are available, and a few of those commonly employed by geographers are discussed here.

The Range

Definition: | The **range** of a set of measurements $X_1, X_2, X_3, \ldots, X_n$ is defined to be the absolute difference between the largest and smallest values.

The simplicity of the range as a measure of dispersion is both an advantage and a disadvantage. The range may be useful if it is desirable to obtain some very quick calculations that can give a rough indication of dispersion. The disadvantage of the range is that it is based only on the two extreme cases. Extremes are likely to be unusual cases. It is, therefore, rare that an extreme value is selected in a sample. Furthermore, the range will ordinarily be greater for large samples than small ones simply because in large samples there is a better chance of including the most extreme observations. Referring again to the travel distance data of Table 4.1, reordered in ascending order of distance, it can be quickly seen that the range is (43 − 1 =) 42 miles. The value of the range may be inflated relative to the mean or the median. One might expect either the mean or median to occur somewhere near the middle of the range (about 21), whereas in fact the mean travel distance is only 11.44 miles, and the median is just 6 miles.

The Quartile Range and Other Divisions

A *conventional descriptive statistic* is one that divides the range of the data into several intervals with approximately equal numbers of observations in each interval. A *quartile* includes 25 percent of the observations in the data set at hand; a *quintile* contains 20 percent of the observations, and so forth. Instead of finding the median, which has half the observations above or below it, we may wish to determine the value of, for example, the *quartile deviation Q*, which is arbitrarily defined as half the distance between the first and third quartiles. Symbolically,

$$Q = \frac{Q_3 - Q_1}{2} \qquad (4.3)$$

where Q_1 and Q_3 represent the first and third quartiles respectively. The value of dividing data into quartiles is that they are less likely to be influenced by one or a few extreme values.

Quintiles refer to fifths of the data set, and *deciles* refer to tenths. The choice of interval is solely a function of what is most useful for accurately describing the data at hand. If we have a large number of observations, the distribution can be divided into deciles by locating values that have one-tenth, two-tenths, or nine-tenths of the observations with lower values.

Quartile, quintile, and decile deviations should not be associated with the construction of a histogram, as was pointed out in the previous chapter. The range of data values in the class intervals of a histogram are usually equal. The range of data values within quartiles, quintiles, and other divisions are not necessarily equal; only the frequency of

Descriptive Statistics 91

observations is as equal as possible.[4] The *interquartile range* is obtained by subtracting the smallest value in any quartile from the largest value in that quartile.

Variability

Numerical measures of central tendency, such as the mean or median, tell us nothing of the range, or **variability,** in the data under study. Figure 4.3 contains two "smoothed" polygons. In both, the mean has the same value, but the curves are quite dif-

Figure 4.3 Two Symmetrical Frequency Distributions with Identical Means but Different Dispersions

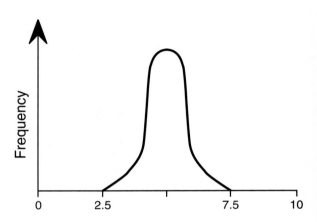

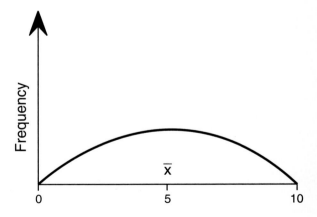

---

[4]If, for example, we wished to find quartiles for a data set with fifty observations, the second quartile (median) would be the average value of the twenty-fifth and twenty-sixth values in the array, such that there were twenty-five observations above and twenty-five observations below. The first quartile would be the value of the thirteenth observation in the array and the third quartile would be the value of the thirty-eighth observation. However, the thirteenth and thirty-eighth observations may belong to only one interval, so there will be twelve observations in two of the intervals and thirteen in the other two intervals.

ferent. Since the figures are symmetrical (50 percent of the area under the curve is to the left of the mean and 50 percent is to the right), the median for the curve is identical to the mean. Both measures of central tendency fail to provide unique descriptions due in part to the **dispersion,** or variability, in the data.

Two distributions are shown in Figure 4.4 and illustrate a situation where the mean and range for one distribution closely approximate the mean and range for a second distribution. The data for average monthly precipitation in Ho Chi Minh City, Vietnam, depict a typical wet-and-dry tropical regime with considerable variability throughout the year, while Salvador, Brazil, demonstrates much less variation, characteristic of the wet tropics. However, neither regime has a great range in precipitation. Here, then, are two distributions with similar means and ranges, but with unequal variability. Figure 4.4 demonstrates the limitation of the range: it fails to establish where data accumulate within

Figure 4.4 Frequency Histograms of Precipitation Data for Ho Chi Minh City, Vietnam, and Salvador, Brazil

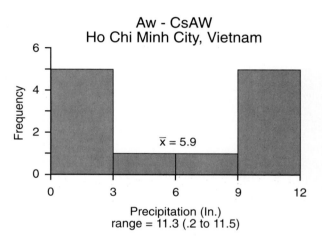

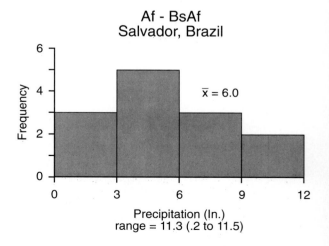

the histogram. In the case of Salvador, the data are centered about the mean; in the case of Ho Chi Minh City, the distribution is bimodal, as there are two precipitation maximums separated by a dry period.

Observations in a distribution may vary widely or they may be very close together. As illustrations of both cases, consider the sample values of average winter (January, February, and March) temperatures in the northern hemisphere in selected marine west coast climate stations (Table 4.2), and the average winter temperatures in selected humid microthermal climate stations (Table 4.3).

A glance at the winter temperatures in the marine west coast climate sample and those of the humid microthermal sample verifies that the humid microthermal data are more variable than those of the marine west coast data. In other words, the humid microthermal data are more scattered about their mean than the marine west coast data. *Scatter* suggests wide distances from the average, and *cluster* suggests that the temperatures are concentrated around the average. Thus the measure of variability that we are seeking should be some indicator of how far, on average, the observations in the distribution are from the mean.

To compare the variability in these data, it is first necessary to compute the means for the two samples. From Equation 4.2, we compute for the humid microthermal sample:

Table 4.2 **Average Winter Monthly Temperatures of Selected Marine West Coast Weather Stations**

Station	Month	Temperature
Aberdeen, Scotland	January	38
	February	38
	March	40
Reykjavik, Iceland	January	32
	February	32
	March	34
Valentia, Ireland	January	44
	February	44
	March	45
Hokitika, New Zealand	July	45
	August	46
	September	50
Portland, Oregon	January	39
	February	42
	March	46

Sources: Corbet (1976:101) and Trewartha, Robinson, and Hammond (1968:112).

Table 4.3 Average Winter Monthly Temperatures of Selected Humid Microthermal Weather Stations

Station	Month	Temperature
Omaha, Nebraska	January	22
	February	25
	March	37
Montreal, Canada	January	13
	February	14
	March	26
Moscow, Russia	January	12
	February	15
	March	23
Shenyang, China	January	8
	February	14
	March	30
Madison, Wisconsin	January	17
	February	20
	March	31

Sources: See Table 4.2.

$$\overline{X} = \frac{\sum_{i=1}^{n} X_i}{n}$$

$[22 + 25 + 37 + 13 + 14 + 26 + 12 + 15 + 23 + 8 + 14 + 30 + 17 + 20 + 31] \div 15 = 20.47$

and for the west coast marine sample,

$[38 + 38 + 40 + 32 + 32 + 34 + 44 + 44 + 45 + 45 + 46 + 50 + 39 + 42 + 46] \div 15 = 41.00$

It is now possible to assess variability around the mean temperature for the sample (see Tables 4.4 and 4.5): Consider the difference between the January temperature in Aberdeen, Scotland, and the mean winter temperature of the sample: 38° − 41° = −3°F; the difference between August temperature in Hokitika, New Zealand, and the mean for the sample: 46° − 41° = 5°F; and so forth. We may refer to this difference between (X_i) and the mean ($\overline{X}$) as a *statistical distance*. However, sometimes X_i is greater than $\overline{X}$ and sometimes it is less than $\overline{X}$, which yields positive and negative values. An important question is, what does one do about these positive and negative numbers? Remember that the mean is a fulcrum, and the sum of the variations, or deviations of items from their mean, must be zero. Tables 4.4 and 4.5 demonstrate this quality. The answer to this problem is to square each deviation $(X_i - \overline{X})^2$ and then sum the squared deviations [i.e., $\Sigma(X_i - \overline{X})^2$]. This yields what is known as the **sum of squares,** or *total squared variation about the mean.* From the sum of squares, it is now possible to derive two valuable statistics, the *variance* and the *standard deviation.*

Table 4.4 Deviations and Squared Deviations from the Mean (20.47°) of Humid Microthermal Climate Sample Data

Temperatures	$(X_i - \bar{X})$	$(X_i - \bar{X})^2$
22	1.53	2.34
25	4.53	20.52
37	16.53	273.24
13	−7.47	55.80
14	−6.47	41.86
26	5.53	30.58
12	−8.47	71.74
15	−5.47	29.92
23	2.53	6.40
8	−12.47	155.50
14	−6.47	41.86
30	9.53	90.82
17	−3.47	12.04
20	−0.47	0.22
31	10.53	110.88
Total	0	943.72

Table 4.5 Deviations and Squared Deviations from the Mean (41°) of West Coast Marine Climate Sample Data

Temperatures	$(X_i - \bar{X})$	$(X_i - \bar{X})^2$
38	−3.0	9.0
38	−3.0	9.0
40	−1.0	1.0
32	−9.0	81.0
32	−9.0	81.0
34	−7.0	49.0
44	3.0	9.0
44	3.0	9.0
45	4.0	16.0
45	4.0	16.0
46	5.0	25.0
50	9.0	81.0
39	−2.0	4.0
42	1.0	1.0
46	5.0	25.0
Total	0	416.0

Variance

Definition: The **variance** is defined as *the sum of the squared deviations from the mean divided by N* for the population parameter, and *divided by n − 1* for the sample statistic.

It is important to bear in mind that sample statistics are *estimates* of the population statistics. Such estimators should be unbiased. Unbiased samples (regardless of sample size) taken from a population with a known mean should give sample means which, when averaged, will yield the population mean. The same may be said of standard deviations. An estimator that does not do so is called *biased*. The sample mean, $\overline{X}$, is an unbiased estimator of the population mean μ. However, the sample variance as computed above is not unbiased. On the average it will underestimate the magnitude of the population variance, σ^2 (sigma squared). In order to overcome this bias, mathematical statisticians have shown that when sums of squares are divided by $n - 1$ rather than by n, the resulting sample variances will be unbiased estimators of the population variance. The rationale for this will be discussed in Chapter 6. The population and sample variances, respectively, are represented symbolically by Equations 4.4 and 4.5:

$$\sigma^2 = \frac{\sum_{i=1}^{N}(X_i - \overline{X})^2}{N} \tag{4.4}$$

$$s^2 = \frac{\sum_{i=1}^{n}(X_i - \overline{X})^2}{n - 1} \tag{4.5}$$

Since the temperature data of Tables 4.2 and 4.3 are taken from only small samples of all the possible weather stations in these climate zones, Equation 4.5 is the appropriate computation here. The procedure shown in Tables 4.4 and 4.5 and Equation 4.5, if done on a hand calculator, might be considered a somewhat cumbersome way of computing the variance. A short form of the variance, Equation 4.6, avoids numerous rounding errors produced by having to subtract each element from the mean and squaring the result.

$$s^2 = \frac{n\sum_{i=1}^{n}X_i^2 - \left(\sum_{i=1}^{n}X_i\right)^2}{n(n-1)} \tag{4.6}$$

To obtain a summation of X_i^2, it is helpful to establish a work sheet column in which each observation, X_i, is squared, unless your calculator has a memory which will accumulate and save values. Using the humid microthermal climate sample data, Box 4.2 demonstrates the procedure used to solve for s^2 with Equation 4.6. For the humid microthermal sample, we obtain

BOX 4.2 **Computation of Sample Variance of Humid Microthermal Climate Sample Data Using Equation 4.6**

(1) X	(2) X^2
22	484
25	625
37	1,369
13	169
14	196
26	676
12	144
15	225
23	529
8	64
14	196
30	900
17	289
20	400
31	961

$$\bar{X} = \frac{\sum_{i=1}^{n} X_i}{n} = \frac{307}{15} = 20.47$$

$$s^2 = \frac{n \sum_{i=1}^{n} X_i^2 - \left(\sum_{i=1}^{n} X_i\right)^2}{n(n-1)}$$

$$n \cdot \Sigma X_i^2 = 108{,}405 \quad (\Sigma X_i)^2 = 94{,}249$$

$$s^2 = \frac{108{,}405 - 94{,}249}{210} = 67.41$$

$$s^2 = \frac{14{,}156}{210} = 67.41$$

For the west coast marine climate sample data,

$$s^2 = \frac{6{,}239}{210} = 29.71$$

From this point on in the text the term *sum of squares* will be encountered with regularity.

Standard Deviation

Variance always represents squared units and therefore is not an appropriate measure of dispersion. In order to express this concept in terms of the original units, simply compute the square root of the variance. The result is a statistic termed the **standard deviation.** Again, the symbolism differs depending on whether the data represent the population or the sample. The symbol for the population standard deviation is the Greek symbol σ (sigma) and the symbol for sample standard deviation is s. Symbolically, sample standard deviation is computed as follows:

$$s = \sqrt{s^2} = \sqrt{\frac{n \sum_{i=1}^{n} X^2 - \left(\sum_{i=1}^{n} X_i\right)^2}{n(n-1)}} \tag{4.7}$$

The standard deviation for the temperature experiment is expressed in the original units of measurement, degrees Fahrenheit. For the humid microthermal data, this value is computed,

$$s = \sqrt{s^2} = \sqrt{67.41} = 8.21°F$$

For the marine west coast data,

$$s = \sqrt{s^2} = \sqrt{29.71} = 5.45°F$$

From this discussion and example it should now be clear that a large standard deviation implies a corresponding (large) degree of variability. An important theorem, stated by the Russian mathematician P. L. Chebyshev, provides a means of using the standard deviation to interpret the dispersion in a data set that is not bell-shaped:

Definition: **Chebyshev's Theorem.** Given a set of data, no matter how they are distributed—population or sample—X_1, X_2, . . . , X_n, it may be shown that a proportion of at least $(1 - 1/k^2)$ of the observations (where the value of k is greater than 1) will lie within k standard deviations of their mean $\overline{X}$, and that the proportion falling beyond the limits of $\overline{X} \pm k$ will be less than $1/k^2$.

Table 4.6 shows that if $k = 3$, then at least $1 - 1/9 = 89$ percent of the observations in the sample lie between $\overline{X} - 3s$ and $\overline{X} + 3s$. Consider the precipitation data for Salvador, Brazil, in Table 4.7 and Figure 4.5.

Descriptive Statistics

Table 4.6 Proportion of Observations Occupying an Interval About the Mean When $k = 1$, 2, and 3

k	$1 - 1/k^2$
1	0
2	3/4
3	8/9

Table 4.7 Average Monthly Precipitation for Ho Chi Minh City, Vietnam, and Salvador, Brazil (in Inches)

	Jan.	Feb.	Mar.	Apr.	May	Jun.	Jul.	Aug.	Sep.	Oct.	Nov.	Dec.
Ho Chi Minh City	0.2	0.5	0.5	2.6	7.7	11.2	9.5	10.9	11.5	10.2	4.8	1.5
Salvador	2.9	2.9	6.4	11.4	11.9	7.7	8.1	4.4	3.3	3.7	5.6	3.9

Source: See Table 4.2.

According to Chebyshev, we can expect a certain *proportion* of the twelve monthly precipitation observations to fall within a specified interval. Table 4.6 gives values of $1 - 1/k^2$, for $k = 1$, $k = 2$, and $k = 3$. At least three-fourths (75 percent) of the observations will occupy an interval of 2 standard deviations on either side of the mean, or, as shown in Figure 4.4, at least nine of the twelve months should fall within the interval 6 ± 2 (3.03), or nine of the twelve months should fall within [0,12]. It may be seen that this condition is easily met, that is, all twelve months fall within the interval [0,12]. In most distributions the proportion of measurements falling within the specified interval will exceed $1 - 1/k^2$.

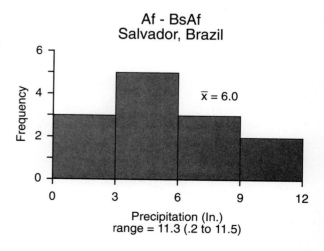

Figure 4.5 Frequency Distribution of Precipitation Data for Salvador, Brazil

When a distribution is mound- or bell-shaped, as in Figure 4.6, one can be less conservative with a statement concerning proportion within a specified interval. Statisticians know well that "real-world" sampling distributions often produce frequency distributions that are normal, a situation that concentrates a relatively high proportion of the measurements near the mean. This is expressed in the so called **empirical rule:**

> The **Empirical Rule:** Given a distribution that is approximately bell-shaped, the interval
>
> 1. $\mu \pm \sigma$ will contain approximately 68 percent of the measurements;
> 2. $\mu \pm 2\sigma$ will contain approximately 95 percent of the measurements.
> 3. $\mu \pm 3\sigma$ will contain approximately 99.7 percent of the measurements.

In summary, a meaningful measure of variability must be some number that can be computed from the terms in any distribution. When computed, this number will reveal the spread of the terms in the distribution. The most desirable statistic of variability should have the following properties:

1. The value of the statistic should not be related to the number of observations. Just because there are many cases in the distribution does not mean that the value of the statistic should also be large.

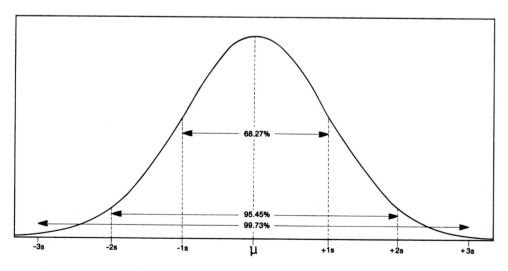

Figure 4.6   Proportion of Area under Normal Curve at ± 1, ± 2, and ± 3 Standard Deviations away from Mean

Descriptive Statistics

   2. The statistic should be independent of the mean. We are interested only in how scattered the values are. The mean alone tells nothing about the variability, and its size should not influence the ideal statistic. The mean may only be used as a point of reference.

Coefficient of Variation

Relative variability between data sets cannot always be shown by comparing standard deviations. A sequence of large income values will probably have a larger standard deviation than a sequence of lower incomes. The standard deviation of annual precipitation in Boston is 7.88 inches, while it is only 4.12 inches in San Diego. However, the distribution of values in Boston is much more uniform about the mean than is the case in San Diego. Squaring and summing larger numbers in Boston dictates a larger value for variability. To compare *relative variability* of the two precipitation regimes, one must take into account the fact that Boston has a much larger value for mean annual precipitation (41.53 inches as opposed to 9.9 inches in San Diego). Standard deviations can be scaled by the mean to produce a measure that can be compared between data sets. This measure is called the **coefficient of variation.** This statistic is expressed symbolically as

$$CV = \frac{s}{\bar{X}} \times 100 \tag{4.8}$$

Standard deviation is expressed in the same units as the mean. The coefficient of variation is thus a statistic that is independent of the unit of measurement. For data sets with low variability, the coefficient of variation approaches zero. Data sets with high variability produce a coefficient of variation that approaches or exceeds 100.

Returning to the temperature data of Tables 4.4 and 4.5, the humid microthermal stations have a mean winter temperature of 20.47°F and a standard deviation of 8.21F; the marine west coast stations have a mean winter temperature of 41.00°F and a standard deviation of 5.45°F. Comparing the two standard deviations, it is evident that the microthermal climate has more variation in winter temperature than the west coast climate. However, the means for the two stations are not from the same population of temperature data.

Substituting the means and standard deviations of the two temperature samples into Equation 4.8, we obtain

$$CV_{hm} = (100 \times 8.21) \div 20.47 = 40.12$$

and

$$CV_{wc} = (100 \times 5.45) \div 41.00 = 13.30$$

This verifies that the variability in winter temperatures is greater in humid microthermal climates than it is in marine west coast climates, assuming that the samples are representative of these climate types.

In another application of the coefficient of variation, data on monthly average temperatures at the Los Angeles International Airport weather station were obtained for the years 1940–1980, and standard deviations were computed. Computing coefficients of variation for each of the twelve months, the results were then graphed (Figure 4.7). It is now possible to see how temperatures vary throughout the year in this representative Mediterranean climate station.

The coefficient is also useful when two sets of data that must be compared are expressed in different units. In such a case the standard deviations could not be directly compared. At the Los Angeles Airport, for example, January rainfall averages 2.60 inches over the same forty years (1940–1980), with a standard deviation also of 2.60 inches. In April, the rainfall averages 0.87 inches, with a standard deviation of 1.059 inches. The coefficients of variation for the two months are 99.8 and 121.7 percent respectively. It can thus be seen that rainfall is much more variable than temperature in both months, and, despite a relatively small variation in April temperature, the April rainfall variation is considerable.

Skewness and Kurtosis

Statistics that measure dispersion essentially describe the "width" of the distribution. However, variances do not tell us anything about the shape of the distribution. Not all data sets produce bell-shaped distributions like that of Figure 4.6. Consider, for example, the data for percent non-English speakers in thirty-eight metropolitan areas in the United States (Appendix B-3). Figure 4.8 is a frequency histogram of these data. Note

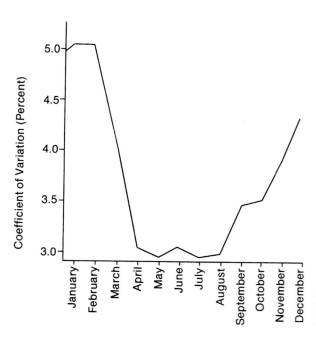

Figure 4.7 Coefficients of Variation in Average Monthly Temperatures at Los Angeles International Airport

Figure 4.8 A Histogram Illustrating Skewness

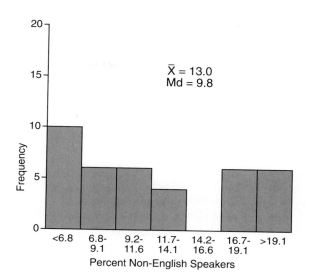

that over half the observations fall in the first three class intervals, while the mean, 13.0, falls in the central class. Thus, the median is below the mean, yielding a skewed distribution. The modes of skewed distributions are displaced to the left if the skew is right, and to the right if the skew is left.

Furthermore, regardless of the presence or absence of skew, most of the data of a sample or a population may be very close to its mean, producing an extremely peaked distribution; or the data may be dispersed evenly so that virtually all the class intervals of the histogram have approximately the same number of observations. This condition is measured by what is termed kurtosis. A condition described as **leptokurtic** occurs in situations where the highest frequencies are in a few class intervals, and a **platykurtic** distribution is one in which the frequencies of all class intervals are nearly identical. A bell-shaped distribution is one with virtually no skew or kurtosis.

Skewness measures the degree of asymmetry about the mean of a distribution. There are several methods for measuring skewness in statistics. One of the most common, derived from the work of mathematician Karl Pearson, takes into account the relative position of the mean and the median in a distribution. Known as Pearson's coefficient of skewness, it may be computed with Equation 4.9:

$$Sk = \frac{3(\overline{X} - Md)}{s} \qquad (4.9)$$

where Sk is the relative measure of skewness, with a positive sign indicating right (positive) skew, and a negative sign identifying left (negative) skew; $\overline{X}$ is the sample mean, Md is the value of the median, and s is the sample standard deviation. Sk will yield a value of zero if perfect symmetry is present. As long as Sk does not exceed ± 3, it may be considered moderate (Croxton, Cowden, and Klein, 1967, p. 202).

Kurtosis measures how peaked or flat the distribution appears, and the most common measure is computed as:

$$Km = \frac{\Sigma(X_i - \overline{X})^4/n}{(s^2)^2} - 3 \qquad (4.10)$$

Negative values of *Km* indicate platykurtic (flat) curves; positive values represent leptokurtic (peaked) curves in which the distribution has longer tails than a normal distribution with the same σ; and a value of 0 represents a **mesokurtic** curve, which is the representative bell-shaped curve of the normal distribution.

In the interest of accuracy, whenever possible, a computer should be used to compute skewness and kurtosis. We assume that many readers will have access to a computer, and most certainly to a calculator. The latter may be used to compute a skew coefficient, although computation of kurtosis is more difficult. Readers who have access to a computer with either SAS or SPSS-X are referred to Appendix A for instructions on how to obtain distribution statistics through these programs.

DISTRIBUTION STATISTICS FOR SPATIAL DISTRIBUTIONS

Since geographers are usually concerned with two-dimensional space, it is often desirable to have techniques to describe the central tendency and dispersion of phenomena that are distributed over both dimensions. These so-called *centrographic* statistics were introduced into the description and analyses of areal distributions late in the 1950s by David Neft and William Warntz in the United States and by Roberto Bachi in Israel. These techniques refer to people, places, and things distributed over the Cartesian coordinate system described in Chapter 2. Again, points in this system are located in terms of *X* and *Y* coordinates, which represent distances from two intersecting, perpendicular axes.

Figure 4.9 shows the City of Baltimore, Maryland, with an arbitrary set of *X* and *Y* coordinates designed to reference points within the study area. The *X* and *Y* scales are conveniently matched to the map's mileage scale. Locations of individual "control points" of minor civil divisions have also been specified in terms of *X* and *Y* coordinates. For instance, Civil Division 126's control point is mathematically located at (7,1).

The Bivariate Mean

As it was possible to describe the mean of *X*, it is also possible to specify the mean location of the points in *X*- and *Y*-coordinate space. That intersection in *X-Y* space is known as the **bivariate mean.** It may be thought of as the fulcrum of physical geographic space. To specify the bivariate mean of the *control points*, or arbitrary points of reference within Baltimore minor civil divisions, the arithmetic means, $\overline{X}$ and $\overline{Y}$, are computed for the coordinates of these divisions, using Equation 4.2. The full technical term for the intersection of these coordinates is the *unweighted bivariate mean center*.

Descriptive Statistics

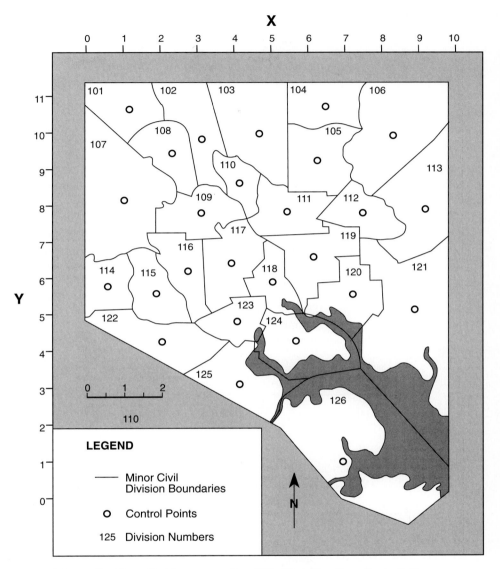

Figure 4.9 **The City of Baltimore on a One-Mile Cartesian Coordinate Grid**

In some cases, the geographic center of an area is of interest in its own right as, for instance, in the case of locating the center of a political jurisdiction. More often, however, one is concerned with distributions of people, places, or things in two-dimensional space. To locate the center of a distribution, a technique is needed that yields the *weighted bivariate mean center,* or *weighted centroid.* To find the coordinates of the weighted centroid, use Equations 4.11 and 4.12:

$$w\bar{X} = \frac{\sum_{i=1}^{n} w_i \cdot X_i}{\sum_{i=1}^{n} w_i} \qquad (4.11)$$

$$w\bar{Y} = \frac{\sum_{i=1}^{n} w_i \cdot Y_i}{\sum_{i=1}^{n} w_i} \qquad (4.12)$$

where w is the phenomenon that represents the value of any characteristic at any control point in the distribution.

One familiar example of weighted centroids is the westward march of population across the United States at ten-year intervals from 1790 to 1990 (Figure 4.10). After 1850 the advance of the population centroid was more rapid as the railroad pushed toward its West Coast destination. During the Great Depression of the 1930s, movement slowed considerably, as few people were financially able to migrate westward. After 1910 a shift southward also began, and in 1980, for the first time in history, the majority of the population lived in the southern half of the nation.

Another use for the weighted centroid is to compare the distributions of several different phenomena in the same area. One example from the literature is shown in Fig-

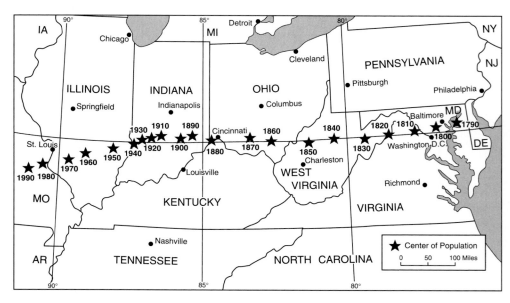

Figure 4.10 Location of Population Centroids in the United States, 1790–1990
Source: U. S. Bureau of the Census (1981:7).

Descriptive Statistics

ure 4.11. Schneider (1967) mapped the mean centers of physicians' offices in Cincinnati in the 1960s, both by medical specialty and over time. In this case, one can see that physicians began to move toward the northern suburbs after 1952. Certain specialties, such as pediatrics, also favored an office location nearer residential areas, particularly in the suburbs. Radiologists, who required an office near a hospital where their practice centered, tended to be clustered near the center of the city. Most centrographic analyses reveal only major shifts in aggregate populations and thus have limited application. In this respect, however, they are no different from single variable (univariate) descriptive statistics.

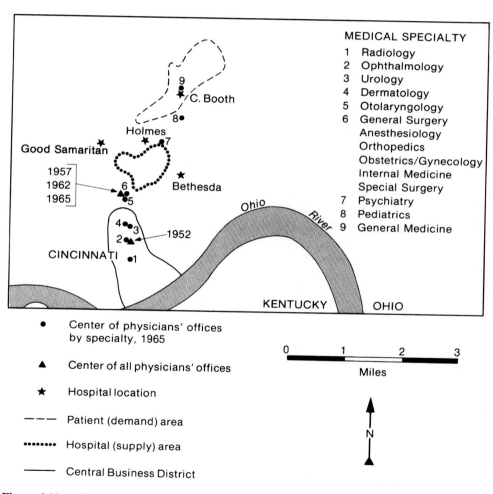

Figure 4.11 Physician Office Location Centroids, by Medical Specialty, Cincinnati, Ohio, 1965

Source: Schneider (1967:38).

Standard Distance

The dispersion concept also has its counterpart in bivariate descriptive statistics. Because distances represent deviations in the geographic sense, it is defined as the areal equivalent of standard deviation. The term used for this is **standard distance.** Just as in univariate statistics the deviations of observations are measured around their mean by the standard deviation (Figure 4.12a), so can the spatial dispersion of a point pattern be described by its standard distance, SD (Figure 4.12b).

The method used for calculating standard distance depends on whether the bivariate mean was weighted or not. If the bivariate mean was unweighted, then standard distance is computed as

$$SD = \sqrt{\frac{\sum_{i=1}^{n}(X_i - \bar{X})^2}{n-1} + \frac{\sum_{i=1}^{n}(Y_i - \bar{Y})^2}{n-1}} \qquad (4.13)$$

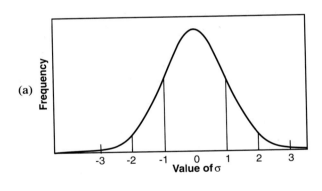

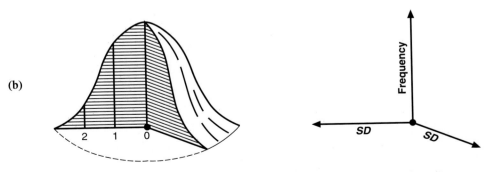

Figure 4.12 Standard Distance: (a) Univariate Standard Deviation, and (b) Bivariate Standard Distance

Descriptive Statistics

These sums of squared deviations are divided by $n - 1$, and the square root of their sum is derived. The procedure required to find *SD,* even using a modern hand calculator, is tedious and error-prone. Once calculated, *SD* can be drawn as a circle of this radius centered on the mean center, as shown in Figure 4.13. As with the standard deviation, standard distance is expressed in terms of the original measurement. In this case, the dispersion is in terms of miles, kilometers, or some other unit of distance.

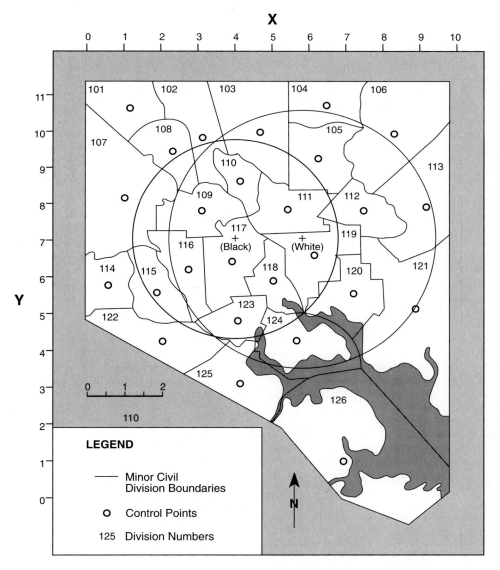

Figure 4.13 **Weighted Centroids and Standard Distances of the White and Black Populations in Baltimore, 1980**

Standard distance, although a convenient measure of dispersion, does not take into account the possibility that the spread about the mean center may be not circular but elliptical. An alternative method of calculation that yields a simplistic elliptical standard distance involves calculation from the separate axes, X and Y. First, compute the standard distance on X values:

$$SD_X = \sqrt{\frac{\sum_{i=1}^{n}(X_i - \bar{X})^2}{n-1}} \quad (4.14)$$

and, then the standard distance on Y values:

$$SD_Y = \sqrt{\frac{\sum_{i=1}^{n}(Y_i - \bar{Y})^2}{n-1}} \quad (4.15)$$

To demonstrate, data for black and white populations in Baltimore minor civil divisions are obtained from census tabulations (Table 4.8). X and Y coordinates are obtained from Figure 4.9. First, the X and Y components of the unweighted bivariate mean for Baltimore are computed. The values of X are summed and divided by 26 (the number of divisions); then the values of Y are summed and divided by 26. The result is $\bar{X} = 4.72$ and $\bar{Y} = 7.00$. To compute standard distance for X and Y, it is necessary to subtract each X coordinate from $\bar{X}$ and each Y coordinate from $\bar{Y}$, squaring the result for each civil division. A summation of $(X - \bar{X})^2$, and $(Y - \bar{Y})^2$, both divided by $n - 1$ (Equations 4.14 and 4.15), yields $SD_X = 2.40$ and $SD_Y = 2.41$. It is not surprising that the X and Y components have nearly identical dispersions because, with the exception of two southern areas (25 and 26), the control points of the areas fall within a shape that approximates a square. The relative dispersion of the civil division control points determines the orientation of the distribution. If the dispersion is not relatively circular, but along an ellipse, it becomes necessary to employ a trigonometric function to compute the width and length of the ellipse.[5]

When the observations are weighted, Equation 4.16 is required to compute a circle of standard distance:

$$SD_W = \sqrt{\frac{\sum_{i=1}^{n} w_i \cdot (X_i - \bar{X})^2}{\sum_{i=1}^{n} w_i} + \frac{\sum_{i=1}^{n} w_i \cdot (Y_i - \bar{Y})^2}{\sum_{i=1}^{n} w_i}} \quad (4.16)$$

---

[5]Ebdon (1985, Chapter 7) demonstrates this more complex method for computing the standard deviational ellipse.

Descriptive Statistics

Table 4.8   X-Y Coordinates, White and Black Population, Baltimore City, Maryland, Regional Planning Districts, 1980

Regional Planning District	X Coordinate	Y Coordinate	White Population	Black Population
101	1.3	10.6	15,831	5,209
102	3.2	9.8	6,670	594
103	4.7	9.9	19,906	906
104	6.4	10.7	16,474	7,927
105	6.3	9.2	9,498	27,287
106	8.3	9.9	39,913	4,992
107	1.1	8.2	4,448	47,220
108	2.4	9.4	1,921	40,754
109	3.2	7.8	112	15,325
110	4.2	8.6	15,874	612
111	5.6	7.8	14,908	30,335
112	7.6	7.8	16,307	3,223
113	9.2	7.9	23,400	13,119
114	0.7	5.8	6,533	9,186
115	2.0	5.5	3,780	19,954
116	2.8	6.2	4,386	36,878
117	4.0	6.4	14,170	69,093
118	5.1	5.9	9,045	6,023
119	6.2	6.5	5,847	64,591
120	7.2	5.5	42,690	3,434
121	8.9	5.1	17,804	1,566
122	2.1	4.3	14,584	2,059
123	4.2	4.9	5,357	1,826
124	5.7	4.3	16,123	920
125	4.2	3.1	5,443	16,332
126	7.0	1.0	14,089	1,746

Source: Baltimore Regional Council of Governments (1982).

Returning to the Baltimore data (Table 4.8), the difference in despersions between white and black populations is demonstrated with this technique. The standard distance of Baltimore's black population is smaller than that of the white population because the black population is more compact about its bivariate mean center than is that of whites.[6]

Computation of $w\overline{X}$ and $w\overline{Y}$ for the white population yields a weighted bivariate mean center of $\overline{X} = 5.87$ and $\overline{Y} = 7.22$. Substituting for the terms in Equation 4.15,

---

[6] A SAS program that computes the weighted values of $w \cdot (X - \overline{X})$, $w \cdot (Y - \overline{Y})$, $w \cdot (X - \overline{X})^2$ and $w \cdot (Y - \overline{Y})^2$ is shown in Appendix A.

$$SD_W = \sqrt{\frac{2{,}384{,}947}{345{,}113} + \frac{2{,}165{,}173}{345{,}113}}$$

$$= \sqrt{6.91 + 6.27} = 3.63$$

Figure 4.13 shows the location of the X and Y coordinates of the weighted centroid for the white population, and the standard distance is a circle of radius 3.63 (coordinate units) miles.

Computation of bivariate center coordinates for the black population yields values of $\overline{X} = 4.15$ and $\overline{Y} = 7.20$. Again, substituting for the terms in Equation 4.16,

$$SD_B = \sqrt{\frac{2{,}106{,}934}{431{,}111} + \frac{1{,}193{,}026}{431{,}111}}$$

$$= \sqrt{4.89 + 2.77} = 2.77$$

Noting the location of the black population mean center and its standard distance on Figure 4.13 shows that the black population is indeed more compact than the white population. The westward displacement of its mean center also reflects the westward dispersion of the black population in the city. In fact, the long axis of a standard deviational ellipse of the black population would be tilted to the northwest.

Comparative Dispersion

Standard distances based on different distributions can also be compared. The ratio of one standard distance to another may be found as a measure of relative dispersion (RD), provided the distributions come from the same population. This measure (Equation 4.17) is computed as

$$RD = \frac{SD_a}{SD_b} \tag{4.17}$$

where a and b are two different distributions. Suppose a standard distance for air pollution dispersion in a county was found to be 2.34 miles, and say the standard distance in the same county for auto ownership was 1.13 miles. The *relative dispersion* of pollution to auto ownership in this county may be computed using Equation 4.17:

$$RD = \frac{2.34}{1.13} = 2.07$$

Now, suppose the ratio of air pollution dispersion to auto ownership in a neighboring county is 5.75 miles to 2.83 miles:

Descriptive Statistics

Table 4.9 Absolute and Relative Measures of Population Dispersion: Selected Countries

Country	Standard Distance (km)	Standard Distance Radius
Australia	615	0.63
United Kingdom	134	0.77
Brazil	697	0.68
Japan	256	1.20
United States	839	0.86
India	538	0.85
China	579	0.52

Source: Taylor (1977:29).

$$RD = \frac{5.75}{2.83} = 2.03$$

Even though the average dispersion for the second county is larger than for the first, the comparison of the two relative measures show that the two counties are markedly similar with respect to their air pollution to auto ownership distribution ratios.

Another method for controlling for areal extent is to divide the standard distance measure by a function of the size of the area being described, such as the approximate radius of a circle drawn around the area. Although this introduces a chance for subtle error, since not everyone will draw the circle exactly the same way, it does allow for a more realistic comparison. Taylor (1977) gives an example where he describes the standard distances of the populations of several countries (Table 4.9) followed by a standardized measure derived by dividing each *SD* by the radius *(r)* of each nation. As a result it is possible to compare these countries directly. Japan has a very dispersed population about its mean center whereas China's population is relatively clustered about its mean center. Many other applications of centrographic techniques may be found in the geographic literature. A particularly interesting set of examples is in Chapter 6 of D. M. Smith's *Patterns in Human Geography*.

KEY TERMS

arithmetic mean 84
array 87
bivariate mean 104
central tendency 83
Chebyshev's theorem 98
coefficient of variation 101
comparative dispersion 112
dispersion 92
empirical rule 100
kurtosis 104
leptokurtic 103
median 87
mesokurtic 104
mode 87
negative skew 88
normal distribution 88
platykurtic 103
positive skew 88
range 90
skewness 103
standard deviation 98
standard distance 108
summation notation 84
sum of squares 94
variability 91
variance 96

REFERENCES

Baltimore Regional Council of Governments. (1982). *Census '80 Population and Housing Characteristics for Regional Planning Districts.* Part A, Table 1. Baltimore: Regional Planning Council.

Corbet, John H. (1976). *Physical Geography Manual.* Dubuque, IA: Kendall/Hunt Publishing Company.

Croxton, F. E., D. J. Cowden, and S. Klein. (1967). *Applied General Statistics.* Englewood Cliffs, NJ: Prentice-Hall, Inc.

Ebdon, David. (1985). *Statistics in Geography,* 2d ed. New York: Basil Blackwell.

Schneider, J. B. (1967). *The Spatial Structure of the Medical Care Process.* Philadelphia, PA: Regional Science Research Institute, Discussion Paper Number 14.

Smith, David M. (1975). *Patterns in Human Geography.* New York: Crane Russak & Company.

Stahl, Sidney M., and James D. Hennes. *Reading and Understanding Applied Statistics.* St. Louis, MO: The C. V. Mosby Company.

Taylor, P. (1977). *Quantitative Methods in Geography.* Atlanta, GA: Houghton-Mifflin.

Trewartha, Glenn T., Arthur H. Robinson, and Edwin H. Hammond. (1968). *Fundamentals of Physical Geography.* New York: McGraw-Hill Book Company.

U. S. Bureau of the Census. (1981). *Statistical Abstract of the United States: 1981,* 102d ed. Washington, DC: U.S. Government Printing Office.

Wayne County Health Department. (1976). *Data 76.* Detroit, MI: Wayne County Health Department, Air Pollution Control Division.

EXERCISES

1. According to Chebychev's Theorem, (at least) how many observations, where $n = 25$, $\bar{X} = 75$, and $s^2 = 100$, lie in the interval 45 to 105?

2. Suppose a sample ($n = 500$) of the distances traveled by commuting students to reach your campus produced a bell-shaped distribution with $\bar{X} = 25$ and $s^2 = 100$. How many students commute between 25 and 45 miles to reach the campus?

3. A frequency distribution of county population densities in your state projects a bell-shaped distribution with a mean of 30.8. If 84 percent of the counties have population densities of *at least* 41.2, what is the standard deviation for your state's population density distribution? (Hint: see Figure 4.6.)

4. Prove that the expression shown below is equal to zero.

$$\sum_{i=1}^{n}(X_i - \bar{X}) = 0$$

5. Prove that

$$\frac{\sum_{i=1}^{n}(X_i - \bar{X})^2}{n} = \frac{\sum_{i=1}^{n} X_i^2 - \frac{\left(\sum_{i=1}^{n} X_i\right)^2}{n}}{n}$$

Descriptive Statistics

6. Calculate the mean, median, variance, and standard deviation for a sample of 30 observations from *one of the twelve months* of the precipitation data for Des Moines, Iowa (Appendix B-1). Construct a histogram for these data, using six class-intervals of equal range. Do these data support the *empirical rule*?

Population for Canadian Cities 20,000 and Over (in Thousands)

112	36	25
96	35	25
84	34	25
78	34	25
67	34	25
62	33	24
57	32	24
56	32	24
55	31	24
54	31	24
53	31	24
51	31	23
50	31	22
48	30	22
47	30	22
47	30	22
47	29	22
45	29	21
45	28	21
44	28	21
44	28	20
43	28	20
41	27	20
40	27	20
39	27	20
38		20

7. Use the data on Canadian cities to construct a *city size distribution*. Use 10 units as class intervals, i.e., 20-29, 30-39, 40-49, etc. Include all cities over 70,000 in the last class interval.

8. Given that $\bar{X} = 36.6$ and $s = 17.8$, within what range (minimum and maximum) would $1.5s$ of Canadian city sizes fall? According to Chebyshev's Theorem, what percentage of cases *should* fall within this interval?

9. Calculate the range, mean, variance, and standard deviation for the temperature data for Los Angeles, California, for the forty-year record in the table at the top of page 116. Plot the histogram for these data on a graph, using six equal class intervals. Do these data support the *empirical rule*?

Average December Temperature (°F) for Los Angeles, California, 1941 through 1980

Year	Temp.	Year	Temp.	Year	Temp.
1941	56.8	1955	55.6	1968	54.3
1942	55.0	1956	58.8	1969	58.8
1943	56.6	1957	60.8	1970	56.1
1944	56.4	1958	62.4	1971	52.7
1945	54.7	1959	59.6	1972	56.8
1946	54.4	1960	54.9	1973	58.3
1947	55.3	1961	54.1	1974	55.8
1948	52.8	1962	56.6	1975	57.7
1949	53.2	1963	57.9	1976	60.1
1950	59.1	1964	55.2	1977	60.9
1951	54.0	1965	55.1	1978	54.3
1952	55.3	1966	57.5	1979	60.6
1953	56.7	1967	54.1	1980	60.2
1954	57.6				

10. Compute the coefficient of variation for rainfall in San Diego, California, and Boston, Massachusetts, based on the information given on page 101.

Suspended Particulate and Sulfur Dioxide Levels, Wayne County, Michigan, 1977

Sampling Station	TSP*	SO_2*
1	64	.009
2	74	.013
3	75	.013
4	74	.014
5	108	.021
6	51	.005
7	57	.008
8	84	.010
9	94	.011
10	66	.009
11	60	.005
12	56	.009
13	50	.008
14	94	.018

*Total suspended particulate (TSP) is given as the annual geometric mean micrograms per cubic meter; sulfur dioxide is given as the annual geometric mean parts per million.

11. Air pollution monitors located at air sampling stations in Wayne County, Michigan, are located as shown on the map. Average annual mean pollutants for 1977 are shown in the table. Using

Descriptive Statistics 117

these data and the procedures set out in this chapter, compute the weighted *mean areal center* and *standard distance* of either pollutant. Make a photocopy of the map of Wayne County. Place a + symbol at the mean areal center, then sketch a circle with radius SD_w around the mean center. Standard distance units are the same as those of the Cartesian coordinates shown at the margins.

12. Obtain a base map of your state, which shows the boundaries of counties or other minor civil divisions. Establish scales for the X and Y coordinates, following the example of the Wayne County, Michigan, problem. Then obtain data for population, or any other variables that are reported at the county (or other) level. Following the procedures outlined in this chapter, locate a weighted bivariate mean center and standard distance for your chosen variable. Place a + symbol at the mean areal center, then sketch a circle with radius SD_w around the mean center.

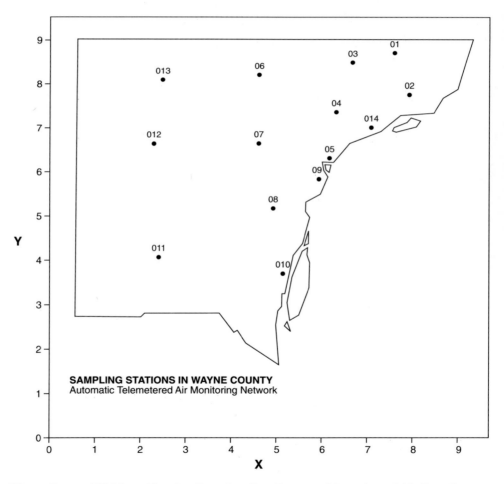

Wayne County, Michigan, Showing Cartesian Coordinates and Locations of Air Sampling Stations

Probability and Probability Distributions

5

The purpose of this chapter is to provide a brief background in probability and its application to data distributions. Because we wish to emphasize applications in this text, we will restrict our discussion of probability to some basic concepts. The theory of probability provides the foundation for the inferential techniques that may be integrated into hypothesis testing, and ultimately, model building.

PROBABILITY OF AN EVENT

We encounter probability casually in everyday communication. For example, we may hear that there is a 40 percent chance of rain tomorrow, or that in baseball there is one chance in five that a certain batter will be intentionally put on base to minimize the damage he can inflict, and we fervently hope that the probability that our automobile will start is 1.0. We sometimes calculate chance and probability in such situations through the process of abstract reasoning. We expect the water to run when we turn on the faucet, and we expect a reaction from the government when we fail to pay our taxes. These expectations are based on inferences from data that we have accumulated over time. We also do not have to draw and count playing cards to arrive at the probability of drawing a heart out of a deck of playing cards. If the deck is standard and complete, we know that the probability of drawing a heart is 13/52. If our deck of cards has been tampered with, a form of bias has been introduced, and our reasonable expectation of structure fails.

Scientists use precise methods to arrive at probabilities. They conduct an experiment repeatedly and keep track of the outcomes. For example, a biogeographer may be interested in nutrient levels of plants under different environmental conditions. A typical experiment might involve measurement of elements such as calcium and phosphorus in plants located in fields selected at random from many locations. One hundred samples taken from the fields weekly over a two-year period might reveal a level of phosphorus ranging from 5 to 8.5 g/m^2, with a mean of 6.6 g/m^2. Upon constructing a histogram, the biogeographer might find that a level of 6.5 to 6.7 occurs 10 times out of the 100 samples, and the researcher can then state that the probability of finding a phosphorus level of 6.5 to 6.7 g/m^2 *in these samples* is .10.

This statement about the ratio of expected outcomes to the total number of observations is what is known as **unconditional probability.** This may be formally stated as

$$P = \frac{\text{Number of events}}{\text{Total possible events}}$$

It should be noted that each experiment is *independent* of all prior experiments. Chapter 2 in the discussion of rules for random sampling emphasized that each potential sample must have an equal probability of being selected. Thus, when one sample case was selected, it was necessary to replace the individual back into the pool, so that the probabilities for selection of all remaining cases remained the same. That is one example of independence—on each incidence of drawing a sample case, the chances for any potential individual to be selected would not be influenced by prior selections. Other examples involve flipping a coin, rolling dice, or drawing playing cards from a deck. Each flip, roll, or draw is independent of previous actions.

Chapter 1 established that geographic research conventionally relies on relatively small samples. For example, one might wish to use soil pH as a surrogate of annual rainfall. Sample data on pH and rainfall are collected at 300 locations in lieu of the literally millions of possible locations that compose the entire population. If conclusions must be drawn about the effect of rainfall upon soil pH from a small sample, it follows that some assurance is expected that an unbiased sample of that size is adequate to make the leap of inference to the population of soils. You shall soon see that you can accept something less than 100 percent assurance.

The classic introduction to the basic concepts of probability involves the proverbial "die." The standard die, of course, has six faces. As it is used here, rolling this die constitutes an **experiment.** If the die is not loaded, when rolled, it has an *equal chance* of turning up any number between 1 and 6. Also, each roll is an independent experiment. Since it is impossible to predict in advance what face will appear on a roll, it is said that this is a *random experiment.* If a roll may be considered as an observation from which a value of 1 to 6 will be received, then any *number* of rolls becomes the *sample* being taken from the infinitely larger *population.* We then use the term *sample space* to specify all possible outcomes of an experiment. If the die is rolled an infinite number of times, the probability of any number between 1 and 6 appearing would be equal to .166; that is, there will be one chance in six of rolling a 1, 2, 3, 4, 5, or 6. As the outcomes of these random experiments are independent and can vary from trial to trial, we refer to the outcome of a random experiment as a **random variable.**

Definition: A random variable X is a variable that assumes its value according to the outcome of a chance experiment.

Composite Outcomes

Very often the primary concern is not with such simple outcomes, but with a *collection* of simple outcomes. For instance, with the roll of the die one might wish to know the probability corresponding to obtaining at least a 4, which is equivalent to the collection of simple outcomes 4, 5, and 6. Or one might be interested in the outcome of a roll that is an even number, which would be equivalent to the set of simple outcomes 2, 4, and 6.

Such a collection of simple outcomes is referred to as a *composite outcome.* In a given trial of an experiment, the composite outcome is said to occur if any one of the simple outcomes which comprise it occurs. Thus, in the die-rolling experiment, the composite outcomes of an even number will have occurred when any of the simple outcomes defining it—2, 4, or 6—occurs.

In order to find the probability of a composite outcome, it is first necessary to invoke the **rule of addition.** That is, the probability of finding one condition *plus* the probability of finding the other condition yields the probability of both. Therefore, the probability of rolling an even value will be the sum of 1/6 + 1/6 + 1/6, the respective probabilities of the simple outcomes 2, 4, and 6. This yields a probability of 3/6 or .5.

Consider Table 5.1, which shows the regional breakdown of low birth-weight cases in the state of Maryland by race in 1986. What would be the probability of drawing at random a birth from the Eastern Shore *or* a birth from the Baltimore Metropolitan Area

Table 5.1 Births and Births of Low Birth Weight, by Race and County Group, Maryland, 1986

Area of Residence	Total Births	*Births of Low Birth Weight*[§]		
		Total	White	Nonwhite
Northwest Counties	4,837	276	244	32
Baltimore Metro Area[†]	21,870	1,349	1,026	323
Baltimore City	13,259	1,609	297	1,312
Nat'l. Capital Area	21,764	1,599	637	962
Southern Counties	3,383	209	126	83
Eastern Shore Counties	4,411	315	193	122
State Total	69,524	5,357	2,523	2,834
Normal Births	64,167		43,457	20,710

[§]Newborns weighing 2,499 grams or less.
[†]Metropolitan area counties, excluding the City of Baltimore.
Source: Maryland Vital Statistics Annual Report (1986:46).

from the total population of Maryland births during 1986? This is a composite situation because births are being drawn from two *mutually exclusive* (spatial) categories. There is no geographic overlap between the Maryland Eastern Shore counties and the Baltimore metropolitan area.

There were 4,411 births in Eastern Shore counties out of the population of 69,524 births in Maryland in 1986. The probability for an Eastern Shore birth is then:

$$P(A) = \frac{4,411}{69,524} = .063$$

There were far more births in the urban Baltimore metropolitan area than in the rural Eastern Shore counties. The probability for a birth in that area is:

$$P(B) = \frac{21,870}{69,524} = .315$$

The addition rule is used for finding this probability. The addition rule can be extended to cover more than two events, as long as they all are mutually exclusive. Births in one place are in no way dependent on births anywhere else. Thus, the probability of a birth occurring in one area or the other $\{P(A \text{ or } B)\}$ is:

$$P(A) + P(B) = .063 + .315 = .378$$

The addition rule applies only to *mutually exclusive events*. Two simple outcomes, by definition, cannot occur at the same time. Two composite outcomes, however, can. Consequently, not only would we want to know the probabilities of the individual composite outcomes, but also the probability of their *joint* occurrence. For example, what is the probability of a farmer growing crop A *and* crop B, or what is the probability of a commuter traveling one segment of the trip home on the expressway *and* another segment of trip on a nearby arterial street? Or, in general, what is the probability of occurrence of outcome A *and* outcome B? Symbolically this is written $P(A \text{ and } B)$.

Logically, the above *joint probabilities*, as they are often called, would be less than the respective probabilities of the individual outcomes. For example, let us say that A represents the probability of drawing at random a Baltimore metropolitan area birth from the population of births, and B represents the probability of drawing a white birth from the population. Note that these events are not mutually exclusive; a Baltimore County birth is from the same population as white births. Therefore, the probability of A and B $\{P(A \text{ and } B)\}$ is computed in the following manner. The probability for drawing a Baltimore metropolitan area birth is:

$$P(A) = \frac{21,870}{69,524} = .315$$

and, the probability for drawing a white birth is:

$$P(B) = \frac{45{,}980}{69{,}524} = .661$$

and the probability of drawing both A and B is:

$$P(A \text{ and } B) = \frac{21{,}870}{69{,}524} \times \frac{45{,}980}{69{,}524} = .208$$

When events are *independent, but not mutually exclusive,* the **rule of multiplication** is invoked. The product of the probabilities is yet another property of probability theory that enables us to calculate the chances for getting both A *and* B events when A and B events are independent. The characteristic of independence means that the probability of event A occurring remains the same regardless of whether B has occurred, and vice versa. Therefore, in the example just given, the location and race of the birth are independent.

There are times when a probability situation will require incorporating both the addition and multiplication rules to calculate the probability ratio. Returning to the die experiment, suppose we wish to know the chance for rolling a 1 and a 6 *in any order* with two rolls of the die. Here the situation is different from previous examples because the *order* of events for successes can be any possible combination. Order of events refers to whether we roll a 6 first and then a 1, or a 1 first and then a 6. In this case there would be only two possible *combinations* or order of events that could occur.

Referring again to Table 5.1, suppose we are interested in calculating the probability of drawing a birth from a northwest Maryland county, or drawing a white, or a nonwhite birth in three random draws from the birth population. The number of successful combinations of sequences or orders possible is six:

NORTHWEST/WHITE/NONWHITE	WHITE/NONWHITE/NORTHWEST
NORTHWEST/NONWHITE/WHITE	NONWHITE/NORTHWEST/WHITE
WHITE/NORTHWEST/NONWHITE	NONWHITE/WHITE/NORTHWEST

To calculate the probability of obtaining a northwest birth, white birth, *and* nonwhite birth in three draws from the 1986 birth population of Maryland, with replacement, the probabilities for each combination must ultimately be summed. Thus, with a northwest/white/nonwhite combination, the probability would be:

$$\{(4{,}837 \div 69{,}524) \times (45{,}980 \div 69{,}524) \times (23{,}544 \div 69{,}524)\}$$
$$= \{.07 \times .661 \times .339\} = .016$$

No matter which of the six combinations came up, the probability would be the same. There are *six* ways, or combinations, to draw *three* specified birth events. The probability for a specified order or combination of drawing a northwest county birth, a white birth, and a nonwhite birth is $(.016)^3$, or 6 times less likely than drawing the three births in *any* order.

Conditional Probability

When composite outcomes overlap (that is, are not mutually exclusive), *conditional probabilities* may be computed; that is, the probability of occurrence of one composite outcome, *given that another composite outcome has occurred*. Recall the discussion on sampling methodology in Chapter 2. A random number table was used as a basis for selecting a sample of locations for an experiment because each potential observation (place) was required to have an equal probability of being selected. To ensure that this state of equal probability holds from one selection to the next, the previously selected places must remain eligible for future selection. If a previously selected location is withdrawn from the pool, then probabilities for the selection of other places increase, as there are fewer possibilities left in the pool. In other words, the probability of selection of the next location is *conditional* on a reduced pool of remaining locations. In cases in which replacement of a draw into the pool is not maintained, *independence* of events does not exist.

What is the probability of a die roll being a 6, *given that the roll is even in value*? If B is the event that a student studies for many hours for an exam on applied statistics and A the event that he or she earns a high mark in the course, then would we expect the probability of A (a high grade) given that B (studying) has occurred to be greater than the probability of A without B? These examples involve situations in which two variables are related so that knowledge of one helps to predict the other.

Returning to Table 5.1, what is the probability of drawing a nonwhite birth of low birth weight, or the probability of drawing a Baltimore City birth (from all births in the state), *or both*? This is a conditional situation. The procedure to follow in this case is first to compute the probability of drawing a nonwhite birth of low birth weight from the population of births (Event A) :

$$P(A) = \frac{2{,}834}{69{,}524} = .041$$

Next compute the probability of drawing a Baltimore City birth from the population of births (Event B) :

$$P(B) = \frac{13{,}259}{69{,}524} = .191$$

But, note that 1,312 nonwhite, low birth-weight births are included in both Baltimore City births *and* in nonwhite ethnic groups' births. Since these 1,312 births have been added into the numerator twice, they must be *subtracted* to eliminate duplication. Thus, the probability that a birth chosen at random from the population is from Baltimore City, or from a nonwhite, low birth-weight group, or both, is

$$P(A \text{ or } B) = P(A) + P(B) - P(A \text{ and } B)$$
$$= .041 + .191 - \frac{1{,}312}{69{,}524} = .213$$

If a birth from Baltimore City had been drawn at random, then *replaced* before drawing another random birth, this would have been an example of independence. But, both Baltimore City births and nonwhite low birth-weight births were considered *simultaneously*. Independence of events does not exist in cases where replacement is not maintained.

The following summarizes the fundamentals of probability:

1. Probability refers to what will happen in the long run.
2. The probability for any given condition or experiment is the ratio of the number of events or outcomes of interest to the total possible events or outcomes, expressed in values from 0 to 1.
3. If we wish to know the probability for one *or* another event or outcome occurring, it is determined by adding the probability ratios for each experiment. The *addition rule,* as previously stated, demands that events A or B be *mutually exclusive;* that is, A and B cannot occur simultaneously in the same experiment.
4. If one wants to know the probability for having A *and* B occur simultaneously, it can be done by multiplying the separate probabilities for A and B. The *multiplication rule,* as just stated, demands that A and B be *independent.* This means that the occurrence or nonoccurrence of one has no influence on the occurrence or nonoccurrence of the other.
5. When events are not independent and it is recognized that the probability of A may be dependent on whether B occurs, the probability ratio is called a *conditional probability.*

PROBABILITY DISTRIBUTIONS

The concept of the *random variable* was introduced earlier. The examples we have seen involving births described situations that had random variables intuitively associated with them. Each draw from the pool of births yields a birth in which the place of residence, condition of birth weight, and race are chance occurrences. After a number of random draws a frequency distribution may be constructed, each column of which corresponds to one of the cells in Table 5.1. That is, there will be a certain frequency of Baltimore City births, a certain frequency of low birth-weight births in Baltimore City, a certain frequency of white, low birth-weight births, and so on. It is then quite simple to convert these frequency distributions into *probability distributions* by replacing the absolute frequencies for each class by its proportional frequency. We may differentiate between variables that are based on characteristics measured on a continuous (metric) scale and those that are measured on noncontinuous (discrete) scales.

Discrete Probability Distributions

In a discrete spatial distribution, each occurrence is separated by a finite distance. Some examples of discrete spatial phenomena are the locations of sink holes, shopping centers, glaciers, parks, airports, and farmsteads. The simplest examples of discrete distributions usually involve a coin flip or die-roll experiment, where the class intervals are nominal, such as "heads" and "tails," or nonfractional, such as 1, 2, 3, 4, 5, and 6.

For example, if the experiment is tossing a coin, the possible outcome would be either heads or tails. The probability of heads equals 0.5, and the probability of tails is likewise 0.5. Figure 5.1 represents a **probability histogram** for the die experiment, where X is an observation (or outcome) and the probability of X, $P(X)$, is equal to 1/6; 16⅔ percent of the area of the histogram falls in each bar. The area contained within the entire histogram represents a probability of 1.0. This comparison of probability and area is purposeful, as we will see later in this chapter.

The probabilities associated with tossing two dice can be calculated as follows. The outcome of throwing dice will produce some number between 2 and 12. Any number on either die has an equal chance of occurring. To roll "2" requires a "1" on each die; to roll "3" requires rolling "1" with one die, and "2" with the other, or vice versa. Hence, there are two ways to roll a "3" and only one way to roll a "2." Table 5.2 lists all the possible combinations.

There are thirty-six possible outcomes for this experiment. The probability for any particular *roll* is equal to the *possible combinations* associated with that roll, divided by the sum of all possible combinations, or 36. For example, there are 6 ways out of 36 possible combinations to roll "7":

$$P(7) = P(1,6) + P(6,1) + P(2,5) + P(5,2) + P(3,4) + P(4,3)$$
$$= 1/36 + 1/36 + 1/36 + 1/36 + 1/36 + 1/36$$
$$= 6/36 = 1/6$$

Again, as with the single-die experiment, the sum of probabilities associated with each roll will equal 1.0. The probability histogram for the two-dice experiment is shown in Figure 5.2. Following on the discussion of distributions in Chapter 3, if the tops of the bars of the histogram are connected with a smooth curve, we will obtain what is conventionally termed as a *probability distribution*.

The Poisson Distribution. Probability applications involve choosing a particular *theoretical* distribution that reflects the process that influences the *observed* geographical pattern and then testing for significance. Comparing actual data to a theoretical function and

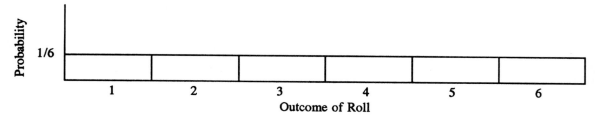

Figure 5.1 Probability Histogram for the Die Experiment When $P(X) = 1 \div 6$

Table 5.2 Possible Outcomes Associated with Rolling Two Balanced Dice

Roll	Possible Combinations					Probability
2	(1,1)					1/36
3	(1,2)	(2,1)				2/36
4	(1,3)	(3,1)	(2,2)			3/36
5	(1,4)	(4,1)	(2,3)	(3,2)		4/36
6	(1,5)	(5,1)	(2,4)	(4,2)	(3,3)	5/36
7	(1,6)	(6,1)	(2,5)	(5,2)	(3,4)	(4,3)	6/36
8	(2,6)	(6,2)	(3,5)	(5,3)	(4,4)	5/36
9	(3,6)	(6,3)	(4,5)	(5,4)		4/36
10	(4,6)	(6,4)	(5,5)			3/36
11	(5,6)	(6,5)				2/36
12	(6,6)					1/36
						Total = 36/36

testing for significance is not a new concept. One such theoretical concept involves *discrete random events* that occur in space and time. This concept is based on the probability of an event happening rarely, if at all, and if it does occur, that the time and place of that occurrence will be independent and random. Place or time *A* will be most likely to exhibit zero occurrences, somewhat less likely to exhibit one occurrence, even less likely to exhibit two occurrences, and so forth. Therefore, a probability distribution of zero, one, two, and so on occurrences will be peaked and skewed, as shown in Figure 5.3. A distribution based on these circumstances is called a **Poisson distribution** named after the French mathematician S. D. Poisson who described it in the 1830s.

This distribution has proven to be especially useful to geographers. As an example, suppose a 2000-year data set were available for the number of meteorites falling each year within a specified region on the moon (Table 5.3). All these meteorites fall at ran-

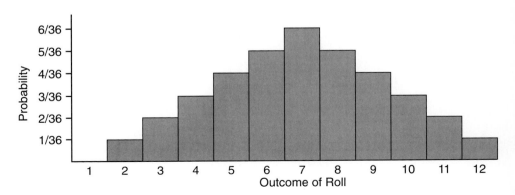

Figure 5.2 Probability Histogram for Two-Dice Experiment

Figure 5.3 A Typical Poisson Probability Distribution

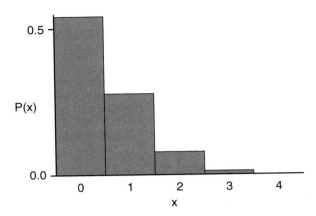

dom through the time period. There is no way of predicting when a meteorite will fall, but we can predict the probability of a given frequency of meteorites expected to fall within any year, assuming that the process that causes falling meteorites does not change. This frequency distribution does not produce a probability distribution that is symmetrical about the mean value of 0.61 sightings per year (1219 impacts in 2000 years). There were zero sightings in 1,093 of the years sampled, the maximum number of possible events per sampling unit of one year is 4 (656 percent greater than the mean), and events below zero are impossible. Hence, the distribution is skewed to the right.

While there is no way of predicting when a meteorite will fall, it may be possible (using the Poisson distribution) to predict the probability of a given number of impacts expected within any year. First, however, let us see if the Poisson distribution produces a distribution similar to that shown in Table 5.3 and if that variable has certain properties.

The number of meteorites (X) sighted in a year is the Poisson variable. In this case X will assume discrete values from 0 to 4. Before a Poisson distribution for these data is developed, however, the variable must meet certain criteria:

1. Its mean must be small relative to the maximum possible number of events per sampling unit. As previously indicated, the mean event is 0.61, and the maximum number of sightings per year is 4.0. The time period in which meteorites

Table 5.3 Frequencies of Meteorite Sightings within a Hypothetical Study Site on the Moon

Number of Meteorites per Year (X)	Number of Years
0	1,093
1	647
2	218
3	32
4	10

are counted must be long enough for a substantial number of impacts to occur; in this case it is one year. A time span of 1 week, for example, would be unrealistic for reporting sightings. Also, the total period of record should be large; in the present example it is 2000 years.
2. An occurrence of the event must be independent of prior occurrences within the sampled unit (1 year). Thus, the impact of one meteorite during the year must not enhance or diminish the probability of other meteorite impacts.
3. Finally, the probability of two or more events (impacts) occurring simultaneously is near zero.

If the occurrence of one event enhances the probability of another such event, a clustered or contiguous distribution is obtained. If the occurrence of one event impedes that of another such event during the sampled unit of one year, a distribution that is uniform over time results. The Poisson distribution can be employed as a model for *randomness*, or independence in time, and, as will soon be seen, in space.

The Poisson probability that in a given year X meteorites will fall is given as:

$$P(X) = \frac{e^{-\lambda} \lambda^x}{X!} \qquad X = 0, 1, 2, \ldots \tag{5.1}$$

where λ (lambda) = density, or mean occurrence per unit of time. In the meteorite example, this is expressed in impacts per year. ! (factorial) is the number of permutations of X. For example, the set $\{1\}$ has 1! permutation, namely, 1; the set $\{1,2\}$ has 2 permutations, namely, 12 and 21, and is represented symbolically as 2! The set $\{1,2,3\}$ may be represented 6 ways; thus $3! = 3 \cdot 2 \cdot 1 = 6$. e is simply a mathematical constant with a value 2.7183, and forms the base for natural or Napierian logarithms, in the same way that ten forms the base for the common logarithms introduced in Chapter 2.

Thus, the Poisson distribution is dependent only on the mean occurrence in time or space (in the present example 0.61). For example, if the Poisson probability for the occurrence of two impacts in a single year is desired, then:

$$P(2) = \frac{2.7183^{-0.61} \times 0.61^2}{2!}$$
$$= .101$$

Hence, for a period of record of 2000 years, $2000 \times .101 = 202$; there should be two sightings in 202 years. The symbol X is often referred to as a "state," but it means a specific number of events, or in our example, sightings. Equation 5.1 calculates what is known as the *Poisson density function*.

It follows that there is a "family" of Poisson probability functions, each member of which is specified by a particular value of λ. Figure 5.4 illustrates probability distributions for $\lambda = 0.2$, $\lambda = 0.5$, and $\lambda = 1.0$, respectively. This diagram shows probabilities only for small values of X, even though Equation 5.1 produces probabilities to all positive integers. The probabilities for values of X exceeding five grow

Figure 5.4 Three Poisson Probability Distributions

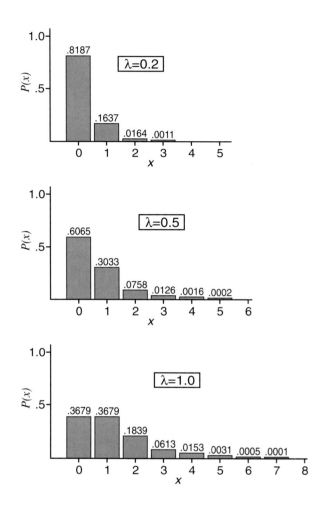

too small to illustrate on such a graphic. You can see by comparing these distributions that as λ increases, the probabilities for smaller values of X decrease, and probabilities for larger values of X increase. The steps in the computation process are shown in Box 5.1.

Now consider geographic distributions of spatially discrete phenomena, such as tornadoes, meteorite impacts, or traffic accidents. Plotting any of these phenomena could result in a map in which the items appeared to be clustered, uniformly spaced, or randomly spaced. In this instance *random spatial distribution* is used in a very special way and is given a very precise meaning. Henceforth, in a random spatial distribution of a set of points in a given area, it is assumed that any point has the same

> **BOX 5.1** **Procedure for Finding Poisson Probabilities and Expected Frequencies**
>
> 1. Establish a table with five columns, as shown below.
> 2. Multiply the individual values of X by their observed frequencies.
> 3. Sum the columns of f_O and $X f_O$.
> 4. Compute $\lambda = \Sigma X f_O \div \Sigma f_O = 1219 \div 2000 = .6045$
>   where f_O is *observed* frequency.
> 5. Compute individual values of $P(X)$ either with Equation 5.1 or with Appendix Table 6. Unless λ is *exactly the same value* as found in this table, computed probabilities will differ from those found in the table under the closest value of λ. The values below are from the table values under $\lambda = 0.6$. The rounding error just mentioned becomes apparent in the column headed $P(X)$, where $X = 2$.
> 6. Compute the values of $f_E = P(X) \cdot \Sigma f_O$
>   where f_E is the frequency *expected* in a Poisson process.
>
> | X | f_O | $X \cdot f_O$ | $P(X)$ | f_E |
> |---|---|---|---|---|
> | 0 | 1,093 | 0 | .5488 | 1,098 |
> | 1 | 647 | 647 | .3293 | 659 |
> | 2 | 218 | 436 | .0988 | 198 |
> | 3 | 32 | 96 | .0198 | 40 |
> | 4 | 10 | 40 | .0030 | 6 |
> | Totals | 2,000 | 1,219 | | |

chance of occurring in any unit area as any other point, and that the placement of each point has not been influenced by that of any other point. The assumptions underlying the Poisson process that leads to a random distribution are (Cliff and Ord, 1981):

- There are no interactions between different subareas, whether inhibitory or attractional;
- There is no possibility of multiple groupings of individuals within each subarea (no point clusters); and
- There is no tendency for neighboring areas to display similar traits (for example, because of common proximity to an important resource).

The mean density refers to the *intensity* of the process or the expected number of individuals *per unit area*. The idea is to compare the actual distribution of items in a study area

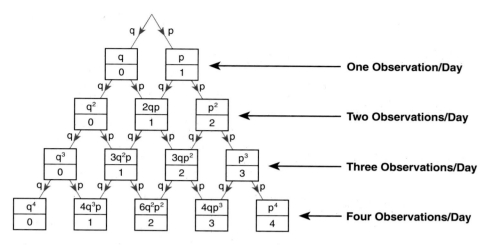

Figure 5.5 Probabilities Associated with One Through Four Dry Observations in a Binomial Process

with that which would be expected if they were distributed at *random*.[1] This technique is called *quadrat analysis* and will be demonstrated in Chapter 9.

The Binomial Distribution. A common probability distribution involves the repetition of events where there are only two possible outcomes. For instance, in certain places in the tropical biome, if we observed the weather at any given random moment of any random day out of any year, about half of the observations would be wet and the other half dry. A common form of shorthand is to symbolize the probability of a wet observation as p and the probability of a dry observation as q. Then, $p + q = 1.0$, as these are the only possible outcomes of all observations.

Now, consider the probability of having different numbers of dry observations in cases where one, two, three, or four daily observations are drawn from the long-run precipitation record. Figure 5.5 illustrates the outcome. Each horizontal line of boxes shows all the possibilities for precipitation with the stated number of random daily observations. The term in the upper half of the box is the probability of having the number of dry observations shown in the lower half of the box.

Thus, if three random observations are drawn from a random day, the probability of having two dry observations is $3qp^2$. The diagram shows why. To get to this box requires two moves (i.e., observations, shown by lines connecting the boxes) of probability p each, and one of probability q. The probability of doing this *once* is $q \times p \times p = qp^2$. But there are three possible alternative ways of achieving this (*ppq, pqp, qpp*, or dry, dry, wet, etc.). The total probability of reaching this box is therefore $qp^2 + qp^2 + qp^2 = 3qp^2$.

---

[1]Some examples are found in Boots and Getis, *Point Pattern Analysis,* and Unwin, *Introductory Spatial Analysis.*

If the terms in the upper halves of the boxes in Figure 5.5 are summarized, a familiar algebraic expression known as the **binomial expansion** is obtained:

$$q + p = (q + p)^1$$
$$q^2 + 2qp + p^2 = (q + p)^2$$
$$q^3 + 3q^2p + 3qp^2 + p^3 = (q + p)^3$$
$$q^4 + 4q^3p + 6q^2p^2 + 4qp^3 + p^4 = (q + p)^4$$

Thus, it is readily seen that the probabilities associated with all possible outcomes of n events (with only two possible outcomes) are given by the terms in the expansion of $(q + p)^n$. Table 5.4 shows how the binomial expansion may be applied to the wet-dry experiment.

The probabilities of *no dry* observations, *one dry* observation, *2, 3, 4, 5,* or *6 dry* observations in a day with six random "draws" may now be computed. Simply apply the expansion of $(q + p)^6$, or $6q^5p + 15q^4p^2 + 20q^3p^3 + 15q^2p^4 + 6qp^5 + p^6$. But if $p = q = 0.5$, this may be restated as follows (with the corresponding number of dry observations noted):

$(0.5)^6$ no dry observations;
$6\,(0.5)^6$ one dry observation;
$15\,(0.5)^6$ two dry observations;
$20\,(0.5)^6$ three dry observations;
$15\,(0.5)^6$ four dry observations;
$6\,(0.5)^6$ five dry observations;
$(0.5)^6$ six dry observations.

Since $(0.5)^6 = 0.015625 = 1/64$, these probabilities may be written as:

1/64   6/64   15/64   20/64   15/64   6/64   1/64,

or   0.016   0.094   0.236   0.312   0.236   0.094   0.016.

This result can be further illustrated by a histogram similar to that of Figure 5.2.

Table 5.4   **Binomial Expansion of Wet-Dry Observation Experiment**

$n = 1$			W		D		
$n = 2$		WW		WD		DD	
				DW			
			WWD		WDD		
$n = 3$		WWW	WDW		DWD		DDD
			DWW		DDW		
		WWWD		WWDD		WDDD	
$n = 4$	WWWW	WWDW		WDWD		DWDD	DDDD
		WDWW		DWWD		DDWD	
		DWWW		WDDW		DDDW	
				DWDW			
				DDWW			

The following are always true about any binomial ($p = q$) distribution based on a maximum q value of 6:

- The distribution will be symmetrical. There is the same likelihood of having six dry observations (and no wet ones) as there is of having no dry observations (and six wet ones);
- The mean will be 3. That is, if we divide the number of dry observations in a six-observation day, we expect an answer of 3;
- The greatest probabilities cluster around the mean. Most six-observation days will yield about half dry, half wet; and
- The greater the deviation from the mean, the smaller the probability of that number occurring. It is highly unlikely that all six observations will be wet ones, or that all six will be dry.

This would be of no use to a weather forecaster. The probability that it would be raining at any given time would remain about 0.5, unless all the other meteorological assumptions that help determine the condition of precipitation are relaxed. But, *over the long run,* for every 1,000 sets of six random daily observations, about 16 will be all dry, 94 will yield five dry and one wet, 236 will yield four dry and two wet, 312 will yield three dry and three wet, 236 will yield four wet and two dry, 94 will yield five wet and one dry, and 16 will be all rainy observations.

Continuous Probability Distributions

Continuous probability distributions involve data that are measured on metric scales. Several examples were given in Chapter 2. The probability model for the frequency distribution of a random variable, as was demonstrated with the dice experiment in the previous section, involves a smooth curve called the *density function.* In the example of the two-dice experiment (Figure 5.2), as in many "real-world" frequency distributions, the curve was approximately bell-shaped. As you observed in Chapter 4, these distributions come in many shapes.

The Normal Distribution. The most important probability distribution in statistics is the **normal distribution.** The normal distribution has already been shown (Figure 4.6), but the applications for this distribution have yet to be demonstrated. The normal distribution, based on metric measurement, forms the basis of a large group of statistical tests known as *parametric statistics,* in which an observed distribution of data values may be compared with a theoretical normal distribution. Several of these techniques are demonstrated in Chapters 6, 7, and 8. It is helpful, however, to understand first how the concept of probability that we have discussed may be combined with the normal curve.

Consider an example, a sample of fifty annual precipitation measurements for Denver, Colorado, between 1921 and 1970 (Table 5.5). A frequency distribution of these data demonstrates that the sample is approximately normally distributed (Figure 5.6). In Denver the amount of precipitation in any given year is more likely to be close to the long-term average than far away from that average. This phenomenon of *central tendency* is common in many biological and physical processes. If, for instance, a sample of the

Table 5.5 Annual Precipitation for Denver, Colorado, 1921–1970 (in Inches)

Year	Precip.	Year	Precip.	Year	Precip.
1921	14.55	1938	24.82	1955	11.58
1922	12.95	1939	7.95	1956	8.99
1923	21.42	1940	13.13	1957	17.71
1924	11.07	1941	21.25	1958	13.96
1925	9.78	1942	21.73	1959	13.74
1926	13.05	1943	10.37	1960	11.26
1927	15.88	1944	13.73	1961	17.66
1928	14.61	1945	13.14	1962	8.01
1929	15.42	1946	12.30	1963	11.84
1930	9.12	1947	16.21	1964	9.95
1931	13.71	1948	14.99	1965	18.49
1932	11.15	1949	17.57	1966	8.07
1933	15.22	1950	14.01	1967	16.03
1934	11.75	1951	15.50	1968	9.82
1935	17.90	1952	10.42	1969	18.32
1936	17.38	1953	12.56	1970	11.11
1937	12.24	1954	6.27		

Source: National Oceanic and Atmospheric Administration (1985).

heights of male college students is taken, extremely short or extremely tall students would occur less frequently than those near the average height for the population of male college students. This common occurrence of bell-shaped frequency distributions has led statisticians to refer to them as normal distributions.

A random sample of only a few years of precipitation data from the Denver record might unintentionally yield several dry or wet years and produce a frequency distribution that is skewed to the left or right. The larger the sample drawn, the more likely that it will be of symmetrical shape, even if the population distribution is slightly skewed. Furthermore, if many samples are drawn and the means of these samples computed, these sample

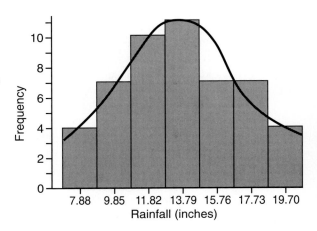

Figure 5.6 Frequency Histogram of Annual Precipitation, Denver, Colorado, 1921–1970

means will tend to be normally distributed. This is what is known as the **central limit theorem**.

Definition: According to the **central limit theorem**, if random samples of n observations are drawn from a population with finite mean, μ, and standard deviation, σ, a frequency distribution of sample means, $\overline{X}$, will be normally distributed with mean equal to μ and standard deviation $\sigma/\sqrt{n}$. The approximation of "normality" will become more and more accurate as n increases.

From this definition, it may be shown that the difference between $\overline{X}$ and μ will approach zero as n increases. In other words, when n is large, the probability density function for the *sample* will approximate the probability density function for the *population*. This result explains why the normal distribution and results derived from it are so commonly used with sample means, even when the population is not normal. The closer to normal the population, the smaller the samples need to be to yield a normal distribution of sample means. On the other hand, some populations require sample sizes well over 100 before the distribution of $\overline{X}$ becomes near normally distributed.

As one example, consider only city populations of United States cities of between 50,000 and 1 million inhabitants. A frequency distribution of these populations is heavily skewed to the right (Figure 5.7a), that is, there are an excessive number of cities with small populations. If several hundred random samples of $n = 25$ cities are drawn, the density function assumes a more normal shape but is still noticeably asymmetrical (Figure 5.7b). Increasing sample sizes to 100 improves symmetry, as shown in Figure 5.7c. Evidently even larger samples would be necessary to approximate normality with any assurance.

Recalling Figure 4.6, the mean, standard deviation, and the shape of the normal distribution are based on an entire population. The interval $\mu \pm \sigma$ contains 68 percent of the area under the normal curve, $\mu \pm 2\sigma$ contains just over 95 percent of the area under the curve, and $\mu \pm 3\sigma$ contains over 99 percent of the area under the curve. *Probability* is the vehicle that enables the statistician to use information in a sample to make inferences about, or describe, the population from which the sample was drawn. Thus, the central limit theorem enables us to reason from the population to the sample. This knowledge, and statistics, may be used to project from the sample to the population. For instance, the *empirical rule* tells us that 68 percent of the observations (in a normal distribution) will lie within one standard deviation. Distance from the mean is thus being measured in terms of standard deviations. Another way of putting this is that distance, in standard deviations from the mean, translates into area under the curve, which also equals the probability that an observation would be found at that distance from the mean.

Cumulative Probability Distribution Function. The bell-shaped density function shown in Figure 4.6 has density under the curve equal to 1.0, or 100 percent. However, when the probability shown under the density function is accumulated from 0 to 100 percent along the Y axis, the resulting curve is an s-shaped curve called a *normal probability*

Figure 5.7 Frequency Distributions of Populations of U.S. Cities between 50,000 and 1 Million

Source: Adapted from Snedecor and Cochran (1989: p. 47).

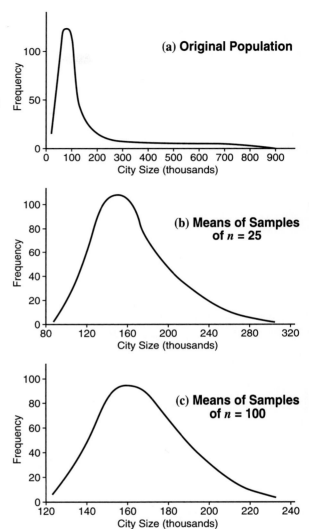

density function (Figure 5.8).[2] Note that 16 percent of the cases have "accumulated" at the mean minus one standard deviation and that 84 percent of the cases accumulate at the mean plus one standard deviation.

---

[2] The mathematical function $f(X)$ for normal probability density is:

$$f(X) = \frac{e^{-(X-\mu)/2\sigma^2}}{\sigma\sqrt{2\pi}}$$

where $(-\infty < X < \infty)$. The symbols e and π are constants, equal to 2.7183 and 3.1416, respectively. The population mean, μ, and the population standard deviation, σ, are employed as variables, as opposed to $\bar{X}$, and s, since small samples are less likely to produce symmetrical curves.

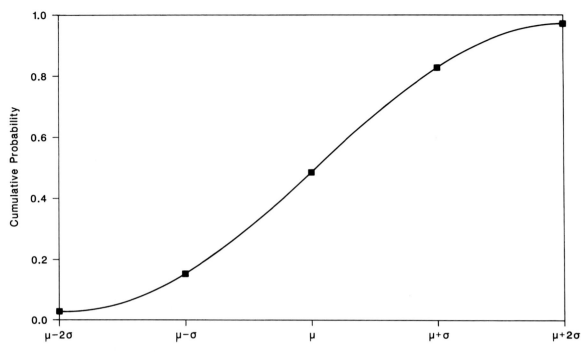

Figure 5.8 **Cumulative Probability Curve of a Normal Distribution**

Given certain properties of the normal distribution, we can make elementary probability statements. For example, the Denver precipitation data yield $\overline{X} = 13.79$ inches, and $s = 3.94$ inches. Therefore, the probability that Denver will receive either less than or more than 13.79 inches of rainfall, based on the period of record, is .50. What is the probability that Denver will receive at least 17.73 inches of precipitation in any given year? Since 17.73 is the mean plus one standard deviation, Figure 5.8 shows a cumulative probability of .84, or 84 percent. This was one of the points used to define the probability distribution function. The probability that Denver will receive more than 17.69 inches of precipitation in any given year is therefore equal to $(1.00 - .84)$, or .16. The probability that Denver will receive between 13.79 and 17.69 inches in any given year is equal to $(.84 - .50)$ or .34.

Standard (z) Scores. Now consider the probability associated with Denver receiving more than 20 inches of annual precipitation in one year. The event, $X = 20$ inches of annual precipitation, falls somewhere under the density function between $\overline{X} \pm 1s$ and $\overline{X} \pm 2s$. Symbolically, this is written $(\overline{X} + 1s < X < \overline{X} + 2s)$.

Since the normal curve depends on the two parameters μ and σ, there are many different normal curves. All standard tables of this distribution are for the distribution with $\mu = 0$ and $\sigma = 1$. Consequently if there is an observation X_i with mean μ and standard deviation σ and you wish to use a cumulative normal frequency distribution table (Ap-

Probability and Probability Distributions

pendix Table 2), you must rescale X_i so that the mean becomes 0 and the standard deviation becomes 1. The rescaled measurement is given by

$$z = \frac{(X_i - \mu)}{\sigma} \tag{5.2}$$

Equation 5.2 is operative only if we are transforming a value X_i relative to its *population* mean. If the transformed value X_i is relative to a sample mean, Equation 5.3 must be used:

$$z = \frac{(X_i - \overline{X})}{s} \tag{5.3}$$

The quantity z goes by various names —*standard normal variate, standard normal deviate,* or, as we will refer to it, *standard score.*

To find the probability of Denver receiving more than 20 inches of annual precipitation, first use Equation 5.3 to compute a z value corresponding to 20 inches of rainfall:

$$z = \frac{(20 - 13.79)}{3.94} = 1.58$$

Thus, 20 inches is 1.58 standard deviations to the right of the mean. In Figure 5.9, the probability of Denver receiving less than 20 inches of annual precipitation appears as an unshaded area under the curve. The shaded area, or *tail,* of the curve represents the probability that *more than* 20 inches of precipitation will be observed in a given year within the period.

The table of cumulative normal frequency distribution (Appendix Table 2) merits some explanation. In the body of the table are probability values that reflect the area under *half* of the normal curve. It is not necessary to show the values corresponding to the other half, since the curve is symmetrical. The values in the left-hand column and top row refer to z values, corresponding to the horizontal axis of Figure 6.3 or 6.6. The top row values are merely extensions of the z column values. To find the probability value associated with $z = 1.96$, locate 1.9 in the column of z, then along that row to the column

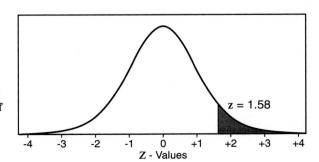

Figure 5.9 Standardized Normal Distribution for Annual Precipitation in Denver, Showing the Probability of 20″ or More of Precipitation

headed by 0.06, and read .4750. This tells us that the area 1.96 standard deviations to the right (*or to the left*) of the mean contains 47.5 percent of the area under that half of the curve:

z	0.00	0.01	0.02	0.03	0.04	0.05	0.06	0.07	0.08	0.09
.	.	.	.	.	.	.	.	.	.	.
1.9	.4713	.4719	.4726	.4732	.4738	.4744	.4750	.4756	.4761	.4767

Figure 5.10 shows the area under the normal curve associated with the probability of Denver receiving between 12.65 and 17.5 inches of rainfall. The z value for 12.65 inches of rainfall is $-.29$; the z value for 17.5 inches of rainfall is .94. From Appendix Table 2, we find a probability of .1141 for $z = -.29$, and a probability of .3264 for $z = .94$. The area to the *left* of the mean precipitation value (negative z) must be added to the area to the *right* of the mean (positive z). Therefore the probability that annual Denver rainfall would be between 12.65 and 17.5 inches is $.1141 + .3264 = .4405$. Box 5.2 sets out this procedure in detail.

Table 5.6 provides useful formulas for finding probabilities related to the normal distribution. Any combination of z values that refer to intervals under the normal curve may be found in this table. For instance, if the problem requires the probability of Denver rainfall outside the interval 12.65 and 17.5 inches, in terms of z scores, that is the interval to the left of .1141 and to the right of .3264. Since the area between those values $= .4405$, then the area remaining under the curve is $1 - 2A$ or $1 - (.1141 + .3264) = .5595$.

The Fractile Diagram. Chapter 2 showed that if a probability distribution is converted to an *ogive*, it assumes an *S*-shape (Figure 5.8). This phenomenon is based on the fact that the normal distribution is bell shaped. If the cumulative probabilities for each observation, arrayed from smallest to largest, are plotted on a sheet of *probability paper*, however, if the distribution is normal, the cumulative distribution assumes the form of a straight line (Figure 5.11). This is what is known as a *fractile diagram*. The advantage of the fractile diagram over the skew coefficient is that with the diagram you can observe which observations deviate most from the mean.

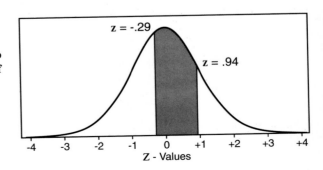

Figure 5.10 Area under the Normal Curve Corresponding to the Probability of Denver Annual Precipitation of Between 12.65" and 17.5"

| BOX 5.2 | **How to Convert z Scores to Probabilities Using the Table of Cumulative Normal Frequency Distribution** |

Problem: What is the probability that Denver would experience an annual rainfall between 12.65 and 17.5 inches of rainfall, based on the data of 1921–1970? During that period, $\overline{X}$ = 13.79 inches, and s = 3.94 inches.

First, convert the X values to z scores (Equation 5.3):

$$z = \frac{(X_i - \overline{X})}{s}$$

$$z = \frac{(12.65 - 13.79)}{3.94}$$

$$= -0.29$$

$$z = \frac{(17.5 - 13.79)}{3.94}$$

$$= 0.94$$

Since the first z score is negative, the probability associated with that value refers to the area to the left of $\overline{X}$. The positive z score indicates area to the right of $\overline{X}$. In Appendix Table 2, find the row in the body of the table corresponding to 0.2 under the column Z and under the column labeled 0.09. The area for this score is .1141. Then find the row in the body of the table corresponding to 0.9 under the column Z and under the column labeled 0.04. The area for this score is .3264. These two values must now be summed:

$$.1141 + .3264 = .4405$$

which is the probability that Denver should experience annual rainfall between 12.65 inches and 17.5 inches.

Table 5.6 Formulas for Finding Probabilities Related to the Normal Distribution

Probability of a Value	Formula*
(1) Lying between 0 and Z	A
(2) Lying between −Z and Z	2A
(3) Lying outside the interval (−Z,Z)	1 − 2A
(4) Less than Z (Z positive)	0.5 + A
(5) Less than Z (Z negative)	0.5 − A
(6) Greater than Z (Z positive)	0.5 − A
(7) Greater than Z (Z negative)	0.5 + A

* A refers to the area under the curve, found in the body of the table.

Source: From Snedecor and Cochran (1989:42).

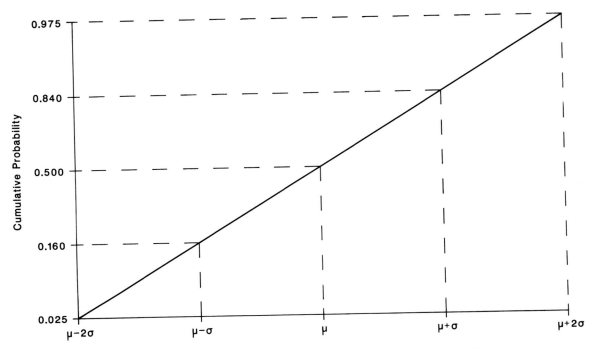

Figure 5.11   **Cumulative Probability of Normally Distributed Data on Probability Graph**

The fifty years of Denver precipitation data of Table 5.5 have been arrayed in Table 5.7. The probability that a precipitation value of 6.27 to 8.4 inches could occur in this sample (4 observations out of 50) is .08; the probability of 8.5 through 10.9 inches (7 observations out of 50) is .14, and so forth. The cumulative probabilities for the upper limits of each interval—8.4, 10.9, 13.4, 15.9, 18.4, 20.9, and 21+ inches of rainfall in this time period—are .08, .22, .50, .74, .90, .92, and 1.0. The cumulative frequency ogive for these data is shown in Figure 5.12a. A few wet years near the end of the period cause the curve to deviate from the cumulative normal density function of Figure 5.8.

A probability graph is structured to increase the scale for probabilities at the lower and upper ends of the scale, so that the gentle slopes in these areas are steepened just enough to equal the slope in the center of the graph. A plot of the same seven points from the Denver precipitation record on a probability graph is shown as Figure 5.12b. The straight line connects the mean (13.79 inches at $P = .5$) with $\pm s$ (9.89 inches at $P = .16$ and 17.69 inches at $P = .84$). That line represents the theoretical normal cumulative frequency distribution for the fifty-year Denver record.

Given that the fifty-year record is only a sample of the long-term Denver rainfall record, it is possible that some elements of the sample distribution, especially those of the six wettest years, were an aberration not balanced by several dry years that might have occurred before or after this fifty-year period. Thus, the record may contain a *sampling error*. The question is, do the points representing the observed data depart from the theoretical normal curve in excess of that attributed to sampling error? To answer this ques-

Table 5.7 Denver Precipitation Values (Inches), 1921–1970, Arrayed in Ascending Order

Rank	Value	Rank	Value
1	6.27	26	13.71
2	7.95	27	13.73
3	8.01	28	13.74
4	8.07	29	13.96
5	8.99	30	14.01
6	9.12	31	14.55
7	9.78	32	14.61
8	9.82	33	14.99
9	9.95	34	15.22
10	10.37	35	15.42
11	10.42	36	15.50
12	11.07	37	15.88
13	11.11	38	16.03
14	11.15	39	16.21
15	11.26	40	17.38
16	11.58	41	17.57
17	11.75	42	17.66
18	11.84	43	17.71
19	12.24	44	17.90
20	12.30	45	18.32
21	12.56	46	18.49
22	12.95	47	21.25
23	13.05	48	21.42
24	13.13	49	21.73
25	13.14	50	24.82

tion, it is necessary to compute values for **confidence bands.** The meaning and application of the term *confidence* will be given in the Chapter 6; for now, it suffices to know that it relates indirectly to sampling error. For any point on the cumulative probability scale, the width of the confidence band is given by

$$\pm [z \sqrt{(pq/n)}] \times 100$$

It is not necessary to compute the values for the confidence band, because they may be taken directly from Appendix Table 3. The left-hand column of that table refers to the sample size, which is 50 in this example. In each row there are two values for combination of pq. The upper value refers to what is called the .95 fractile, and the lower value coincides with the .99 fractile. In most applications, the 95 percent confidence limit is the band of choice. The column pq is also an error term, and it is conventional procedure to select $p = .95$. The intersection of this column with row $N = 50$ contains the value 6.0 percent. This value is in terms of cumulative probability. Thus, on the probability graph, at several points within the range of precipitation values, a distance of 6 percent should be measured above and below the line that represents the theoretical cumulative normal distribution function. The bands curve away from the line because distances at the extremes

Figure 5.12 Cumulative Probability Curve for Denver Precipitation Data, 1921–1970: (a) on an Arithmetic Graph, and (b) on a Probability Graph

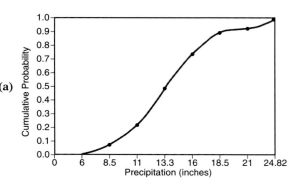

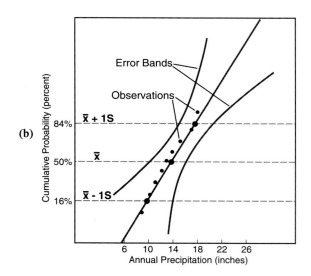

of the probability scale are stretched out by comparison with distance near the middle of the probability range (Figure 5.12b).

Recall that the dots on Figure 5.12b represent the upper limits on seven class intervals of the Denver data. Thus, the intervals with upper limits of 13.5, 16, and 21 inches have more (if the dots appear above the line) or fewer (if the dots appear below the line) observations than would be expected if the distribution were bell-shaped. In theory, 68 percent of the observations in a normal distribution should fall within ± 1 standard deviation of the mean. The mean of the fifty-year record is 13.79 inches and the standard deviation is 3.90 inches. A count of the frequency of values that fall between 9.89 inches and 17.69 inches reveals that nearly 70 percent of the observations fall within that interval.

This chapter has introduced the fundamentals of probability that will help you understand inferential statistics. In the following two chapters, attention is turned to testing hypotheses about the means and variances of population characteristic based on sample data. You should bear in mind that both *probability theory* and the concept of *random sampling* are essential components that will be required each time that a statistical experiment is conducted.

KEY TERMS

binomial distribution 132
binomial expansion 133
central limit theorem 136
conditional probability 124
confidence bands 143
continuous probability distribution 134
cumulative probability distribution 136
discrete probability distributions 125
experiment 120
fractile diagram 140
normal distribution 134
Poisson distribution 127
probability histogram 126
random variable 120
rule of addition 121
rule of multiplication 123
standard score 138
unconditional probability 120

REFERENCES

Boots, B. N., and A. Getis. (1988). *Point Pattern Analysis.* Newbury Park, CA: Sage Publications.
Cliff, A. D., and J. K. Ord. (1981). *Spatial Processes.* London: Dion.
Maryland Center for Health Statistics. (1986). *Annual Vital Statistics Report Maryland, 1986.* Baltimore: Maryland Department of Health and Mental Hygiene.
National Oceanic and Atmospheric Administration. (1985). *Local Climatological Summary: Denver, Colorado.* Washington, DC: U.S. Department of Commerce.
Snedecor, G. W., and W. G. Cochran. (1989). *Statistical Methods,* 8th ed. Ames, IA: Iowa State University Press.
Unwin, D. (1981). *Introductory Spatial Analysis.* London: Methuen.
U.S. Bureau of the Census. (1991). *State and Metropolitan Area Data Book, 1991.* Washington, DC: U.S. Government Printing Office.
U.S. News and World Report. (1989). "A Smoking Gun for Smokers," November 13, p. 89.

EXERCISES

1. If we desire to roll a 2 when we roll one die, what are our chances for success?
    a. 1/2 or .50   **b.** 1/6 or .17   **c.** 2/6 or .33

2. What would be the probability of randomly drawing a 6 from a box of nonrepeated numbers that are written on separate pieces of paper?
    a. 6/30 or .2   **b.** One chance in 6   **c.** More information is needed.

3. Write out the formula for finding the probability of any set of equally likely events.

4. What would be the probability of drawing an ace or a king from a deck of fifty-two cards? What *rule* do we apply to compute this probability?

Questions 5–12 refer to Table 5.1.

5. What is the probability of drawing at random a normal birth from the birth population of Maryland in 1986?

6. What is the probability of drawing at random a birth with low birth weight from the birth population of Maryland in 1986?

7. What is the joint probability of drawing at random a birth from the Baltimore Metropolitan Area, Baltimore City, and the National Capital Area from the birth population of Maryland in 1986?

8. What is the probability of drawing at random a white birth or a nonwhite birth from the Eastern Shore from the birth population of Maryland in 1986?

9. What is the joint probability of drawing at random from *all* low-birth-weight births a low-birth-weight birth that was from the southern counties *and* a low birth-weight birth that was nonwhite?

10. What is the probability of drawing at random from all Maryland births a low-birth-weight birth in the Eastern Shore region, or a low-birth-weight birth that was nonwhite, or both, from Table 5.1?

11. On a sheet of graph paper, construct a probability histogram of total low-birth-weight births for the six Maryland regions. Scale the vertical axis from "0" at the bottom to .30 at the top. Along the horizontal axis, label the bars by their regional names.

12. Extend Figure 5.5 by considering the probability of drawing five daily observations from the long-run precipitation record. Compute the probabilities when $p = q = 0.5$.

13. Using the nomenclature of Table 5.4, show the combinations possible with two dice (substituting a black die for W and a white die for D) for rolling a 4.

14. What is the probability that a z-score lies between -2.50 and -0.49?

15. The heights of a large sample of buildings in central business districts in the eastern United States were found to be approximately normally distributed with mean 67.56 feet and standard deviation 25.7 feet. What proportion of the buildings have heights less than 52 feet?

16. Using the information in Question 15, what height is exceeded by approximately 5 percent of the buildings?

17. Assume that you have a sample of annual precipitation data that has a mean of 25 inches and a standard deviation of 2.5 inches. Compute a value of z that corresponds to the likelihood of a year in which 20 inches of precipitation was recorded. Compute z scores with $s = 1.5$ and $s = 5.0$. How and why do the computed z scores differ as s changes?

18. The following table shows a random sample of U.S. metropolitan county data on percent change in money income per capita, 1979–1987. These data are normally distributed and some descriptive statistics for the data in that table are as follows:

$$n = 35 \quad \overline{X} = 58.25 \quad s = 12.55$$

From this information, what is the probability that a county's per capita income would have changed by 25 percent?

Random Sample of U.S. Metropolitan County Data on Percent Change in Money Income Per Capita, 1979–1987

County	Percent	County	Percent
Montgomery (New York)	64.2	Kankakee (Michigan)	48.3
Madison (Georgia)	68.7	Cascade (Montana)	45.3
Travis (Texas)	63.3	Wood (West Virginia)	53.3
Ascension (Louisiana)	35.3	Pueblo (Colorado)	34.2
Broome (New York)	70.1	Peach (Georgia)	75.5
Burleigh (North Dakota)	48.1	Scott (Kentucky)	68.1
Chittenden (Vermont)	90.5	Chatauqua (New York)	57.5
Kendall (Illinois)	50.1	Anderson (S. Carolina)	67.2
Union (Ohio)	65.3	Thurston (Washington)	46.6
Clark (Ohio)	60.6	Bristol (Rhode Island)	87.0
Douglas (Colorado)	56.2	Dona Ana (New Mexico)	61.5
Henderson (Kentucky)	53.6	Bullitt (Kentucky)	59.7
Martin (Florida)	60.8	El Paso (Colorado)	63.1
Bradford (Florida)	64.0	Howard (Indiana)	52.8
Fort Bend (Texas)	37.4	Jasper (Missouri)	58.6
Boyd (Kentucky)	46.1	Jackson (Oregon)	50.3
Rankin (Mississippi)	57.7	Whatcom (Washington)	49.6
Onslow (N. Carolina)	68.1		

Source: U. S. Bureau of the Census (1991).

19. What is the probability that a county's per capita income would have changed between 49.6 and 65.3 percent?

20. If you wished to know the probability that there was a corn farm in Winnebago County, Illinois, in 1989, what data would you require to compute that probability?

Four of five people who die of lung disease each year in the United States are smokers. The following table shows the death rate from lung disease by state in 1986. States (including the District of Columbia) have been sorted in descending order by their lung disease death rate (per 100,000 population).

Death Rate from Lung Disease by State, United States, 1986

State	Rate	State	Rate
Wyoming	49.1	**U.S. Median**	**29.5**
Nevada	48.3	Utah	29.4
Colorado	42.6	Maryland	29.2
Montana	41.1	Pennsylvania	29.1
New Mexico	39.9	Oklahoma	29.0
Kentucky	39.6	Rhode Island	28.9
West Virginia	39.4	Iowa	28.4
Idaho	39.0	Florida	27.6
Arizona	37.9	Illinois	27.5
Washington	37.5	South Carolina	27.5
Maine	36.6	North Carolina	27.4
Alaska	35.4	Massachusetts	26.5
Oregon	34.9	Arkansas	25.6
Indiana	34.4	New York	25.6
Vermont	34.1	South Dakota	25.6
Ohio	32.7	New Jersey	25.3
California	32.5	Texas	25.3
Tennessee	32.2	Nebraska	25.0
Michigan	31.5	Louisiana	24.9
Missouri	31.1	Mississippi	24.7
Georgia	31.0	Wisconsin	24.0
New Hampshire	30.9	Minnesota	23.7

21. What is the probability that a state death rate equal to that of your state would be drawn at random from the population? ($\mu = 30.9$; $\sigma = 6.74$)

22. What is the probability that a state drawn at random would yield a rate *above* 30 per 100,000?

23. Compute standard normal deviates (z) for the *range* of the data.

24. Using Equation 4.8, compute the value of Sk for these data. Is this distribution skewed? If so, in which direction?

Inferential Statistics and Hypotheses Involving Means

6

There are several purposes for statistics. One of the most common uses—shown in Chapter 4—is to reduce large quantities of data to manageable and understandable form. It is difficult at best to digest 100 soil pH values obtained at four different forest sites, for instance, but if means and standard deviations are computed for pH at the four sites, a biogeographer can readily describe differences in pH at the four sites. As we pointed out in an earlier chapter, a *statistic* is a measure computed from a sample. The geographer usually takes the means and standard deviations of samples as epitomes or summaries of the *populations* from which the samples were obtained.

Another purpose of statistics is to help researchers make reliable decisions, or **inferences,** from sample data by carrying out statistical experiments. No inference from sample to population would be necessary if the entire population could be directly measured. Usually it is not economically feasible to measure some characteristic for every individual. An *inference* is a proposition or generalization derived by reasoning from other propositions, or from empirical evidence (sample data). In statistics, several inferences may be drawn from tests of *statistical hypotheses*. For example, suppose the biogeographer finds that the mean pH in the sample of 100 measurements was 6.2. If the pH of the soil in the biome from which the samples were taken is expected to be 7.0, is the difference between the sample mean, 6.2, and the theoretical expectation, 7.0, *significant*? Is the difference of 0.8 small enough to warrant saying that the result was due to some deviant pH values drawn *by* *chance* in the sample? Is it large enough to lead us to believe that it is a *significant* departure from chance expectation? As scientists, researchers must be skeptics. Scientists assume that all results are chance results until demonstrated otherwise. The core of our approach to data gathered through experiments is to *establish chance expectation as our hypothesis and to try to fit the sample data to the chance model.*

In the soil pH experiment, if the sample mean pH is "close enough" to the expected value, we say that the mean of the sample data is "not significantly different" from that which was expected, and we attribute the difference to chance. If the mean is "sufficiently far" from the expected value, it is said that the mean is "significantly different," and cannot be attributed to chance alone. Inferences may be made *from samples to populations or from one sample to another.* When the biogeographer says that the pH sample mean is 6.2, it may be inferred that because $\overline{X} = 6.2$, $\mu = 6.2$, or near enough to 6.2 in the population from which the sample was drawn.

The researcher may be only indirectly interested in the population parameter. An economic geographer may, for instance, be studying the female labor force, where sample proportions of women in the labor force are obtained from two regions of a country or state. The hypothesis is that the proportion of women in the labor force of the two regions is not significantly different, and that any differences found in the two regions are due to chance. The economic geographer is interested in testing this hypothesis only in regions A and B. The experiment is carried out and statistical results are obtained that support the hypothesis—that there is no significant difference. The *inference* is made from the statistical evidence of a difference between region A, on the one hand, and region B, on the other hand, that the hypothetical proposition is correct in regions A and B. Thus, it is possible for one's interest to be limited to specific places.

Inferential statistics also allows assessment of the *probability of obtaining significant differences* of means, medians, variances, standard deviations, and so on from sample data under assumed population conditions. This will be demonstrated in the examples that follow. First, however, some fundamental terms are defined and the process of hypothesis formation and testing is explained.

BASIC CONCEPTS FOR INFERENTIAL STATISTICS

For beginning students, statistics is much like a foreign language. It is filled with jargon, including such terms as *confidence interval, null hypothesis, critical region,* and *Type I error.* All these terms, and more, will become clear with practice as you learn how to set up data-collection experiments, establish hypotheses, and carry out the tests for these hypotheses. You have already learned the basics of collecting sample data from Chapter 2, techniques for displaying these data in Chapter 3, and procedures for summarizing data distributions in Chapter 4. Assuming that the purpose for collecting sample data is to test a formal hypothesis, the following section explains the steps of this procedure.

Formulating and Testing a Hypothesis

Table 6.1 sets out six steps in the process of testing a hypothesis. The first step is to establish the hypothesis. The hypothesis is a prediction of how the statistics used in analyzing the data of an experiment will turn out. It is stated in such a way that it may be analyzed with statistical tools. It must be a very precise statement. To illustrate, the first example will involve a simple human geographic problem.

As part of a larger analysis involving the expenditure of tax funds in the United States, data have been collected that give the percent of the population that is over sixty-four years of age, by county, for 1984 (Table 6.2). A sample of thirty-eight counties has been selected at random from over 3,000 counties in the nation. We know that the mean proportion of county population over sixty-four years of age (μ) in the nation is 11.7 percent. We assume that our sample (n) of thirty-eight counties accurately reflects the population mean.

There are two ways to state a statistical hypothesis. The first is termed the **null hypothesis,** which is symbolized as H_0. This is what may be called the *chance hypothesis.* This form of hypothesis basically states that there is no significant relationship between the statistics that are being tested. In terms of the elderly population, some null hypotheses might simply state H_0: $\bar{X}_{SOUTH} = \bar{X}_{WEST}$, or H_0: $\bar{X} = \mu$, or H_0: $\bar{X}_{MIDWEST} - \bar{X}_{NORTHEAST} = 0$. The null hypothesis is a succinct way to express the testing of sample data against *chance expectation,* which is that the sample mean, standard deviation, and so on do not differ from, say, the population mean, standard deviation, and so on or that two sample statistics are not significantly different from each other.

Even though the null hypothesis states that the expected difference is zero, that does not literally mean 0.0. When we draw samples from the population of a variable (X), every element of that population (X_i) has an equal probability of selection, regardless of the magnitude of the values in that variable. Since the mean is influenced by extremely deviant values, the *chance selection* of two or three extreme sample values could pull the sample mean away from the population mean, such that $\bar{X} - \mu$ yields either a positive or negative value. This is what is termed *chance variation,* or *random error,* and the central limit theorem takes this into account.

Standard Error. A certain amount of fluctuation is expected around $\bar{X}_A - \bar{X}_B = 0$. The average amount of fluctuation differs with sample size. Recall that the central limit theorem stated that a distribution of sample means will on average be identical to the population mean. However, the theorem does not tell us how far, or different, any single sample

Table 6.1 The Six Steps in Hypothesis Testing

Step 1. Formulate the null (or alternative) hypothesis.
Step 2. Specify the appropriate sample statistic and its sampling distribution.
Step 3. Select a level of significance (α).
Step 4. Construct a decision rule.
Step 5. Compute the value of the test statistic.
Step 6. Make the decision.

Table 6.2 Percent of Population 65 Years of Age and Older, Randomly Sampled Counties, United States, 1984

County, State	Percent
Midwest	
Isanti, Michigan	10.6
Oconto, Wisconsin	15.5
La Porte, Indiana	11.6
Coles, Illinois	12.7
Jackson, Iowa	14.3
Ward, South Dakota	9.4
Jefferson, Wisconsin	12.8
Adams, Ohio	13.2
Clay, Minnesota	11.2
Dubuque, Iowa	11.9
Dodge, Wisconsin	16.9
Genesee, Michigan	8.9
Northeast	
Randolph, West Virginia	14.3
Franklin, Massachusetts	14.7
James City, Virginia	9.3
Windham, Connecticut	12.7
Caledonia, Vermont	13.3
Essex, New York	15.5
Caroline, Maryland	14.5
Bedford, Pennsylvania	13.8
South	
Chesterfield, South Carolina	11.3
Lenoir, North Carolina	11.4
Cooke, Texas	15.0
Yazoo, Mississippi	14.2
De Kalb, Alabama	14.2
Lauderdale, Mississippi	12.8
Jackson, Mississippi	7.0
Franklin, North Carolina	14.1
West	
La Paz, Arizona	10.8
Elko, Nevada	8.6
Yamhill, Oregon	13.1
Grant, Washington	11.0
McKinley, New Mexico	5.2
Deschutes, Oregon	12.6
Washoe, Nevada	9.5
Canyon, Idaho	12.4
Fairbanks, Alaska	5.9
Kauai, Hawaii	11.5

Source: U. S. Bureau of the Census (1988).

mean will be from the population mean. In order to assess this, it is necessary to formulate a summary measure, termed the **standard error,** that describes the distribution of the sample means around their average value. This measure represents the same sort of information as the standard deviation described in Chapter 4, except that in this use it refers to the distribution of sample means rather than individual cases in a single sample. The symbol for standard error is $\hat{\sigma}$ (pronounced "sigma hat").

The concept of standard error was discovered by researchers who generated all possible samples of size n from a population and calculated the mean of their sampling distribution. They then calculated the distribution of the individual sample means. This was a tedious procedure. Thanks to the perseverance of these researchers, we now accept that the standard error is equal to the ratio of the population standard deviation to the square root of the sample size:

$$\hat{\sigma} = \frac{\sigma}{\sqrt{n}} \qquad (6.1)$$

In other words, it is possible to use the distribution of the individual observations in the population to calculate the standard error rather than use the sample means from all possible samples. However, the population standard deviation is usually unknown. Therefore, we conventionally replace Equation 6.1 by an alternative formula, which uses the only information that we have, the sample standard deviation:

$$\hat{\sigma} = \frac{s}{\sqrt{n}} \qquad (6.2)$$

It may seem strange and arbitrary that we may calculate this summary measure based on the ratio of the observed standard deviation from a single sample of data to the square root of the number of observations in it. It is defensible, however, because of certain distributional properties of the sampling distribution of the mean. If certain conditions hold, the distribution of the sample means tends toward normality and the mean of the sampling distribution equals the population mean on the average. This will be amplified in the section on specification of the sample statistic.

From the sample of county populations over sixty-four years of age (Table 6.2), the mean of that sample is expected to differ from the known population mean of 11.7 percent, but only because the sample is small relative to the number of counties in the nation. The probability of drawing thirty-eight values that are within only a few tenths of a percent of the population mean is low. Thus, there must be some way to account for chance selection of deviant values in testing a null hypothesis, and the standard error provides the necessary leeway. In the tests that follow, the standard error will be an integral element in the testing of hypotheses. For each form of statistical test, there is a corresponding expression for the computation of the standard error.

The second way of stating a hypothesis, symbolized as H_1, is called the **alternate,** or *research,* hypothesis. The alternate hypothesis states that there *is* a significant difference between the statistics. Symbolically, this is designated as $H_1: \overline{X} \neq \mu$, or $H_1: \overline{X} > \mu$ or

$H_1: \overline{X} < \mu$. *The null hypothesis is the one that is tested directly; the alternate hypothesis is supported when the null hypothesis is rejected as being unlikely.*

The need for two hypotheses arises out of a logical necessity: the basis of the null hypothesis is to avoid what is known as a *Type I error*. A **Type I error** is the rejection of a null hypothesis when it is actually true. Statisticians symbolize the probability of making a Type I error as α. If a critical value of $\alpha = .05$ is established, H_0 will be erroneously rejected approximately 5 percent of the time. In order to avoid this type of error, α could be set at a much lower level, say .001, signifying a risk of making a Type I error only about one time in every thousand. However, unless there is a compelling reason to be extremely conservative about making a Type I error, the .05 level is routinely adopted as a rejection level in geographic experiments. This is because the .05 level corresponds closely to two standard deviations from the mean of a normal probability distribution. It is considered a reasonably good gamble. Some researchers prefer the .01 level of significance. This is quite a high level of certainty; however, one chance in a hundred is too stringent, and some significant results may be discarded simply because of variation in a sample.

For example, suppose that a null hypothesis about a sample of soil pH in the eastern United States were actually true; that is, there is no difference between the sample mean and a mean pH that is not harmful to the environment. If a Type I error were made and the null hypothesis falsely rejected, there could be serious political and/or economic consequences. It could, for instance, lead to the spending of millions of federal dollars enforcing pollution regulations when in fact the regulations were too stringent. In situations such as these, it makes sense to set a more conservative level of significance (for example $\alpha = .01$). However, the lower we set α, the greater is the likelihood that we will make a *Type II error*.

In a **Type II error,** we fail to reject the null hypothesis when it is actually false. The symbol β is used to signify this error. The Type II error is far more common than its counterpart, Type I. For example, suppose the federal agency charged with enforcing pollution laws adopted the .01 level of significance as the basis of rejecting the null hypothesis about pH. Subsequent studies are conducted in which the result obtained would have occurred by chance only 2 percent of the time. The null hypothesis would thus not be rejected and funding would not be allocated for regulation. It is clear, then, that the lower the rejection level, the less the likelihood of a Type I error and the greater the likelihood of a Type II error. Conversely, the higher the rejection level, the greater the likelihood of a Type I error and the smaller the likelihood of a Type II error.

Specifying the Sample Statistic

Every statistical test is based on a test distribution. From the previous chapter, you are already familiar with the standard cumulative normal frequency distribution. There are several other distributions that geographers might use in statistical modeling—including the t, χ^2, F, and Poisson distributions. Which one is to be employed depends on the type of data, the size of the sample, and the characteristics of the population from which the sample data are drawn. For the testing of hypotheses about means, both the Z test and the t test are demonstrated in this chapter.

Tests of significance are divided into two major groups—parametric tests and nonparametric tests. A *parameter,* as stated in Chapter 4, is a population value. A U.S. census publication, *County and City Data Book,* lists the proportion of the population over sixty-four years of age for every county in the nation. The mean of those 3,000+ proportions is a parameter, μ. Similarly, the variance (σ^2) and the standard deviation (σ) of a population, or any other population measure, is a parameter. A *statistic,* conversely, is a measure computed from a sample. The random sample of counties of Table 6.2 yielded a mean ($\overline{X}$), variance (s^2), and standard deviation (s), among other statistics, which may be interpreted as estimates of the parameters for the proportion of the population over sixty-four.

Whenever statistical tests are employed, certain assumptions are made. Tests in which the hypotheses pertain to population parameters are called **parametric tests.** The first assumption with a parametric test is that the samples used to test hypotheses are drawn from populations that are normally distributed. In the case of the population over sixty-four, the assumption is made that county proportions of that age group are symmetrically distributed about the population mean of 11.7 percent. If the populations from which samples are selected are not normal, then statistical tests that depend on the normality assumption are violated. As a result, the conclusions drawn from sample statistics may be erroneous. When in doubt about the normality of a population, or when one knows that the population is not normally distributed, one should use a *nonparametric test* that does not make the normality assumption. Tests of this type are described in Chapter 9.

Another assumption of parametric statistics is known as the *homogeneity of variance assumption.* In analysis of variance, as you will see in Chapter 7, the variances within the subsamples are assumed to be homogeneous from subsample to subsample, within the bounds of sampling variation. If the variances are not homogeneous, the F test used in the analysis of variance cannot properly be used as a test statistic. The result is that an F test, which states that there is no significant differences among sample variances, may not produce a significant difference, when in reality there are significant differences between the means of samples.

The assumptions of normality and homogeneity have both been examined thoroughly by statisticians, and the evidence so far is that the importance of these assumptions is overrated. According to Kerlinger (1964, p. 259–260):

> Unless there is good evidence to believe that populations are rather seriously non-normal and that variances are heterogeneous, it is usually unwise to use a nonparametric statistical test in place of a parametric one. The reason for this is that parametric tests are almost always more powerful than nonparametric tests. (The power of a statistical test is the probability that the null hypothesis will be rejected when it is actually false.) . . . nonparametric tests are often quick and easy to use and are excellent for preliminary, *if not always definitive, tests.* [italics added]

A third assumption is that the data to be analyzed are measured on the continuous metric scale, interval or ratio. Parametric tests like the t and F tests depend on this assumption, but many nonparametric tests do not. Rank data, or frequencies of nominal classifications, for example, should be analyzed with nonparametric tests.

When testing hypotheses about means, two parametric tests may be employed if the assumptions just explained are not violated. If the null hypothesis is $\bar{X} = \mu$, and the population standard deviation σ is known, or if the sample is very large, then a Z test may be used. If σ is unknown, or if the sample size is under 30, a t test should be used. The reasons for this rule of thumb will become evident when the actual procedures for these tests are described.

Selecting the Level of Significance

Suppose it is necessary to perform a statistical test of the null hypothesis that there is no difference between the mean of the sample of county populations in Table 6.2 and the United States county mean. Suppose further that the outcome leads to an acceptance that there is no difference and that the two means are equal. The conclusion, as stated above, is to *accept the null hypothesis*.

Recall again our discussion of the *central limit theorem* from Chapter 5. A distribution of sample means approximates a normal distribution. However, even if the population distribution is skewed, the distribution of sample means will show *less* skew, the skew being inversely related to the size of the sample. As n approaches infinity, this distribution approaches normal (Figure 6.1). The kurtosis of the distribution of sample means may also be closer to normal than is the kurtosis of the actual population from which the samples were drawn. Based on this, sample means may be considered to be

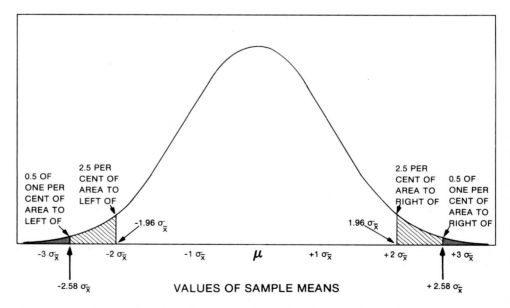

Figure 6.1 **Expected Distribution of Sample Arithmetic Means from a Normally Distributed Population**

normally distributed unless there is reason to believe that the population distribution from which they were taken is known to be markedly non-normal.

To state that one expects a sample statistic to deviate significantly only so many times out of 100, for example, is to establish a *significance level* (α). This is the probability of making a Type I error. A rather long-standing convention is to stipulate that a Type I error will be made only 5 times out of 100. Symbolically, this specification is stated as $\alpha = .05$. The smaller the α, the more deviant a sample statistic will have to be for us to reject H_0 (and thereby to commit a Type I error).

The α level establishes the size of the rejection region of whatever test distribution we are using. Figure 6.2 illustrates the location of these regions in a symmetrical distribution. A test statistic is computed according to specifications, and if that statistic diverges sufficiently from the population parameter (or another test statistic), the test statistic will fall into one of the *tails* of the test distribution. Where that cutoff point, called the **critical value,** occurs is a result of choosing a particular significance level. The tails of the curve are called the **rejection region,** meaning that the null hypothesis is rejected. The body of the curve (within the *confidence limits*) is termed the **acceptance region.** The confidence limits are related to the standard error, as will be demonstrated below.

One must be concerned with both tails of a distribution if a null hypothesis is stated. It is impossible to accept the null hypothesis if, for example, $\mu > \bar{X}$ or $\bar{X} > \mu$. Thus, the null hypothesis requires what is termed a **two-tailed test.** In most research experiments, however, the researcher has a reasonable expectation of the results, and it may be expect-

Figure 6.2 Confidence Intervals and Critical Regions in Two-Tailed and One-Tailed Significance Tests

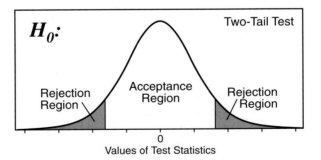

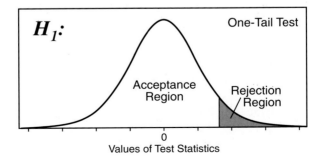

ed that one statistic will exceed another. For example you should expect soil pH to be lower in areas subjected to acid rain than in comparable areas where the atmosphere contained little or no sulphur pollution. One might expect rural Midwestern counties to have a higher proportion of elderly population than counties in other regions of the United States.

Whenever a directional alternative hypothesis, say $\bar{X} > \mu$, can be justified from prior empirical testing, then a **one-tailed test** must be applied. With a one-tailed test, the proportion of the distribution associated with the pre-established α level can all be located in one tail, which increases the power of the experiment (i.e., makes it more likely to find a significant difference when, in fact, one exists). Which tail is designated is a function of the expected sign of the test statistic. For instance, a negative t value refers to the left tail, while a positive t value refers to the right tail. The decision to establish a one-tailed test, however, should be based on an understanding of the process under study and not as a method of influencing the outcome of the test. The choice of significance level should thus be stated before the test statistic is computed.

Whenever a two-tailed test is in effect, the stipulated value of α must be split between the two tails of the test distribution. In Figure 6.2, the two shaded tails of the distribution are labeled $\alpha/2$. If $\alpha = .05$, then each tail contains .025 of the area under the curve.

COMPUTING TEST STATISTICS

Steps 5 and 6 of the hypothesis-testing procedure involve the computing of the test statistic and rendering a decision on the hypothesis based on the outcome of the test. The test statistic is a measure that is computed from the data of the sample. There are many possible values that the test statistic may assume, depending on the nature of the sample and the assumptions about the population. As will be demonstrated, the test statistic serves as a decision maker, since the decision to reject or not to reject the null hypothesis depends on the magnitude of the test statistic. The general form of a parametric test statistic usually appears as follows:

$$\text{test statistic} = \frac{\text{computed statistic} - \text{hypothesized parameter}}{\text{standard error of the computed statistic}}$$

The key to statistical inference is the choice of the test distribution. As stated above, the choice of either the Z test or the t test as a test distribution is determined by certain conditions. We will begin with a relatively rare situation in which a Z test should be used.

Significance of the Difference between $\bar{X}$ and μ When σ Is Known

To illustrate this test, let us return to the data of Table 6.2. Suppose it is known that the national proportion of population over sixty-four years of age is 11.7 percent. Furthermore, suppose it is also known that the standard deviation, σ, is 4.0 percent. The sample data of Table 6.2 yield a mean, $\overline{X}$, of 12.04 percent. A null hypothesis is that the difference between the sample and population mean is insignificant. This magnitude of difference is thus expected on the basis of sampling variation. Having no reason to choose a lower limit, we set $\alpha = .05$. As this is a two-tailed test, each tail of the test distribution will contain .025 of the area under the curve. Under the conditions stated earlier, the appropriate test distribution is Z for this experiment.

You are already familiar with Appendix Table 2, which is used with a Z test. You should also remember that the two tails of this distribution, because they contain 5 percent of the area under the curve, may be represented by a Z value of 1.96 (Figure 6.1). To refresh your memory, recall that the body of Appendix Table 2 contains areas that correspond to standard deviations from the mean. A P value of .4750 represents all but 2.5 percent of the area under half the normal curve (the table represents only half the curve). Finding .4750 in the body of the table, you may then see that this corresponds to $Z = 1.96$.

The equation for the Z significance test is

$$Z = \frac{(\overline{X} - \mu)}{\hat{\sigma}_{\overline{X}}} \qquad (6.3)$$

where $\overline{X}$ is computed from the sample, μ is the hypothesized parameter, and $\hat{\sigma}_{\overline{X}}$ is the standard error of $\overline{X}$, which is computed by the expression $\sigma \div \sqrt{n}$. If many random samples of size n of county population proportions over sixty-four years were drawn, the expected standard deviation of the distribution of sample means would be $\sigma \div \sqrt{n}$.

Substituting the values for $\overline{X}$, μ, and $\hat{\sigma}_{\overline{X}}$ from the elderly population example, you may now solve for Equation 6.3:

$$Z = \frac{12.04 - 11.7}{4.0 / \sqrt{38}} = .52$$

You will recall from Chapter 5 that Appendix Table 2 was employed to find the probability that any given value X_i would depart from its distribution mean μ, with given σ. Now this table may be used to determine the probability that a sample mean differs significantly from the population mean with a given standard error. A portion of that table showing the referent row appears as follows:

Z	0.00	0.01	0.02	0.03	0.04	0.05	0.06	0.07	0.08	0.09
.5	.1915	.1950	.1985	.2019	.2054	.2088	.2123	.2157	.2190	.2224

If all the assumptions of this test are met, the probability that a sample mean of 11.7 percent (given $\sigma = 4.0$ percent) would differ from the population mean, 12.04, is .1985. The null hypothesis stated that there was no significant difference between the sample mean and population mean. The critical region for this test begins at $P = .4750$. Thus, the Z value based on the deviation of sample and population means (.52) is far below the critical value (1.96) established as the point where the null hypothesis could no longer be accepted. Figure 6.3 illustrates the position of $Z = .52$ relative to the critical

Figure 6.3 The Value of Z for $\overline{X} = 11.7$, $\mu = 12.04$, and Standard Error = .65, Relative to its Critical Z Value, 1.96

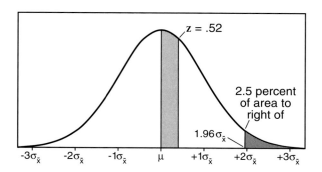

value indicating the rejection region of the normal test distribution. If the sample mean had diverged far enough from the population mean so that a Z value of say, 2.0, had been the result of the Z test, you could not have accepted H_0.

Confidence Limits for μ When σ Is Known

The question might arise, how far from the population mean, μ, would the sample mean have to deviate to cause one not to accept H_0? In other words, what are the limits of $\overline{X}$ within which researchers may be confident that their sample mean could deviate from μ solely due to sampling variation? With a random sample of size n, where n is large enough that $\overline{X}$ can be assumed to be normally distributed, one can calculate these upper and lower limits for $\overline{X}$. From knowledge of the normal distribution, regardless of where $\overline{X}$ is located relative to μ, approximately 95 percent of the possible values of $\overline{X}$ that constitute the distribution are within two standard deviations of the mean. The lower and upper limits for $\overline{X}$ are $\mu - 1.96\sigma_{\overline{X}}$ and $\mu + 1.96\sigma_{\overline{X}}$, so that the interval, $\mu \pm 1.96\sigma_{\overline{X}}$, will contain approximately 95 percent of the possible values of $\overline{X}$. This is what is termed the *confidence interval* for μ. Given any sample mean, $\overline{X}$, and Z value, the lower and upper confidence limits may be computed by

$$\overline{X} \pm Z_{(1-\alpha)}\sigma_{\overline{X}} \tag{6.4}$$

where $\sigma_{\overline{X}} = \sigma \div \sqrt{n}$.

From the elderly population example, $n = 38$ counties, $\overline{X} = 12.04$, and $\sigma = 4.0$. An approximate 95 percent confidence interval for μ may then be computed:

$$\mu = 12.04 \pm (1.96)(4.0 \div \sqrt{38})$$
$$= 12.04 \pm 1.27$$
$$= 10.77 \text{ percent and } 13.31 \text{ percent}$$

You cannot be *sure* that the population mean falls within the limits just given, but you may be 95 percent confident that it does. The remaining 5 percent of the time, you may expect, due to sampling variation, that the true mean lies outside these limits.

Inferential Statistics and Hypotheses Involving Means

In this example, where $Z = 1.96$, the confidence interval is interpreted as follows: In repeated sampling approximately 95 percent of the intervals constructed by Equation 6.4 will include the population mean. The total area under the curve of $\overline{X}$ *beyond* the confidence limits (α) is $\mu \pm 1.96\sigma_{\overline{X}}$. The area *within* the confidence limits is then $1 - \alpha$. The following is the probabilistic interpretation of Equation 6.4:

> In repeated sampling, from a normally distributed population, $(1 - \alpha) \times 100$ percent of all intervals of the form $\overline{X} \pm Z_{(1-\alpha)}\sigma_{\overline{X}}$ will in the long run include the population mean, μ.

As stated earlier, you may use any reasonable confidence interval; however, the most frequently used values are .95 and .99.

Determination of Sample Size for Estimating Means. In Chapter 2, the procedure for drawing samples was discussed, but the question of how large a sample to take was postponed because the explanation involves the concepts of confidence interval and standard error. The size of sample selected for an experiment is important because to take a larger sample than is required to achieve the desired results is inefficient, and samples that are too small often lead to results that are of no practical use. What follows is a technique for determining the sample size required for estimating a population mean.

The concept of the confidence interval was just demonstrated, showing that the width of this interval is determined by $\pm Z \hat{\sigma}_{\overline{X}}$. Notice that increasing the magnitude of Z produces a wider interval. If the value of Z is inflexible, however, the interval width may be controlled only by reducing the standard error, which is done by increasing the sample size. Thus, the size of sample is determined by the size of σ, the desired degree of reliability, and the desired interval width. Assuming that one is drawing an *independent* sample from a population of sufficiently large size, the minimum sample size required is expressed by

$$n = \frac{Z^2 \sigma^2}{d^2} \tag{6.5}$$

where $d = Z\hat{\sigma}_{\overline{X}}$.[1]

---

[1] Populations that number fewer than several hundred are large enough to warrant ignoring a problem that arises from estimating sample size from a "small finite population." In the unlikely event that the population is smaller than about 500, it is suggested that you apply a finite *population correction,* in which case

$$d = Z \frac{\sigma}{\sqrt{n}} \sqrt{\frac{N-n}{N-1}}$$

which, when solved for *n*, yields

$$n = \frac{NZ^2\sigma^2}{d^2(N-1) + Z^2\sigma^2}$$

These equations require a knowledge of σ, but it is rare that we know the value of the population standard deviation, and it must be estimated. If an estimate cannot be made from a previous or similar study of the phenomenon of interest, it will be necessary to collect a small preliminary sample and to compute s as an estimate of σ. Or, if you can be reasonably sure that the population from which the sample is to be selected is approximately normally distributed, you may assume that the range (R) of the variable is about equal to 6 standard deviations, and compute σ ≈ $R/6$. This approach obviously requires some knowledge of the smallest and largest value of the variable in the population.

As an example, consider once again the elderly population experiment. What minimum size sample should be selected from among the over 2,000 nonmetropolitan counties in the United States? First, you must choose the desired width of the confidence interval, the level of confidence desired, and the magnitude of the population variance in the population over sixty-four.

Assume that you could accept an interval width of about 4 percent (within about 2 percent of the true value of the mean in either direction). Also assume a conventional α level of .05. Further, assume that a preliminary sample revealed that the population standard deviation is about 6 percent. You now have the necessary information to compute the sample size: $Z = 1.96$, σ = 6.0, and $d = 2$. Employing Equation 6.3, we obtain

$$n = \frac{(1.96)^2 (6)^2}{(2)^2} = 35.6$$

Rounding up to the next whole integer yields a sample size of 36. This is the minimum sample size that is necessary to compute a sample mean that is reliable in the terms that have just been discussed.

The *t* Distribution and *t* Tests for Differences in Means

As stated before, population parameters are rarely available, particularly the standard deviation, σ. A test of the difference between a sample mean and the population mean almost always involves an unknown population variance, forcing one to estimate that variance from the data of the sample before a significance test can be applied.

An estimate of σ may be obtained from Equation 4.6, the sample standard deviation. Now an estimate of the standard error may be computed using Equation 6.2. Returning to the sample data of U.S. county proportions of population over sixty-four years of age, the sample standard deviation, s, was 2.66 percent. Substituting this into Equation 6.2 yields

$$\hat{\sigma}_{\bar{X}} = \frac{s}{\sqrt{n}} = \frac{2.66}{\sqrt{38}} = .43$$

Since $\hat{\sigma}_{\bar{X}}$ is only an estimate of the standard error of the population mean, the Z test is inappropriate. A test distribution for this purpose was invented by W. S. Gossett (writing under the pseudonym Student) in 1908 and modified by R. A. Fisher almost twenty years later. This distribution has since been known as the **Student *t* distribution.** It is an important procedure in the statistics of small samples. As a test of differences between sample and population means, it is expressed symbolically as

$$t = \frac{\overline{X} - \mu}{\hat{\sigma}_{\overline{X}}} \qquad (6.6)$$

Thus, t is the deviation of a sample mean from its population mean measured in units of the mean's estimated standard error. A table of the distribution of t is found in Appendix Table 4.

The t distribution is a family of curves. Figure 6.4 shows the distribution of t expected with a relatively small sample. Values are dispersed beyond $t = 4$. The larger the sample, the less dispersed is the distribution. For increasingly small samples, t is more dispersed than the normal distribution, reflecting the greater likelihood of departure of sample means from the population mean. The distribution of t required for a particular experiment must take into account the number of **degrees of freedom** (ν) available. This term refers to the latitude of variance a statistical problem has. To illustrate, refer to the elderly population data of Table 6.2. The thirty-eight values sum to 457.7, and the mean is 12.04 percent. Recall from Chapter 4 that if you subtract the sample mean from each county percentage, the sum of those thirty-eight deviations should equal zero. Also note that these deviations are not independent. Once it has been established that the deviations are taken from the mean, the values of only thirty-seven deviations are free to vary. As soon as thirty-seven deviations are known, the thirty-eighth is completely determined. To generalize, for any given sample on which a single restriction has been placed, the number of degrees of freedom is $n - 1$. In the population experiment, $n = 38$; therefore, $\nu = 38 - 1 = 37$. In other words, by computing a mean, one degree of freedom has been removed.

Common sense tells us that if n is small, chance alone might produce a considerable difference between s and σ. A glance at Appendix Table 4 reveals that as degrees of freedom (ν) shrink, the value of t increases at all levels of α, to allow for a broader acceptance region for H_0. As ν increases, t decreases and approaches Z. Note, for example that at $\nu = \infty$, where $\alpha = .05$, both t and $Z = 1.96$. The t statistic will vary much more than the Z statistic with small samples. For example, consider the means and standard deviations

Figure 6.4 A Distribution of t Values for a Small Sample

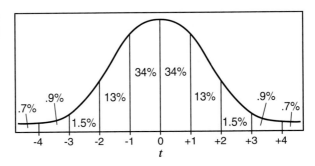

of four random samples (n = 5, 10, 15, and 20) drawn from the Denver precipitation record, 1921–1970 (Table 6.3). As the sample size increases, the value for s decreases, approaching σ, and, if each value of $\overline{X}$ were contrasted with μ using Equation 6.6, computed values of t would approach Z for the same sample size.

To illustrate the use of the t statistic with a known population mean (μ), a sample mean ($\overline{X}$), and unknown population standard deviation (σ), reconsider the county proportions of elderly population in the United States (Table 6.2). Since σ is unknown, the standard deviation of the sample data is computed as an estimate of σ, yielding $s = 2.66$. Then with Equation 6.2, the estimate of the standard error is calculated as .44. Using the .05 α level as a criterion, test the null hypothesis that $\overline{X} = 12.04$ percent is the mean of a random sample from a population with $\mu = 11.7$ percent. Substitute into Equation 6.5, and solve for t:

$$t = \frac{\overline{X} - \mu}{\hat{\sigma}_{\overline{X}}} = \frac{12.04 - 11.7}{.44} = .77$$

From the t table of Appendix Table 4, it appears that, for $\nu \approx 37$, P lies between .6 and .7, and the hypothesis may not be rejected. The interpretation is that there is a high probability (~.65) that a difference in means of only .34 with a standard error of .44 and

Table 6.3 Samples of Annual Precipitation for Denver, Colorado, 1921 through 1970 (Inches), n = 5, 10, 15, and 20

$n = 5$	$n = 10$	$n = 15$	$n = 20$
13.05	11.15	17.66	17.90
21.25	10.37	21.25	9.78
6.27	18.32	17.57	11.07
21.73	6.27	15.50	16.21
7.95	18.49	13.73	10.37
$\overline{X} = 14.05$	11.75	15.22	13.14
$s = 7.24$	16.21	6.27	17.57
	14.55	11.15	21.42
	21.42	10.42	15.42
	21.73	11.75	7.95
	$\overline{X} = 15.03$	13.14	8.07
	$s = 5.10$	12.30	14.01
		14.99	14.29
		12.56	8.01
		21.73	15.50
		$\overline{X} = 13.78$	12.56
		$s = 4.48$	13.13
			14.99
			21.25
			10.42
			$\overline{X} = 13.51$
			$s = 4.17$

Inferential Statistics and Hypotheses Involving Means

37 degrees of freedom is expected *by sampling variation alone*. The *t* distribution table in the Appendix is designed for two-tail tests, and should be referenced at the stated α level with the null hypothesis in mind. The row of the table corresponding with $\nu = 40$ appears as follows:

Level of Significance (*P*)

ν	70	.60	.50	.40	.30	.25	.20	.10	.05	.025. . .
40	.388	.529	.681	.851	1.050	1.167	1.303	1.684	2.021	2.329

Each row of the table relates to a *t* distribution with a unique number of degrees of freedom. The *t* distribution, like the normal distribution, is symmetrical. The critical values shown are sign-free. If the computed value of *t* is positive, as it is in this case, interpret the table value as positive also. Since $\alpha = .05$ was initially established, refer to that column in the table. The critical value ($\alpha = .05$) for the elderly population experiment is 2.021.

Confidence Limits of μ Based on $\bar{X}$ with Standard Error Estimated from Sample

In the elderly population experiment, we concluded that the sample mean was indeed the mean of a random sample selected from a population of U.S. counties with a mean of 11.7 percent over sixty-four years of age. From a knowledge of the sample alone, what can be said about the confidence limits within which μ may be expected to occur? Again, compute lower and upper limits for μ. Assuming a willingness to err no more than 5 times in 100, one value of μ will be obtained that cuts off the *lower* 2.5 percent tail of the distribution of sample means around μ, and another value of μ that cuts off the *upper* 2.5 percent of the distribution of sample means around μ.

Both values may be computed from Equation 6.7, in which you substitute the already computed values of $\bar{X}, \hat{\sigma}_{\bar{X}}$ and the *t* value for the appropriate confidence limits:

$$\mu = \bar{X} \pm t\hat{\sigma}_{\bar{X}} \quad (6.7)$$
$$\mu_1 = 12.04 \pm 2.021\,(.44)$$
$$\mu_2 = 12.04 \pm .89$$
$$= 11.15 \text{ and } 12.93 \text{ percent}$$

Again, you cannot be *sure* that the true population mean for the over sixty-four population falls within these limits, but you can be 95 percent confident that it does, given a mean that is based on a random, independent sample from a normally distributed population. In other words, if 100 samples of data were drawn from this record, and each of the 100 sample means were substituted into Equation 6.7, one would expect that the confidence limits computed would include the population value 95 times out of 100 and to exclude the population value 5 times in 100.

If you refer back to the computation of confidence limits for this experiment with known σ, you may now wonder why the confidence interval was wider in the previous computation until you examine the values that produced these intervals. The *population* standard deviation was 4.0, whereas the *sample* standard deviation was 2.66. With a larger value of standard deviation, a larger value of standard error is also obtained (.65 in the former example and .44 here). Despite the fact that the value of t corresponding to $\alpha = .05$ is larger than that for its counterpart in the Z test ($2.021t$ vs. $1.96Z$), the larger standard error, as it should have, pushed the confidence limits farther apart. Figure 6.5 illustrates this situation.

The *t* Test for the Difference between Independent Sample Means

The t test can also be used to compare two independent samples where at least one sample has a sample size under 30 and where you may assume that the two samples are

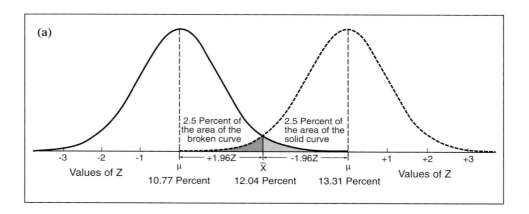

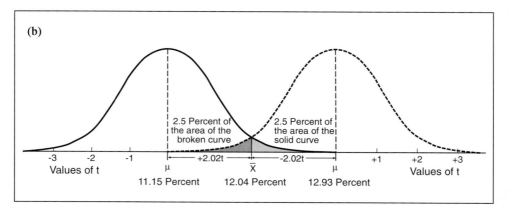

Figure 6.5 The 95 Percent Confidence Limits for μ for the Over 64 Population, with (a) $\sigma = 4.0$, $n = 38$, $Z = 1.96$, and (b) $s = 2.66$, $\nu = 37$, $t = 2.02$

from the same population in regard to their variances. The following problem illustrates a situation of this kind. A geographer is interested in comparing certain attributes of two drainage basins, one of 297 square miles and the other of 289 square miles. One characteristic of interest to the researcher is the annual discharge rate of the basins (measured in cubic feet per second, or cfs). To control for basin size, each discharge rate is divided by its basin area. A graduate assistant earlier collected data on (n_1) fourteen years of annual discharge rates from the first basin. Although the individual yearly discharge rates for this basin are not available, the sample mean flow per unit area ($\overline{X}_1$) for the basin has been reported by a reliable source to be 2.05 cfs/area, and its standard deviation (s_1) for the fourteen years has been reported to be .404. Using a secondary data source, the researcher collects a second random sample of ($n_2 = 9$ years) annual discharge rates from an adjacent basin. Based on other local research results, the variances of discharge rates from the basins are assumed to be virtually equal. From the second set of sample data, a mean ($\overline{X}_2 = 1.64$ cfs/area) and standard deviation ($s = .31$) are computed. A null hypothesis is formulated, which states that there is no difference in discharge rates per unit area between the two basins. Since one measure of difference between the basins is their mean discharge rate per unit area, the null hypothesis may be stated symbolically:

$$H_0: (\mu_1 - \mu_2) - (\overline{X}_1 - \overline{X}_2) = 0$$

Using a Type I error (α) of .05 as a criterion, the statistical procedure involves determining the probability of t, where t is the ratio of $\overline{X}_1 - \overline{X}_2$ to an estimate of the *standard error of the difference* between the two sample means. The values of σ for the populations of discharge rates are unknown, so we must again make an estimate of the standard deviations of the basin discharge rates, and again we do this with the information given by the two samples. The following procedure applies when the sample sizes are unequal.

As a first step, compute the *sum of squares (SS)* for each sample with either Equation 6.8 or 6.9:

$$SS = \sum_{i=1}^{n} (X_i - \overline{X})^2 \tag{6.8}$$

$$SS = \sum_{i=1}^{n} X_i^2 - \frac{\sum_{i=1}^{n}(X_i)^2}{n} \tag{6.9}$$

For the basin experiment, s_1 has been given and s_2 has been computed from the sample data. From Chapter 4, recall that a sample standard deviation is derived from its variance. Thus, the sum of squares for these samples may be computed directly with Equation 6.10:

$$SS = \nu\,(s^2) \tag{6.10}$$

Compute SS_1 and SS_2:

$$SS_1 = 13(.163) = 2.12$$
$$SS_2 = 8(.096) = .77$$

From these values, an estimated value of the standard deviation may be computed from the so-called *pooled variance*:

$$s_{1+2}^2 = \frac{SS_1 + SS_2}{(n_1 + n_2 - 2)} \tag{6.11}$$

Substituting the values obtained from using Equation 6.10 to obtain the sums of squares, compute

$$s_{1+2}^2 = \frac{2.12 + .77}{21} = .14$$

From this, the standard error of the difference between the two means may now be computed:

$$\hat{\sigma}_{\bar{x}_1 - \bar{x}_2} = \sqrt{s_{1+2}^2 \left(\frac{1}{n_1} + \frac{1}{n_2}\right)} \tag{6.12}$$

Or, *if the two samples are of the same size* ($n_1 = n_2$),

$$\hat{\sigma}_{\bar{x}_1 - \bar{x}_2} = \sqrt{\frac{s_1^2}{n_1} + \frac{s_2^2}{n_2}} \tag{6.13}$$

Again, substituting the values of $\hat{\sigma}_{1+2}$, n_1, and n_2 from the discharge experiment,

$$\hat{\sigma}_{\bar{x}_1 - \bar{x}_2} = \sqrt{.14\left(\frac{1}{14} + \frac{1}{9}\right)} = .16$$

Finally, the desired significance ratio may be obtained using Equation 6.14:

$$t = \frac{(\bar{X}_1 - \bar{X}_2)}{\hat{\sigma}_{\bar{x}_1 - \bar{x}_2}} \tag{6.14}$$

Substituting for the terms of the equation,

$$t = \frac{2.05 - 1.64}{.16} = 2.56$$

As this is a two-tailed test, consult Appendix Table 4, referring to $v = n_1 + n_2 - 2$ and $\alpha = .05$. Note that one degree of freedom was lost when the sum of squares for the first sample was computed about $\overline{X}_1$ and another degree was lost when the second sum of squares was computed about $\overline{X}_2$.[2] The probability that a $t = 2.56$ with 21 degrees of freedom would have happened by chance (sampling variation) is between .02 and .01. The critical value of t that defines the tail of the distribution is 2.08. The difference in means is too great to attribute to chance, and one cannot accept the null hypothesis that the discharge samples from the two drainage basins came from the same population.

Confidence Limits of $\mu_1 - \mu_2$

When you have concluded that a significant difference exists between $\overline{X}_1$ and $\overline{X}_2$ you may compute the confidence limits of the difference in means, $\mu_1 - \mu_2$. This is obtained by solving the equation

$$\overline{X}_1 - \overline{X}_2 = (\mu_1 - \mu_2) \pm t \hat{\sigma}_{\overline{X}_1 - \overline{X}_2} \tag{6.15}$$

for $\mu_1 - \mu_2$. The value of t depends upon (1) the α level chosen, and (2) the degrees of freedom, $v = n_1 - 1 + n_2 - 1$.

To illustrate the use of Equation 6.15, recall the discharge experiment. Remember that $\hat{\sigma}_{\overline{X}_1 - \overline{X}_2} = .16$. To obtain the $(1 - \alpha)$ 95 percent confidence limits of $\mu_1 - \mu_2$, consult Appendix Table 4, for $\alpha = .05$ and $v = 21$, $t = 2.08$. Substituting into Equation 6.15:

$$2.05 - 1.64 = \mu_1 - \mu_2 \pm (2.08)(.16)$$
$$\mu_1 - \mu_2 = .41 \pm .33$$
$$= .08 \text{ and } .74 \text{ cfs/area}$$

A One-Tailed t Test

An example of a one-tailed t test may be demonstrated with a hypothesis based on females in the labor force in Maryland minor civil divisions. Because there are more available job opportunities in metropolitan areas, one might expect that urban women would be more likely to be gainfully employed than rural-based women. In the population of 324 Maryland minor civil divisions, it is possible to define without ambiguity those that are in metropolitan areas and those that are rural. The variable of interest is proportion of females in the labor force. In an example of a *stratified sampling procedure,* twenty predominantly urban civil divisions are selected at random, and a sample

---

[2]Some textbooks stipulate that the degrees of freedom for this t test must be computed from a rather complicated equation:

$$df = \frac{(s_1^2 + s_2^2)^2}{\frac{s_1^2}{(n_1 - 1)} + \frac{s_2^2}{(n_2 - 1)}}$$

However, this tends to approximate the simpler computation, unless you are comparing samples from different populations, both of which have extremely large variances.

of fifteen predominantly rural civil divisions is chosen (Appendix B-2). This is a test of the null hypothesis, $\overline{X}_{URBAN} - \overline{X}_{RURAL} \neq 0$. Box 6.1 demonstrates the procedure for performing a one-tailed test of the alternative to the null hypothesis stated above.

For this experiment the degrees of freedom, ν, equals 30. Two degrees of freedom were lost when the variances were computed about $\overline{X}_1$ and $\overline{X}_2$. The assumption here is that $\overline{X}_2 > \overline{X}_1$, and the sampling error is therefore all in one tail of the distribution. As a

---

BOX 6.1

Procedure for Test of Alternative Hypothesis, Two Sample Means Test, $n_1 = 12$, $n_2 = 20$, σ Not Known

1. Formulate a hypothesis.

   $H_1: \overline{X}_2 - \overline{X}_1 > 0$. The mean proportion of females in the labor force in urban civil divisions in Maryland in 1980 significantly exceeded that of rural civil divisions.

   Sample data:

Rural	Urban	
38.89	42.89	
35.18	38.77	
30.25	39.60	
44.16	37.03	
49.33	37.42	
34.35	43.48	
41.64	39.70	
36.64	39.53	
38.84	31.33	
36.55	40.13	
33.39	40.57	
33.27	36.57	$\overline{X}_2 = 38.61$
$\overline{X}_1 = 37.71$	39.72	$s_2 = 3.37$
$s_1 = 5.3$	31.13	
	44.01	
	35.80	
	38.87	
	39.54	
	36.10	
	39.91	

2. Specify the sample statistic.

   Since these are small, independent samples, and population parameters are unknown or unspecified (but presumed to be normally distributed), a t test is specified (Equation 6.14):

   $$t = \frac{(\overline{X}_1 - \overline{X}_2)}{\hat{\sigma}_{\overline{x}_1 - \overline{x}_2}}$$

3. Select the level of significance.
   $\alpha = .05$. Accept risk that Type I error could be made 5 times in 100.
4. Construct a decision rule.
   Accept H_1 if $|t| > 1.697$ ($v = 30$, $\alpha = .05$—one tailed test). Since the table is constructed for a two-tailed test, and this is a one-tailed test, we must look under the column labeled $P = .10$ for the critical value.
5. Compute test statistic (s_1^2 is rural variance, s_2^2 is urban variance).
   First, compute s_{1+2}^2 (the pooled variance—Equation 6.11):

$$s_{1+2}^2 = \frac{v_1 s_1^2 + v_2 s_2^2}{(n_1 + n_2 - 2)}$$

$$= \frac{(11)28.09 + (19)11.36}{30} = 17.49$$

Then, compute $\hat{\sigma}_{\bar{X}_1 - \bar{X}_2}$ (the standard error—Equation 6.13):

$$\hat{\sigma}_{\bar{X}_1 - \bar{X}_2} = \sqrt{17.49\left(\frac{1}{11} + \frac{1}{9}\right)} = 1.88$$

Substitute for terms in Equation 6.14:

$$t = \frac{37.71 - 38.61}{1.88} = -0.48$$

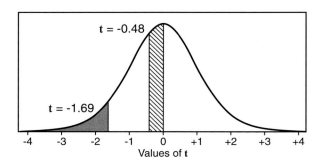

6. Make the decision.
   Probability that this magnitude of difference of means could occur by chance is greater than .9. Reject alternative hypothesis. Urban civil divisions did not have a significantly higher mean proportion of females in labor force than rural civil divisions in Maryland in 1980.

result, the value of t in Appendix Table 4 corresponding to $\alpha = .05$ must be taken from the column $P = .10$. Whereas the urban proportion of the female labor force does exceed the rural proportion in this sample, with a standard error of 1.73, the difference is significant. The t value, -0.48, would occur through sampling variation in more than 90 out of 100 samples of $n = 30$.

A Paired Comparisons t Test

In the previous experiments involving the difference between two sample means, the sampling was independent and random. Frequently, we come across a situation where it is necessary to assess the change in some phenomenon at a set of locations, or to assess the effectiveness of a treatment or experimental procedure on related observations. When inherent pairing exists between the pairs of elements in two samples, the samples are not independent. The order in which the pairs of values are drawn is unimportant. The concern is with changes that have occurred over time in the place samples. Such a test is known as a *paired comparisons* test.

Related or paired observations may occur in many situations. Villages in a Third World country, for example, might be targeted for the dissemination of family planning information. The rate of birth control use in these villages may be measured before and after a campaign supporting family planning measures. Crop production rates for farms (or counties) before and after the application of fertilizers or pesticides may be compared, and so forth. Instead of performing the analysis with individual observations, one uses the *difference* between individual pairs of observations as the variable of interest.

Instead of comparing differences of means of independent samples, compute the mean of the differences between pairs of values on each observation, here symbolized as $\bar{d}$:

$$\bar{d} = \frac{\sum_{i=1}^{n} d_i}{n} \qquad (6.16)$$

The variance for this statistic is given in Equation 6.17:

$$s_d^2 = \frac{n \sum_{i=1}^{n} d_i^2 - \left(\sum_{i=1}^{n} d_i \right)^2}{n(n-1)} \qquad (6.17)$$

where n is the number of pairs of samples. The standard error of the mean of $\bar{d}$ is computed as

$$\hat{\sigma} = \frac{s_d}{\sqrt{n}} \qquad (6.18)$$

The t test for a paired comparisons experiment involves finding the ratio of the mean difference between columns to the standard error of that mean:

Inferential Statistics and Hypotheses Involving Means

$$t = \frac{\bar{d} - 0}{\hat{\sigma}_{\bar{d}}} \qquad (6.19)$$

For example, suppose you are analyzing the use of fertilizer in Third World countries. You might expect that use of fertilizer in these nations would increase over the twelve-year period 1975 to 1987. Data are available for average annual fertilizer use for some 200 countries in 1975–1977 and 1985–1987 (World Resources Institute, 1990). The question is whether or not the change in amount of fertilizer used is significant. To test the null hypothesis that $\bar{d} = 0$, a random sample of twenty-four developing countries is selected and data on annual fertilizer use in the 1970s and 1980s are obtained. Box 6.2 illustrates the hypothesis-testing procedure for this problem.

BOX 6.2

Procedure for Paired Comparisons t Test: Average Annual Fertilizer Use, 1975–77 and 1985–87, in 24 Developing Countries

1. Formulate a hypothesis.
   H_0: $\bar{d}$ $(X_1 - X_2) = 0$ (two-tailed test).
   The observed differences in fertilizer use constitute a random sample from a normally distributed population of differences, and are assumed, under the null hypothesis, to be insignificantly different from zero.
   Sample data:

Average Annual Fertilizer Use (kg/ha of cropland) in Sample ($n = 24$) of Developing Countries, 1975–77 and 1985–87

Country	1975–77	1985–87	d	d^2
Angola	4	4	0	0
Botswana	2	0	−2	4
Burkina Faso	3	5	2	4
Central African Republic	1	1	0	0
Chad	2	2	0	0
Djibouti	0	0	0	0
Egypt	188	347	159	25,281
Ethiopia	2	4	2	4
Gambia	9	21	12	144
Guinea	1	0	−1	1
Kenya	22	46	24	576
Mali	5	15	10	100
Mozambique	4	2	−2	4
Sierra Leone	1	2	1	1
Togo	2	7	5	25
Zaire	2	1	−1	1
Honduras	14	19	5	25
Bangladesh	29	68	39	1,521

Average Annual Fertilizer Use (kg/ha of cropland) in Sample ($n = 24$) of Developing Countries, 1975–77 and 1985–87 (continued)

Country	1975–77	1985–87	d	d^2
Indonesia	27	100	73	5,329
Iran	23	63	40	1,600
Kampachea	0	0	0	0
Oman	12	96	84	7,056
Yemen	7	13	6	36
Papua New Guinea	20	30	10	100

2. Specify the sample statistic.
   A random sample of twenty-four paired observations calls for a paired comparisons t test (Equation 6.19):

$$t = \frac{\bar{d} - 0}{\hat{\sigma}_{\bar{d}}}$$

3. Select the level of significance:
   $\alpha = .05$. Accept risk that Type I error could be made 5 times in 100.

4. Construct a decision rule.
   Accept H_0 if $|t| \leq 2.069$ ($df = 23$, $\alpha = .05$) two-tailed test.

5. Compute test statistic, using Equation 6.16. The average difference between paired observations (average annual fertilizer use) is

$$\bar{d} = \frac{\sum d_i}{n} = \frac{466}{24} = 19.42$$

The pooled variance is computed with Equation 6.17:

$$s_d^2 = \frac{n \sum d_i^2 - \left(\sum d_i\right)^2}{n(n-1)}$$

$$\frac{24\,(41{,}812) - 217{,}156}{24\,(23)} = \frac{786{,}332}{552} = 1{,}424.51$$

The standard error, $\hat{\sigma}_{\bar{d}}$, is computed with Equation 6.18:

$$\hat{\sigma}_{\bar{d}} = \frac{s_d}{\sqrt{n}} = \frac{37.74}{\sqrt{24}} = 7.70$$

Substituting for the terms in Equation 6.19:

$$t = \frac{\bar{d}}{\hat{\sigma}_{\bar{d}}} = \frac{19.42}{7.70} = 2.52$$

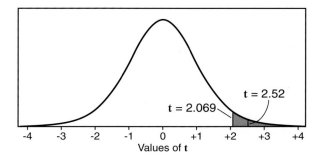

6. Make the decision.
   Probability that this magnitude of average difference in means could occur by chance is about .02. Reject H_0. Average annual fertilizer use in developing countries was significantly higher in 1985–87 than it was in 1975–77. The critical value of t for a two-tailed test with $\alpha = .05$, for $\nu = 24$, is 2.069.

■ ══ ■

The difference in fertilizer use in the twenty-four countries turns out to be significant. However, had we made a Type I error, that is, rejecting a true null hypothesis, the result might have somehow influenced agricultural policy toward developing countries. If the object at the outset was to avoid a Type I error, the α level could have been set to .01. As a result we would have accepted H_0 since the critical value for $\alpha = .01$ and $\nu = 23$ was 2.807.

No attempt has been made in this chapter to contrast "large-number methods" and "small-number methods." When σ is known, the normal curve is appropriate for samples of any size, large or small. When σ is not known, and the population of this characteristic is known or assumed not to suffer from extreme skew or kurtosis, and when s is employed in its place, the t distribution (a "small-number method") is the appropriate distribution. As n increases, the t distribution approaches the normal distribution, so that for samples in excess of thirty observations, the normal distribution is sometimes applied.

In all experiments of inferential testing, the interpretation of results should be weighed carefully in light of the theory being tested or process being evaluated. The examples used in this chapter were selected to illustrate techniques for hypothesis testing under specified assumptions, namely to ascertain whether statistically significant differences exist between means. Bear in mind that the results of individual statistical experiments by themselves do not necessarily change or influence the larger picture of geographic theory that they are designed to test.

KEY TERMS

acceptance region 157
alpha (α) level 157
alternative hypothesis 153
confidence limits 143
critical value 157
degrees of freedom 163
hypothesis formulation, steps 151
independent sample means 161
inference 149
null hypothesis 151
one-tailed test 158
paired comparisons test 172
parametric test 155
rejection region 157
research hypothesis 153
sample size 161
significance level 156
standard error 151
student t distribution 162
two-tailed test 157
Type I and II Errors 154

REFERENCES

Kerlinger, Fred N. (1964). *Foundations of Behavioral Research.* New York: Holt, Rinehart and Winston.

National Oceanic and Atmospheric Administration. (1989). *Local Climatological Data, Des Moines, Iowa.* Asheville, NC: National Climatic Center.

U.S. Bureau of the Census. (1980). *Census of Population and Housing: Minor Civil Divisions, Maryland.* Washington, DC: U.S. Government Printing Office.

U.S. Bureau of the Census. (1988). *County and City Data Book, 1988.* Washington, DC: U.S. Government Printing Office.

World Resources Institute. (1990). *World Resources 1990–91.* New York: Oxford University Press.

EXERCISES

1. In the examples below, state whether or not an error has been made, and if so which type (I or II).
   a. H_0: $\bar{X} = \mu$, $\alpha = .05$, two-tailed test. Obtained $P = .03$. Statistical decision: H_0 is false. Actual status of H_0: True.
   b. H_0: $\bar{X}_1 = \bar{X}_2$, $\alpha = .05$, two-tailed test. Obtained $P = .04$. Statistical decision: H_0 is false. Actual status of H_0: False.
   c. H_0: $\bar{X} = \mu$, $\alpha = .025$, two-tailed test. Obtained $P = .10$. Statistical decision: fail to reject H_0. Actual status of H_0: False.
   d. H_0: $\bar{X}_1 = \bar{X}_2$, $\alpha = .01$, two-tailed test. Obtained $P = .006$. Statistical decision: Reject H_0. Actual status of H_0: False.

2. Explain the meaning of the following terms or concepts:
   a. α level
   b. null hypothesis
   c. alternative hypothesis
   d. rejection region

3. Under what condition must we use α/2?

4. What is the difference between a *one-tailed test* and a *two-tailed test*?

5. In terms of a Type I error, what is the consequence of selecting an α level of .05?

6. Since it is possible to test a statistical hypothesis with any size sample, why do we prefer larger sample sizes?

7. Compute sample sizes n given the following sets of parameters:
   a. $Z = 1.96$, $\sigma = 5.0$, $d = 3$
   b. $Z = 1.64$, $\sigma = 5.0$, $d = 2$
   c. $Z = 1.96$, $\sigma = 3.0$, $d = 5$

8. The 95 percent confidence limits for μ as obtained in a sample of precipitation values were 3.91 inches and 4.67 inches. Is it correct to say that 95 times out of 100 the population mean, μ, falls inside the interval from 3.91 inches to 4.67 inches? If not, what would the correct statement be?

9. Compute 99 percent confidence limits for $(\mu_1 - \mu_2)$ based on the results of the example in Box 6.1.

10. Draw a sample of 20–25 values at random from the population of female labor force data of Appendix B-2. Given $\mu = 37.36$ percent and $\sigma = 5.64$ percent, use the six steps of the hypothesis-testing procedure to compare your sample mean and μ.

11. Two samples of female labor force data are drawn at random from two strata of the population of minor civil divisions in Maryland. One stratum represents predominantly black (B) urban residential areas; the other stratum represents predominantly white (W) urban residential areas. The means and standard deviations of these samples are:

$$n_B = 24;\ n_W = 26;\ \overline{X}_B = 36.09;\ s_B = 6.37;\ \overline{X}_W = 39.11;\ \text{and}\ s_W = 5.99$$

Test the hypothesis that there was no difference in the means of percent female labor force in black and white urban districts in Maryland in 1980. Follow the six steps of the hypothesis-testing process.

12. Why must you use the t distribution rather than the normal (Z) distribution in the means test of problem 10?

13. A random sample of ten mean yearly flow discharge readings (measured at the head of the Potomac estuary) are given in the following table:

Sample Data of Mean Annual Discharge

Year	cfs
1966	7.1
1972	21.4
1944	9.6
1976	12.3
1957	9.0
1969	5.5
1937	18.3
1956	10.3
1946	9.1
1931	10.9

The population mean for the entire discharge record at this site is 11.27 cfs, with $\sigma = 3.27$ cfs. Using the six steps for testing a hypothesis of equality of means, compare $\overline{X}$ and μ for this problem.

14. A second random sample of ten flow discharge readings are: 12.4, 10.9, 8.3, 9.8, 8.4, 7.1, 11.3, 9.1, 5.5, and 12.6 (cfs). Assume we do not know σ, and carry out a test of the difference of the sample mean of this experiment and that of problem 13.

15. Compute confidence limits for $(\mu_1 - \mu_2)$ based on the results obtained in Problem 13.

16. Were rents in the city of Baltimore higher than rents in suburban Baltimore County in 1980? The following table contains the average monthly contract rent for renter-occupied housing units in twenty-five census tracts, selected at random from both jurisdictions.

Baltimore City ($ Rent)	Baltimore County ($ Rent)
147	205
183	257
185	213
136	187
177	199
164	389
86	177
128	204
151	213
171	205
185	167
153	161
148	255
174	198
177	250
152	284
178	175
71	230
139	183
147	167
207	247
177	249
283	293
147	155
172	284

Carry out the six steps for testing a hypothesis of equality of independent sample means, where $n_1 = n_2$.

17. Two independent random samples of annual precipitation for Denver, Colorado, have been selected. The first sample contains total annual precipitation values (in inches) drawn from the record of 1921–1940, and the second sample contains total annual precipitation values drawn from the record of 1950–1970. Test H_1: The sample drawn from the later period is significantly drier than that drawn from the early period. The means, standard deviations, variances, and sample sizes for the two samples are as follows:

Early period: $\bar{X} = 13.55$; $s = 2.158$; $s^2 = 4.659$; $n = 7$
Late period: $\bar{X} = 10.35$; $s = 2.831$; $s^2 = 8.017$; $n = 7$

18. Read the following news article from the April 18, 1993, issue of the *Washington Post*. Describe a statistical experiment about some aspect of this story, and state a null hypothesis for that experiment.

Oil-Hungry Asia Relying More on Middle East

By William Branigin
Washington Post Foreign Service

HONOLULU—Despite intensive oil exploration by Asian countries, soaring demand for petroleum is far outstripping regional supplies and creating heavy dependence on Middle Eastern producers, according to energy and security analysts here and in Asia.

The growing Asian reliance on Persian Gulf oil, combined with a perception of U.S. military disengagement from Asia, carries major economic and political implications for the region and its relations with the United States, the analysts said. It appears likely in the coming years to foster closer political ties between Asia and the Middle East and lead to diminished U.S. influence in the region, more assertive and independent military policies by some Asian states and greater regional development of nuclear power, the sources said.

These perceptions coincide with concerns among U.S. military planners that cutbacks in defense spending will reduce the Hawaii-based Pacific Command's ability to project power over an area that covers half the earth's surface, includes 60 percent of its population and accounts for 36 percent of America's global trade.

"We feel we must maintain a credible forward military presence," a Pacific Command intelligence officer said.

A senior U.S. official expressed concern that Washington has failed to appreciate that cutbacks could "create the impression of a power vacuum" in the region and lead to a crisis—with greater expenses in the long run—as Asian countries move to fill the perceived void.

Among the Pacific Command's major responsibilities is patrolling an area through which ships annually carry an estimated $400 billion worth of Persian Gulf oil, some of it destined for the United States. But it is the Asia-Pacific region, with its booming economic growth rates, that is becoming most dependent on Persian Gulf crude.

Almost singlehandedly, the region has been holding up the world oil market in recent years, energy analysts say. Last year, demand for petroleum products grew by 6.3 percent in the region, compared with less than 1 percent in the rest of the world, according to the Program on Resources of Honolulu's East-West Center.

"Internal oil production in Asia has stabilized at about 7 million barrels a day, but demand is exploding," said Fereidun Fesheraki, an oil expert who heads the resources program, an energy think tank. "This region is carrying the world oil market now."

If not for Asia's growth in oil demand of 862,000 barrels a day last year, the world's consumption would have declined by more than 200,000 barrels a day, a recent study by the resources program says. It projects demand for oil products in the region at 19.1 million barrels a day by 2000, compared with 12.8 million barrels a day in 1990.

The failure of Asia's oil production to keep up with this sharply rising, boom-driven consumption partly reflects what Fesheraki called the "highly exaggerated" potential of regional reserves.

Exploration in the South China Sea, the East China Sea and the Bo Hai Bay has failed to produce the anticipated results, he said. "Nobody knows for sure" what these areas contain, he added, "but there is no clear geological evidence" that they hold vast oil reserves.

The area of the disputed Spratly Islands might, if developed, yield 50,000 to 100,000 barrels a day, Fesheraki said, but "it's not going to be the big prize that everybody's looking for."

Last year, about half the oil consumed in the region was imported from outside, two-thirds of that from the Persian Gulf, Fesheraki said.

By 2000, this import dependency is projected to rise to 64 percent, with more than 90 percent of it from the Persian Gulf.

"In terms of security of supply, the Asia-Pacific region is in a more precarious situation than the United States or Europe," the resources program's study says.

"This is a region that cannot take care of itself," Fesheraki said.

The likely result is a "political and economic tilt" by Asian countries toward their Middle Eastern suppliers, he said. "The special relationship between Asia and the Middle East cannot be stopped."

Of particular concern is the prospect of greater sales of weapons, and possibly nuclear technology, by China to Middle Eastern states such as Iran, despite U.S. protests.

"China is poised to become a massive importer of Middle Eastern crude," Fesheraki said. With its oil consumption rising fast—demand grew nearly 10 percent last year alone—China is expected to be a net oil importer by the end of the decade. It now produces about 3 million barrels a day, with net exports of less than 300,000 barrels a day.

The demand for oil grew even faster elsewhere in the region last year, climbing 28 percent in Vietnam, 21 percent in the Philippines, 20 percent in South Korea and 11 percent in Thailand. In South Korea, gasoline use has increased 30 percent a year for the last three years, compared with less than 1 percent in the United States and other Western industrialized countries.

In India, "oil production is falling and demand is going through the roof," Fesheraki said. He predicted that India would be the region's "biggest new player" in the oil market, with imports of 1 million barrels a day by 2000.

Indonesia, an OPEC member that exports about 700,000 barrels of crude a day, is also projected to become a net importer within a

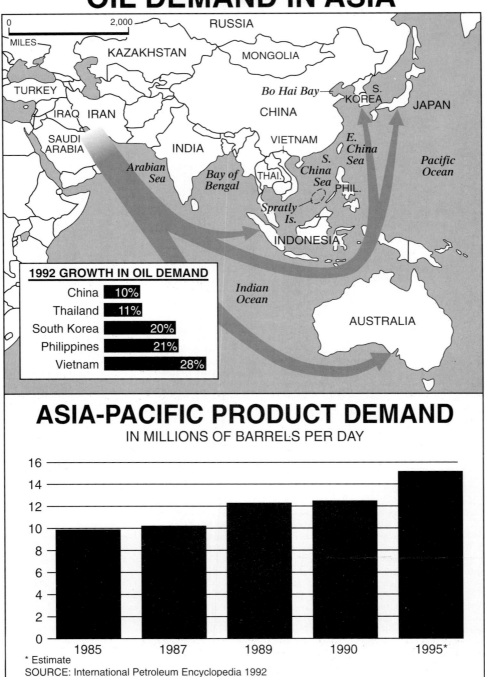

decade. "That's one of the reasons that the government of Indonesia is hellbent on increasing utilization of natural gas, geothermal energy and coal and is talking about a massive investment in nuclear power plants," said a diplomat in Jakarta. A Japanese consulting firm is studying the feasibility of building nuclear power plants on the crowded island of Java, home to 70 percent of Indonesia's 184 million people.

According to Andrew Mack, a professor at Australian National University, Indonesia is contemplating acquisition of up to a dozen nuclear reactors, sparking concern by international environmental groups.

Mack said a desire for energy self-sufficiency also underlies "Japan's absurd obsession with fast-breeder reactors." A recent shipment of plutonium to Japan for use in such reactors generated controversy around the globe and stirred protests by antinuclear activists.

Japan significantly reduced its dependence on Persian Gulf crude after the first "oil shock" of 1973, but the percentage of imports from the gulf has crept back up in recent years, said Richard Leaver, an oil expert at Australian National University.

"There is a problem here of imbalance between regional supply and regional demand that is going to get worse in the short run," Leaver said. Hopes that the former Soviet republics could provide major alternative sources of oil for Asia have not panned out, partly because high production costs and political instability have discouraged large investments by foreign companies, he said.

One consequence of the region's rising consumption is that oil prices—now less than $18 a barrel for Arab light crude—are likely to rise, Fesheraki said. He forecast real price levels of $23 to $25 a barrel by the mid-1990s and $25 to $27 by 2000.

Oil today is relatively cheap for some Asian countries, notably Japan, whose currencies have appreciated against the dollar in recent years. In yen-denominated prices, for example, oil is cheaper today in real terms than it was before the 1973 oil crisis, when it sold for less than $3 a barrel, Fesheraki said.

For the United States, the high growth in Asian oil consumption presents a major opportunity to sell oil-processing technology and equipment. Over the next 10 years, Fesheraki said, Asia needs to invest an estimated $100 billion to meet requirements for new oil refining capacity.

One-Way Analysis of Variance   7

To this point you have encountered only techniques for comparing two samples, but many experiments involve groups of observations. For instance, suppose a geomorphologist has data on twenty-five randomly sampled drainage basins. Each basin in the sample has different slope angles and different mean areas. Suppose the geomorphologist wishes to compare mean areas based on different classes of slope angles. The purpose of breaking the original sample into subsamples, or *replicates,* is to determine the variability of mean area within the subsamples. The criterion for the subsamples could be slope angle, say 0–7°, 7.1–14°, and above 14°.

The analysis of variance (abbreviated **ANOVA**) may be considered as an extension of the difference of means test to more than two populations. ANOVA involves partitioning a sample data set into three or more subsets for purposes of testing a hypothesis that means of the populations of the subsets are equal. ANOVA does this by breaking down the total variance of any sample data set into component sources of variance. In the case of the geomorphologist's drainage basins, the variance in basins has been broken down in one way, by slope angle. If the variation within subsamples is large compared to the differences between subsamples, the differences between subsamples will be difficult to detect. The variance in a sample may be disaggregated any number of ways, depending on the nature of the experiment and the size of the sample.

Suppose the drainage basins are selected from different geological areas, such as glaciated and nonglaciated. If the geomorphologist wished, for example, to compare differences in basin areas by slope and by geological area simultaneously, the variance in basin area would be subdivided two ways, and the resulting test is a *two-way ANOVA*. Associated with each component (slope angle, geological area, etc.) is a specific source of variation, so that in the analysis it is possible to determine the magnitude of the contributions of each of these sources of variation to the total variation. The examples in this chapter are restricted to what is called *one-way analysis of variance*.

COMPONENTS OF VARIANCE

As they were for the means tests in Chapter 6, the prerequisites for ANOVA are that the populations be normally distributed, that the subsamples have homogeneous variances, and that they be selected at random. These conditions refer to the ideal case, though small departures from them are not likely to influence the validity of the results. Variation can creep into the geomorphologist's experiment in several subtle ways: inadvertent differences in the way the slopes were measured, variation in the composition of the soil and rock from one slope to the next, and so on. These sources of variation, or **components of variance,** all combine to produce what is known as **experimental,** or **random, error.** This is the variation that is not accounted for by real differences in slope angle or geologic ages between the samples. Some of this variation is probably unavoidable. The main thing in setting up an experiment is to avoid introducing any systematic error into the statistical analysis by what is known as *randomizing* the observations. If each measurement in the field is sequentially numbered, then the elements of each slope angle class may be selected with a random number table. This way, the various sources of experimental error are mixed, or *confounded,* over the subsamples rather than being concentrated in one or two.

Systematic variance is the variation in some phenomenon that is due to some known or unknown influences that "explain" differences in the phenomenon at different times or places. Any environmental or human factors that influence the probability of occurrence of events in a predictable way are systematic influences. The mean size of morphological units in drainage basins may be *systematically* greater where the slopes are relatively flat than in basins with steeper slopes. The infant mortality rates of populations in Third World countries will tend to be *systematically* higher than the infant mortality rates of populations in industrialized countries. In certain climate regions, the amount of precipitation on the leeward side of a mountain range will be *systematically* lower than the amount of precipitation on the windward side. There may be several causes of systematic variance in any situation, and geographers seek to distinguish between those causes in which they are interested from those in which they are not interested.

In the ANOVA test there are two components of variance. The first is the variance within each replicate; the second is the variance between the replicates. If the subsamples had been drawn from the same population, then their mean values would most likely still have varied from each other. In Chapter 4, we discussed variance that reflected the differences between individual elements in a variable and the mean for that variable. The important consideration is to see whether or not the variation *between* subsample means is significantly greater than that *within* the subsamples themselves. A greater variation between the samples than within the samples would suggest that the two samples had been selected from different populations.

To illustrate the analysis of variance, consider a simple experiment. One index of poverty in a population is its rate of low birth weight births. (A low birth weight is defined as less than 2,500 grams.) Recall from Table 5.1 that in the state of Maryland annual births with low birth weights occurred more frequently in less affluent cities and counties and less frequently in more affluent counties. It then follows that low birth weights might be expected to occur more frequently in low-income neighborhoods than in high-income neighborhoods. The null hypothesis for an analysis of variance states that all group means are equal. For the sake of illustration, suppose three average annual family income brackets are arbitrarily selected: below $15,000, $15,000 to $25,000, and over $25,000. A random sample of census tracts is drawn from the population, and this sample is disaggregated into subsamples corresponding to the three income brackets. The null hypothesis and its alternative are

$$H_0: \mu_1 = \mu_2 = \mu_3$$
$$H_1: \text{at least one mean is different}$$

Each subsample is thus considered as representative of a population of tracts with those average annual incomes.

If these population means are equal, the **between-group variance** is equal to zero. If H_0 is true and the assumption of homogeneous variances and normally distributed populations are met, the population distributions of each group of low birth weight rates will be neatly superimposed upon each other (Figure 7.1a). The group population means will be equal and the distributions will be centered at the same point (the common mean). When H_0 is false, the cause may be that one of the group population means is different

Figure 7.1 The Distribution of Populations of Low Birth Weight Births in Three Regions When (a) H_0 Is True, and (b) When H_0 Is False Because None of the Population Means Are Equal

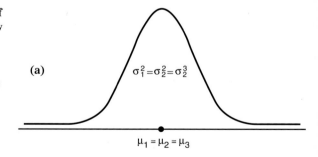
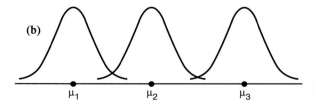

from the other two, or all three census tract group population means are significantly different (Figure 7.1b).

This is not to say that all variance attributed to error is without cause, only that the cause or causes, if there are any, have not yet been discovered. Error variance is the variance left over in a statistical analysis after *all known sources* of systematic variance have been removed from the analysis. In the low birth weight experiment, there may be confounding reasons for low birth weights other than poverty. With this caveat, we proceed to analyze some urban birth data. This experiment is a variation on the family income design described above.

Table 7.1 shows the average annual rates of low birth weight per 1,000 live births of a random sample of thirty-five census tracts in the District of Columbia for the years 1981 through 1985. The sample has been disaggregated geographically instead of by income group. The tracts have been partitioned into northwest, northeast, and southern sectors of the city. The rationale for this division is that there are several socioeconomic and racial differences between the northwest, northeast, and southern quadrants. The southwestern quadrant has few residential neighborhoods, and is thus combined with the southeast.

Statisticians have developed a formalized procedure for analysis of variance which is contained within an ANOVA table. This lists the sources of variation, a column of sums of squares resulting from the various sources, degrees of freedom associated with each, a column called mean square variance, which is nothing more than the sample-based estimates of the variances, and a test statistic (F), which will be explained below. The table for a one-way analysis appears as follows:

Source of Variation	Sum of Squares	Degrees of Freedom	Mean Square Variance	F Test
Between groups	SS_B	$k - 1$	MS_B	
Within groups	SS_W	$N - k$	MS_W	MS_B/MS_W
Total variation	SS_T	$N - 1$		

The Total Sum of Squares

The total variance of all observations (all replicates of all subsamples) is given by SS_T. In the context of this experiment, the total sum of squares is the sum of the squares of the de-

Table 7.1 Average Annual Rate (per 1,000 Live Births) of Low Birth Weight Births in Randomly Selected Census Tracts, Washington, D.C., 1981–1985, Partitioned by Quadrant of City

Northwest			Northeast		South
139	23	94	53	127	141
88	40	49	128	128	96
177	83	103	184	84	105
156	57	77	130	111	205
56	52	44	73	135	130
129	164	193	101		134

Data Source: Government of the District of Columbia (1981–1985).

viations of each observation (tract annual average low birth weight rate) from the mean of all the observations taken together (the so-called *grand mean*). Symbolically, this is

$$SS_T = \sum_{i=1}^{N} (X_i - \bar{\bar{X}})^2 \tag{7.1}$$

where the symbol N indicates the entire sample, as opposed to n_j, which indicates a group sample, where j is the group (in the low birth weight experiment, $j = 1, 2,$ or 3). Notice that Equation 7.1 is the numerator in the variance equation from Chapter 4 (Equation 4.4). For computational purposes, it is more convenient to use Equation 7.2 to compute SS_T directly from sample data:

$$SS_T = \sum_{i=1}^{N} X_i^2 - N\bar{\bar{X}}^2 \tag{7.2}$$

The grand mean for low birth weight rates in Washington ($\Sigma X/N$) is $3801.39 \div 35 = 108.61$. Substituting for the terms in Equation 7.2, we may obtain the total sum of squares for the low birth weight problem:

$$SS_T = 486{,}625.94 - 35\,(108.61)^2 = 73{,}752.57$$

The total number of observations, N, is equal to the number of replicates per sample times the number of samples, or $N = n \times k$.

Between-Groups Sum of Squares

The sum of squares may now be partitioned into *systematic* (between-groups sum of squares) variation and *error* (within-group sum of squares) variation. Obtaining the between-groups sum of squares calls for performing certain computations *between* the groups. We first find the square of the difference between group means and the grand mean ($\bar{\bar{X}}$), multiplying each by group n, then pooling the two values. Symbolically, this is stated as

$$SS_B = n_1(\bar{X}_1 - \bar{\bar{X}})^2 + n_2(\bar{X}_2 - \bar{\bar{X}})^2 + \ldots + n_k(\bar{X}_k - \bar{\bar{X}})^2$$
$$= \sum_{j=1}^{k} [n_j\,(\bar{X}_j - \bar{\bar{X}})^2] \tag{7.3}$$

The subscript k refers to the number of groups into which the sample has been partitioned. The subscript j refers to the index of the group, just as the subscript i is the index value of the individual data value, and Σ indicates a summation over the k subsample means.

The between-groups sum of squares may be computed directly from the sample data with Equation 7.4:

$$SS_B = \frac{\left(\sum_{i=1}^{n_1} X_i\right)^2}{n_1} + \frac{\left(\sum_{i=1}^{n_2} X_i\right)^2}{n_2} + \ldots + \frac{\left(\sum_{i=1}^{n_k} X_i\right)^2}{n_k} - \frac{\left(\sum_{i=1}^{N} X_i\right)^2}{N}$$

$$= \sum_{j=1}^{k} \left[\frac{\left(\sum_{i=1}^{n_j} X\right)^2}{n_j} - \frac{\left(\sum_{i=1}^{N} X_i\right)^2}{N} \right] \qquad (7.4)$$

For the low birth weight experiment, we first compute group means, obtaining for the northwest, northeast, and southern groups, 95.98, 114.89, and 135.0, respectively. Substituting for the terms of Equation 7.3, we obtain

$$SS_B = 18\,(95.98 - 108.61)^2 + 11\,(114.89 - 108.61)^2 + 6\,(135.0 - 108.61)^2$$
$$= 2{,}872.68 + 433.94 + 4{,}176.74 = 7{,}483.36$$

Within-Group Sum of Squares

The next step involves computing within each group the sum of the squared deviations of the individual observations from their mean, then summing these group terms. This sum is subtracted from the sum of the squared values for the entire sample. This component of variation is called the *within-group sum of squares* and may be designated SS_W. Equation 7.5 enables us to compute the within-group sum of squares directly from the sample data:

$$SS_W = \sum X^2 - \frac{\left(\sum_{i=1}^{n_1} X_i\right)^2}{n_1} + \frac{\left(\sum_{i=1}^{n_2} X_i\right)^2}{n_2} + \ldots + \frac{\left(\sum_{i=1}^{n_k} X_i\right)^2}{n_k}$$

$$= \sum X^2 - \sum_{j=1}^{k} \left[\frac{\left(\sum_{i=1}^{n_j} X_i\right)^2}{n_j} \right] \qquad (7.5)$$

Substituting for the terms of the equation with the low birth-weight data, we obtain

$$SSW = 486{,}625.9 - 420{,}356.69 = 66{,}269.21$$

This quantity is also sometimes referred to as the **residual,** or **error, sum of squares.** As a check on our arithmetic, we may confirm that $SS_T = SS_B + SS_W$:

$$SS_T = 7{,}483.36 + 66{,}269.21 = 73{,}752.57$$

The ANOVA Table

The sums of squares just computed may now be used to obtain two estimates of the common population variance, $\hat{\sigma}^2$. Recall from Chapter 6 that the symbol ^ over a term indicates that it is an estimated value. When divided by their respective degrees of freedom, both the within- and between-groups sum of squares will yield independent and unbiased estimates of $\hat{\sigma}^2$. The objective is to compare the estimated variance between subsamples with the estimated variance within subsamples. This is done by computing the ratio of SS_B/SS_W. The null hypothesis states that SS_B/SS_W does not differ more than might be accounted for by chance (sampling variation). The estimated variance *within subsamples* becomes the measure of *chance variation,* since the variation of the items within each subsample is not affected by differences between subsample means.

The variance estimate may also be called the **mean square variation,** since it is computed by dividing each component sum of squares by its degrees of freedom. The degrees of freedom associated with the *total sum of squares* is the number of sample values less 1 ($N - 1$). One degree of freedom is lost because of computing the squares about the grand mean. The degrees of freedom associated with the *between-groups sum of squares* is the number of groups less one ($k - 1$). One degree of freedom is lost because variance is computed about the grand mean. The degrees of freedom associated with the *within-group sum of squares* is the number of sample values less the number of groups ($N - k$). One degree of freedom is lost for each group or category because variances are computed about each group mean. Table 7.2 shows the conventional format for presenting the summary statistics for an ANOVA, using the low birth weight experiment as an example.

The rationale in ANOVA may be clearer if you consider the extreme situation where all subsamples are identical. Then, the means and variances of the replicates would be the same, and the error measure would vanish, there being *no* variance unaccounted for.

Table 7.2 **Summary Statistics for Analysis of Variance of Sample of Low Birth Weight Births in Washington, D.C., 1981–1985**

Source of Variation	Sum of Squares	v	Estimated Variance
Between groups	7,483.37	2	3,741.69
Within groups	66,269.15	32	2,070.91
Total	73,752.52	34	

THE *F* DISTRIBUTION AND THE *F* TEST

Statisticians such as R. A. Fisher have demonstrated that variances of random and independent samples from a normally distributed population are neither normal nor symmetrical. Their distribution follows a curve that is skewed to the right, the exact shape of which depends on σ^2 and the number of degrees of freedom in the sample. The distribution of ratio values, which ranges from 0 to infinity, forms an asymmetrical distribution known as the *F* distribution, named in honor of R. A. Fisher.

A distribution of the kind just described is obtained as follows. Suppose we drew a pair of samples from a normally distributed population and computed their variances. If the two variances were exactly equal, the ratio of between-groups estimated variance (σ^2_B) to within-group variance (σ^2_W) would be 1.0. Since the population from which these samples are drawn is assumed to be normal, sample variances should be clustered around 1.0.

Unlike the *t* distribution, the shape of the *F* distribution is determined by *two* values of ν, computed as $n_B - 1$ and $n_W - 1$. Figure 7.2 shows three representative *F* distributions, based on numerator and denominator ν, respectively, of 1 and 40, 6 and 28, and 28 and 6. A test of the difference between sample variances is thus based on a distribution

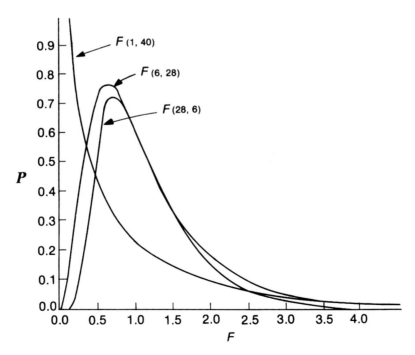

Figure 7.2   Probability Density Functions of *F* for Three Combinations of Numerator-Denominator Degrees of Freedom

that is also skewed to the right. Degrees of freedom for the two components of variance are not pooled because we are not interested in the *difference* between two variances, but rather the ratio of these variances. Any table that gives values of the estimate of variability in samples would have to have both ν and α as arguments.

Once the appropriate F distribution has been determined, the null hypothesis that there is no difference in the variation between groups and in the variation within groups may be tested. The altenative hypothesis, H_1, is that $\sigma_B^2 > \sigma_W^2$. The magnitude of this ratio that will cause acceptance of the alternative hypothesis of greater between-groups variance depends on the significance level chosen. The significance level chosen determines the critical value of F, the value that, like the examples of Chapter 6, separates the acceptance region from the rejection region.

From the ANOVA table, the variance ratio, F, is computed by dividing the estimated variance between groups by the estimated variance within groups. For the low birth weight example, we compute

$$F = \frac{\hat{\sigma}_B^2}{\hat{\sigma}_W^2} = \frac{3,741.69}{2,070.91} = 1.81$$

$$\text{numerator } \nu = 3 - 1 = 2$$
$$\text{denominator } \nu = 35 - 3 = 32$$

Having chosen a critical region ($\alpha = .05$), we can now reject or accept the hypothesis. To reach this decision, compare the variance ratio just computed with the critical value of F (Appendix Table 7). The critical value of F is found at the intersection of $\nu_1 = 2$ and $\nu_2 = 32$:

ν_2	1	2	3	4	5...
.	.	.	.	.	....
32	4.15	3.30	2.90	2.67	2.51...

Figure 7.3 shows the location of the critical value under an F distribution with 2 and 32 degrees of freedom, respectively. With a computed variance ratio of only 1.81 in the low birth-weight experiment, we cannot accept the null hypothesis that these variances come from different populations. In terms of the sample of low birth weights, the variation between the three regions is not sufficient to allow us to conclude that the regions as they are defined make a significant difference in mean rates of low birth weight babies. If the ratio of the two variances had exceeded the critical value for the stated significance level and associated degrees of freedom, we would conclude that the computed variance ratio did not represent a rare event brought about by chance, but instead reflected the fact that something other than chance caused the divergence in variances.

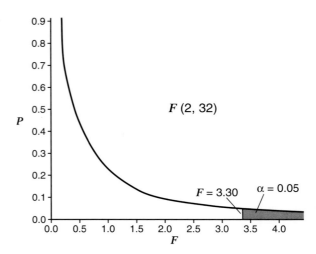

Figure 7.3 Probability Density Function of F for 2 and 32 Degrees of Freedom, Showing Critical Value for $\alpha = .05$

A second example of a one-way ANOVA is demonstrated using wheat yields on experimental farm plots with differing soils. It is well known that soil texture can influence crop yield. Suppose one is interested in determining if a new strain of wheat recently developed in another part of the world will produce different yields on different soil types. Assume wheat is planted in experimental plots with sand, clay, and loam soils, respectively. ANOVA is appropriate if the population of wheat yields is normally distributed. Box 7.1 follows the ANOVA procedure from hypothesis to decision.

The subsamples of wheat yields are based on output from the sand, clay, and loam soil plots. Over four times as much between-groups variance as within-group variance is found in the analysis. The null hypothesis may safely be rejected, since the computed variance ratio exceeds the limits provided for an experiment with $\alpha = .05$, and with 2 and 24 degrees of freedom. The computed value falls in the tail of the distribution. The probability that a difference this great could occur by sampling variation (chance) is .0240, which is well below the .05 level allowed by the hypothesis.

The analysis of variance was designed primarily for controlled experiments in which the sample subjects are considered homogeneous. In psychology and medicine, for instance, a sample of patients might be part of an experiment in which some of the patients were given a specified treatment, while the remaining patients were not. This type of experiment is subject to much more control than the examples illustrated in this chapter. For most of the research undertaken by geographers, a lack of control over the experiments conducted suggests that we consider the results of analysis of variance to be approximate rather than exact.

BOX 7.1 Procedure for ANOVA on Wheat Yield Experiment

1. State hypothesis and level of significance.

$H_0: \mu_1 = \mu_2 = \mu_3$. The soils subsamples come from the same population of soils with respect to their wheat yields. $\alpha = .05$. Test distribution: F. Decision rule: Accept H_0 if computed variance ratio does not exceed the critical value, numerator $\nu = 2$, denominator $\nu = 24$.

2. Random samples of wheat yields (bushels per unit area): $n_1 = n_2 = n_3 = 9$.

Sandy Soil	Clay Soil	Loam Soil
18	21	21
20	23	17
16	14	24
21	24	23
22	21	25
18	20	22
19	19	23
17	21	26
21	23	23

3. Compute SS_T: (Equation 7.3)

$$SS_T = 11{,}912 - 11{,}697.93 = 214.07$$

4. Compute SS_B: (Equation 7.4)

$$SS_B = \{(172^2 \div 9) + (186^2 \div 9) + (204^2 \div 9)\} - \{(562)^2 \div 27\}$$
$$= (3{,}287.11 + 3{,}844.0 + 4{,}624.0) - 11{,}697.93 = 57.19$$

5. Compute SS_W: (Equation 7.5)

$$SS_W = 11{,}912 - 11{,}755.11 = 156.89$$

6. Construct ANOVA Table:

Summary Statistics for Analysis of Variance of Sample of Wheat Yields on Different Soil Plots

Source of Variation	Sum of Squares	ν	Estimated Variance
Between groups	57.19	2	28.60
Within groups	156.89	24	6.54
Total	214.08	26	

7. Consult F table for critical value, $\alpha = .05$, $\nu_1 = 2$, $\nu_2 = 24$. Find *3.40*.

8. Decision:

Reject H_0. Computed variance ratio this high (4.37) unlikely to arise by chance alone. We cannot conclude that yield variances on the three different soil plots belong to the same population. Soil texture makes a difference in this experiment. Accept H_1.

KEY TERMS

ANOVA 183
between-groups sum of squares 187
between-group variance 185
components of variance 184
error sum of squares 189

error variance 186
experimental error 184
F-test 191
mean square variation 189
random error 184
residual sum of squares 189

systematic variance 184
total sum of squares 186
within-group sum of squares 188

REFERENCES

Government of the District of Columbia. (1981–1985). *Vital Statistics Summary.* Washington, DC: Department of Human Services.

U.S. Bureau of the Census. (1986). *State and Metropolitan Area Data Book, 1986.* Washington, DC: U.S. Government Printing Office.

EXERCISES

1. Large-scale drainage basins in the midwestern United States gradually erode into shapes that differ depending partly on their geomorphic structure. Basins that were formed in the Illinoisan or Wisconsin glaciations evolved differently from unglaciated basins. Basin shape is defined by the ratio of the perimeter of a basin to the perimeter of a completely circular basin with the same area. We may ask if there is some component of variance in drainage basin shape that might be attributed to geomorphic history. Perform a one-way ANOVA of the shape indices of fifty-two loess-mantled second-order drainage basins in Illinois to assess the null hypothesis that drainage basin shape is *independent* of glacial history, or $\mu_1 = \mu_2 = \mu_3$. Use $\alpha = .05$.

Shape Ratios of Fifty-Two Drainage Basins

Unglaciated		Illinoisan			Wisconsin
1.30	1.58	1.61	1.28	1.27	1.35
1.24	1.32	1.25	1.32	1.21	1.35
1.24	1.66	1.18	1.27	1.42	1.26
1.25	1.33	1.45	1.31	1.24	1.33
1.17	1.25	1.26	1.22	1.40	
1.49	1.35	1.34	1.24	1.53	
	1.35	1.16	1.22	1.52	
		1.47	1.21	1.22	
		1.18	1.17	1.20	
		1.08	1.22	1.26	
		1.25	1.42	1.30	
		1.49	1.29		

2. From Appendix B-2, draw a random sample of thirty-five Maryland civil divisions. For each sample observation, record the county code (CC) and percent female employment.

The following counties (codes) are predominantly rural/small-town populations in the mountainous western section of the state:

01, 23, and 43;

the following counties (codes) are predominantly rural/small-town populations in the eastern section of the state:

09, 11, 15, 19, 29, 35, 37, 39, 41, 45, and 47;

and the following counties (codes) are predominately urban counties in the central part of the state:

03, 05, 13, 17, 21, 25, 27, 31, 33, and 51.

Partition the sample of percent female employment measures according to region. If necessary, increase the size of the sample in order to have a minimum of three observations in each group. Use one-way analysis of variance to test the hypothesis H_0: that the mean percentages of female employment in the three regions are equal. Use $\alpha = .05$.

3. The data in the following table constitute a random sample of mean personal incomes for United States metropolitan areas for the year 1983, partitioned into the four major U.S. regions. Use one-way ANOVA to test the null hypothesis that the regional means of incomes are equal. (Hint: Computations on hand calculators will be less tedious if you replace the comma in each income value with a decimal point. This will reduce sums of squares by a factor of 1,000.)

Mean Per Capita Incomes, by Region, United States, 1983

Northeast SMSA	Income ($)	Midwest SMSA	Income ($)
Boston, MA	11,252	Akron, OH	9,566
Charleston, WV	9,049	Aurora, IL	10,432
Erie, PA	8,464	Cleveland, OH	10,193
Hagerstown, MD	8,361	Lansing, MI	9,470
Jersey City, NJ	8,543	Madison, WI	10,500
New Haven, CT	10,563	Rochester, MN	11,072
Philadelphia, PA	9,968	Steubenville, OH	8,150
Pittsfield, MA	9,575	Topeka, KS	10,424
Scranton, PA	7,999		
South SMSA	**Income ($)**	**West SMSA**	**Income ($)**
Ft. Lauderdale, FL	11,813	Dallas, TX	11,472
Gainesville, FL	8,218	El Paso, TX	6,798
Jackson, MI	8,527	Laredo, TX	4,816
Montgomery, AL	8,255	Medford, OR	8,168
Roanoke, VA	9,877	Odessa, TX	9,778
		Santa Fe, NM	10,233
		Visalia, CA	7,142
		Yakima, WA	8,020

Source: U.S. Bureau of the Census (1986).

Bivariate Correlation and Linear Regression

8

Prior chapters have laid the foundation for your basic understanding of inferential statistics. You should now be familiar with the procedures for testing hypotheses about means and variances of single variables. In Chapter 7, methodology for partitioning a sample according to some criterion, which could account for significant statistical differences in the variances of the partitioned groups, was discussed. However, *univariate* statistics severely limit the assessment of geographical relationships. In this chapter, a procedure for describing the relationship between two variables measured for a common sample of individual observations will be introduced. This technique in turn lays the foundation for understanding *multivariate* statistics, a methodology that dominates the present-day quantitative literature.

Two variables are *functionally related* when variations in the values of one are systematically associated with variations in the values of the other. Some bivariate relationships of interest to geographers are well known, such as that between proportion of tree cover and rainfall, between education and income, or between land value and distance from a commercial hub. Relationships between two variables measured on the interval or ratio scales, such as transportation cost and distance, may be graphed on orthogonal axes (Figure 8.1). We may observe the relationship between these two variables by plotting pairs of values, using the scales of the two axes. This yields what is known as a **scatter plot.** For example, in Figure 8.1, one given cost, p, is associated with a given distance, d.

Figure 8.1 The Relationship between Distance and Transportation Cost

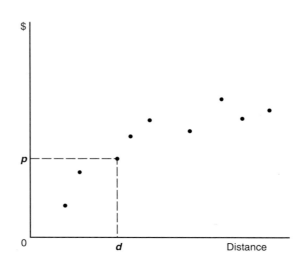

The systematic positive relationship between distance and transport cost is immediately apparent.

Relationships have both *direction* and *strength*. The direction of a relationship can be described as positive, negative, or indeterminate. A **positive relationship** means that as the values of one variable, Y, increase, values of a second variable, X, also increase. Consider education and income as an example of a positive relationship. Increases in years of schooling generally lead to increases in income. Also, warm air can "hold" more water vapor than cool air, so the relationship between temperature and absolute humidity is positive.

A **negative, or inverse, relationship** indicates that as values of one variable increase, values of the other decrease. For example, land values fall systematically with increased distance from downtown, or a commercial center, and atmospheric pressure is inversely related to altitude. As you will soon discover, an algebraic formula that estimates the relationship between variables Y and X can be developed that goes beyond simple specification of a negative or positive relationship. When, however, you are only interested in measuring the strength of the relationship between two metric variables, you may employ *correlation analysis*.

CORRELATION ANALYSIS

In this section the method for measuring the degree of a relationship between two interval-scale variables, Y and X, will be demonstrated. This may be done with either the population or a sample. If, as is usually the case, it is not possible or desirable to use the entire population, a sample of locations (points or areas) at random from within the study area of interest can be selected. Data can then be obtained on the characteristics of interest for each of these locations. The assumptions of parametric correlation analysis are that the variation in Y is independent of the variation in X, and that both samples are drawn from normally distributed populations. There is also a *nonparametric* form of correlation, which you will find in Chapter 9.

The expectation is usually that there is some correlation between variables in geographic space. Geographers use the term **areal association** for this relationship. If areal association exists between the two variables, the scatter plot will appear as an elongated pattern of points, such as Figures 8.1 and 8.2. The latter figure depicts an inverse relationship, where an increase in use of contraceptives is correlated with a drop in the fertility rate in each of thirty-nine countries.

Areal association may be expressed by the **Pearson product-moment coefficient of correlation, r.** The correlation coefficient is independent of the units of the original data and expresses the degree of association between Y and X. This coefficient ranges from -1.0 to 1.0. A value of -1.0 indicates a perfect *negative* correlation between Y and X; a value of 1.0 indicates a perfect *positive* correlation. A perfect correlation would be one in which a single straight line would intersect all of the points in the scatter plot. A scatter plot in which the dots form virtually a circular pattern depicts a nonexistent relationship between two variables. This pattern yields a correlation coefficient of 0.0. Correlation coefficients may be used to test a null hypothesis that no correlation exists in the populations from which the samples Y and X were selected, or $\rho = 0$.

The correlation coefficient is the ratio of the **covariance**—the extent to which Y and X vary together about their common means—to the product of the standard deviations of Y and X. In Chapter 5, the concept of the single variable probability distribution was explained. The student should recognize Figure 8.3 as a *bivariate normal probability surface,* which expresses the same information as a univariate distribution, with the exception that it lies in two dimensions. The volume under any part of the surface may be interpreted as the probability of an individual pairing of runoff and precipitation occurring at that location under the surface. It is important to note here that if the joint distribution is bivariate normal, the variances of Y and X will be equal. This property is referred

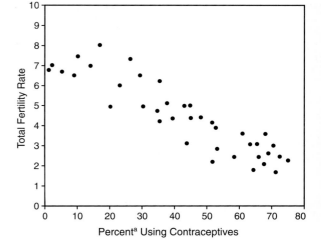

Figure 8.2 Contraceptive Prevalence and Fertility Rates in Thirty-nine Countries, 1988

[a]Percent of married women of childbearing age.

Source: Mauldin and Segal (1988), Figure 3.

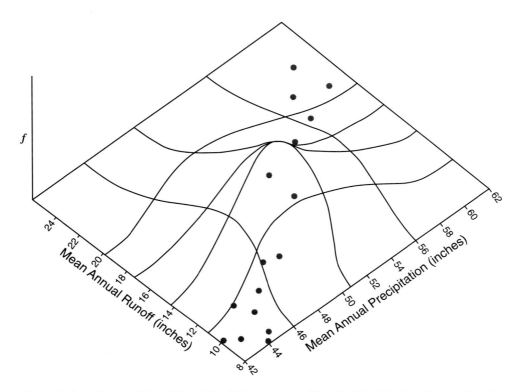

Figure 8.3 Scatter Plot of Runoff and Precipitation, Showing Distribution Expected under the Null Hypothesis

to as **homoscedasticity** and is analogous to the assumption of equal variances made in analysis of variance.

The covariance between two variables may be calculated with Equation 8.1:

$$COV_{1,2} = \frac{\sum_{i=1}^{n} Y_i X_i - \frac{\sum_{i=1}^{n} Y_i \cdot \sum_{i=1}^{n} X_i}{n}}{n-1} \tag{8.1}$$

For example, consider the data in Table 8.1. Mean annual runoff and mean annual precipitation have been recorded for each of sixteen sites. The scatter plot of the sixteen pairs of Y and X values is shown in Figure 8.3. Under the null hypothesis, the *expected* distribution of paired observations should resemble a symmetrical mound, as illustrated in this diagram. If the null hypothesis is true, any tendency toward a correlation coefficient greater than zero is attributable to sampling variation. The *actual* distribution in this example clearly departs from the symmetrical mound, with an apparent positive correlation.

Table 8.1

(Y) Mean Annual Runoff (Inches), (X) Mean Annual Precipitation (Inches), Squares of Y and X and Cross-Product Y · X

	Y	X	Y^2	X^2	$Y \cdot X$
	20.4	60.5	416.16	3660.25	1234.20
	23.0	60.0	529.00	3600.00	1380.00
	17.5	52.9	306.25	2798.41	925.75
	19.3	58.0	372.49	3364.00	1119.40
	21.0	58.5	441.00	3422.25	1228.50
	9.9	42.0	98.01	1764.00	415.80
	11.6	44.5	134.56	1980.25	516.20
	9.5	43.0	90.25	1849.00	408.50
	15.5	53.0	240.25	2809.00	821.50
	11.5	46.5	132.25	2162.25	534.75
	18.5	56.0	342.25	3136.00	1036.00
	12.7	49.0	161.29	2401.00	622.30
	8.2	44.0	67.24	1936.00	360.80
	8.6	44.5	73.96	1980.25	382.70
	10.6	45.0	112.36	2025.00	477.00
	13.1	48.0	171.61	2304.00	625.80
Sums:	230.9	805.4	3688.93	41191.66	12092.20

Source: Hahn (1977), p. 221.

As a first step toward obtaining the correlation coefficient from sample data, the covariance is computed:

$$COV_{1,2} = \frac{12{,}092.2 - \frac{230.9 \cdot 805.4}{16}}{15} = \frac{469.27}{15} = 31.28$$

The numerical value of the covariance can be considerably greater than unity which, as a measure of association, is difficult to interpret. Standardizing it by dividing by the product of the two standard deviations yields the correlation coefficient:

$$r = \frac{COV_{1,2}}{s_1 \cdot s_2} \tag{8.2}$$

Substituting numerical values from the precipitation experiment for the terms in this equation yields

$$r = \frac{31.28}{4.877 \cdot 6.582} = .97$$

A formula for computing r without first deriving means and standard deviations is given by Equation 8.3:

$$r = \frac{n\sum_{i=1}^{n} Y_i X_i - \left(\sum_{i=1}^{n} Y_i\right)\left(\sum_{i=1}^{n} X_i\right)}{\sqrt{n\sum_{i=1}^{n} Y_i^2 - \left(\sum_{i=1}^{n} Y_i\right)^2} \sqrt{n\sum_{i=1}^{n} X_i^2 - \left(\sum_{i=1}^{n} X_i\right)^2}} \quad (8.3)$$

As you can see, it is necessary to obtain the sums of Y, X, Y^2, X^2, and the cross-product $Y \cdot X$. Substituting the required values from Table 8.1 into Equation 8.3, we compute:

$$r = \frac{16(12{,}092.2) - (805.4)(230.9)}{\sqrt{16(41{,}191.66) - (805.4)^2} \sqrt{16(3{,}688.93) - (230.9)^2}}$$

$$= \frac{7508.34}{7704.04} = .97$$

A second experiment involves an economic theme—the relationship between education and unemployment within a metropolitan area. Using a specific example, a null hypothesis states: H_0: There is no correlation between the average percent of persons in Baltimore, Maryland, metropolitan area census tracts who have graduated from high school, and the unemployment rate as a percent of total civilian labor force in these tracts in 1990 ($\rho = 0$). Twenty Baltimore area census tracts were selected at random. Tract averages for unemployment and high school graduation were then collected from the 1990 census. The data and computation of the product-moment correlation coefficient are shown in Box 8.1. The correlation between these two samples is .48.

BOX 8.1 Computation of the Product-Moment Correlation Coefficient for the Unemployment and Education Experiment

Data and computations: (Y) Unemployment rate as a percent of total civilian labor force, (X) high school graduates (percent), random sample of Baltimore, Maryland, census tracts, 1990. $n = 20$.

Unemployment Rate (percent) Y	High School Graduates (percent) X	Y^2	X^2	$Y \cdot X$
2.4	79.78	5.76	6364.85	191.472
2.0	93.60	4.00	8760.96	187.200
3.2	72.60	10.24	5270.76	232.320
4.3	84.00	18.49	7056.00	361.200
2.5	81.10	6.25	6577.21	202.750
1.9	82.90	3.61	6872.41	157.510
0.6	91.60	0.36	8390.56	54.960
4.9	57.50	24.01	3306.25	281.750

Bivariate Correlation and Linear Regression

Unemployment Rate (percent) Y	High School Graduates (percent) X	Y^2	X^2	$Y \cdot X$
1.9	73.60	3.61	5416.96	139.840
2.6	62.00	6.76	3844.00	161.200
3.2	85.50	10.24	7310.25	273.600
2.7	65.90	7.29	4342.81	177.930
4.7	41.00	22.09	1681.00	192.700
2.2	61.30	4.84	3757.69	134.860
5.3	69.60	28.09	4844.16	368.880
6.5	50.20	42.25	2520.04	326.300
7.0	60.80	49.00	3696.64	425.600
1.6	46.90	2.56	2199.61	75.040
3.0	80.70	9.00	6512.49	242.100
5.0	64.80	25.00	4199.04	324.000

Data source: U.S. Bureau of the Census (1992).

Intermediate statistics:

$$\Sigma Y = 67.5 \qquad \bar{Y} = 3.4 \qquad \Sigma Y^2 = 283.45$$
$$\Sigma X = 1{,}405.4 \qquad \bar{X} = 70.3 \qquad \Sigma X^2 = 102{,}923.7$$
$$\Sigma Y \cdot X = 4{,}511.2$$

Computation of r, using Equation 8.3:

$$r = \frac{n \sum_{i=1}^{n} Y_i X_i - \left(\sum_{i=1}^{n} Y_i\right)\left(\sum_{i=1}^{n} X_i\right)}{\sqrt{n \sum_{i=1}^{n} Y_i^2 - \left(\sum_{i=1}^{n} Y_i\right)^2} \sqrt{n \sum_{i=1}^{n} X_i^2 - \left(\sum_{i=1}^{n} X_i\right)^2}}$$

$$= \frac{20(4{,}511.2) - 67.5 \cdot 1{,}405.4}{\sqrt{20(283.45) - 67.5^2} \cdot \sqrt{20(102{,}923.7) - 1{,}405.4^2}}$$

$$= \frac{4{,}640.5}{9{,}629.7} = .48$$

We may now ascertain whether the magnitude of r is large enough to be significant. In effect this is a null hypothesis that there is no correlation in the population between the variables, or H_0: $\rho = 0$, against the alternative H_1: $\rho \neq 0$. Under the null hypothesis, the values of r obtained from different samples would tend to vary around zero, some below and some above. In terms of this experiment, we must test whether the value of $r = -.48$

is far enough removed from zero to allow rejection of H_0. When the variables are approximately normally distributed, the null hypothesis may be tested with Student's t distribution, with $n - 2$ degrees of freedom (Equation 8.4):

$$t = r\sqrt{\frac{n-2}{1-r^2}} \tag{8.4}$$

One degree of freedom is lost by taking the variation about the means, $Y_i - \overline{Y}$, and $X_i - \overline{X}$, and the other results from only $n - 1$ cases being needed to explain all variations, the last case being redundant. For this example, we compute

$$t = .48\sqrt{\frac{20-2}{1-.23}} = 2.32$$

As demonstrated in Chapter 6, this value of t may be evaluated against the critical value in Appendix Table 4. This is a two-tailed test, since we are concerned only with the magnitude of r, regardless of the sign. Degrees of freedom in this case are $n - 2 = 18$, and if α is set at .05, the critical values of t from Appendix Table 4 are ± 2.10. Since the computed value of t in the unemployment experiment exceeds the critical value, we cannot accept H_0 and cautiously conclude that the correlation between the two variables is significant. The probability (P) that a value of t as large, or larger, than 2.32 could occur by sampling variation (chance) is just under .05.

If the alternative hypothesis, which states the direction (positive or negative) of the correlation, is used, the t distribution must be applied as a one-tailed test. In the unemployment example, experience leads us to expect the correlation to be negative ($H_1: r < 0$). The null hypothesis can only be rejected if the value of r is negative *and* the computed value of t exceeds the critical value at the specified level of α.

Expecting this relationship to hold at a different geographic scale is an example of an *ecological fallacy*. For instance, a sample of twenty metropolitan areas was drawn at random from across the nation, and metropolitan area average data for high school graduation and unemployment were collected (U.S. Bureau of the Census, 1986). Using the same hypothesis, $\rho = 0$, correlation analysis was applied. This time, the correlation coefficient was $-.229$, which is not significantly different from zero ($t = -.998$, $P = .33$). Figure 8.4 shows the scatter plot for the latter experiment.

You cannot use a metropolitan-level relationship to predict what will happen at the national level, and you cannot project downward from a nation-scale relationship to predict association at the metropolitan scale. Even given a *bona fide* cause-and-effect relationship, measurable association varies differently at different scales. In this case, there are large socioeconomic differences among American regions. Certain attributes that affect unemployment and education vary differently between regions than within their individual metropolises. Thus, it should not surprise us to find a lower degree of correlation between these characteristics at the national scale. It should be pointed out that there is often *less* variance in human characteristics in *larger* spatial units than in small units.

To illustrate this point, recall the coefficient of variation from Chapter 4. The coefficient of variation enables us to compare the relative variation between two variables by

Figure 8.4 The Relationship between Percent Unemployment and Percent High School Graduates in a Sample of Twenty U.S. Metropolitan Areas, 1980

Source: Authors.

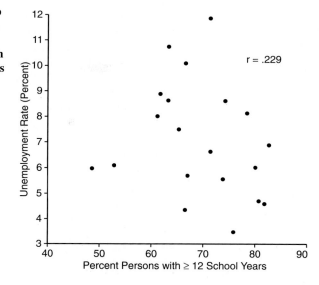

dividing standard deviations by respective means. Meade, Florin, and Gesler (1988, p. 226) stated that the coefficient of variation for mortality rates from leukemia was 7.0 percent comparing states at the national level, and 20.9 percent comparing counties within the state of Wisconsin. Similarly, the coefficient of variation of mortality rates for cancer of the nasopharynx is 188.9 percent for Wisconsin at county level, and 24.4 percent for the United States at state level. Thus, it is wise to interpret correlation coefficients with care, especially if different geographic scales are involved.

Correlation and Causation

Measures of strength of association should not be thought of as something that proves **causation,** but only as a measure of covariation in the measurements. One or more of the following situations may obtain:

1. *Variation in either characteristic may be caused (directly or indirectly) by variation in the other.* The variable that is supposed to be the cause of variations in the other is usually called the independent variable. Thus, because distance between places is thought to affect spatial interaction, rather than vice versa, the amount of movement between places would be made the dependent variable. It is also logical to assume that runoff is the result of precipitation, not the reverse. A measure of association in itself does not imply that Y is caused by X, any more than X is caused by Y. Knowledge of the system itself suggests the direction of cause and effect.
2. *Covariation of the two variables may be due to a common cause or causes affecting each variable in the same way, or in opposite ways.* If we should find that there is areal association between days of sunshine and per capita federal in-

come tax payments, it should not hastily be concluded that it takes a sunny day to entice a person to pay their income taxes! Nor is it necessarily true that making large tax payments ensures beautiful weather. It is possible, however, that in places where the average income is high, the per capita income tax might be high, and these people will have migrated there because of the amenities found in warm, sunny climates.

3. *The causal relationship between the two characteristics may be a result of interdependent relationships.* For example, placing a high tax on automobile use may depress automobile sales, but only among low-income persons who tend to locate their residences on public transportation routes. Upper-income persons (with large families) are more likely to live farther from the congested city center and will continue to purchase automobiles simply because they have no alternative method of travel. Thus, the elasticity of demand for a good or service, the period of time under observation, and any number of other variables can affect changes in tax rates, prices, movement, and so on. In fact, virtually every process that the geographer observes and analyzes is multivariate in nature, and bivariate relationships yield only limited explanations.

4. *The association may be due to chance alone.* Even though there may be no relationship whatsoever between the variables in the population from which the sample is drawn, it is possible that a sample of observations will produce what appears to be a high degree of areal association by chance alone. Thus, we may conclude that in a given group of commuters there was a positive relationship between the distance of their journey to work and their divorce rate. Yet it is difficult to develop a theory as to why this positive relationship should be so; the probability that another sample would yield quite different results is high.

5. *The strength of the association may be affected by one or two extreme values of either variable.* Figure 8.5 is a scatter plot of the relationship of female stomach cancer rates and female pancreatic cancer rates in Québec (Thouez et al., 1991). Most of the paired observations cluster near the origin of the graph, but one observation is located at rates above 300 on both axes. Such a situation may alter both the slope and the strength of the correlation. In this example, if the one extreme case were removed, the slope would change, and the value of the correlation coefficient would certainly be lower. One might remove this outlier and re-

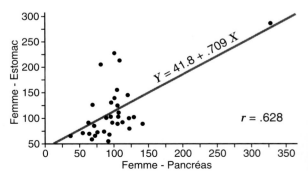

Figure 8.5 A Statistical Relationship Illustrating the Effect of One Observation on the Correlation Coefficient

compute the statistics. The researcher should report the results of both experiments along with a short explanation of what was done to produce the different statistics.

6. *The strength of the association may be affected by a lack of independence among observations.* Observations from areal units next to each other may be influenced by each other. Examples of lack of independence include the way the real estate value of one piece of property affects that of the property around it. Crop yields may be affected by common local weather, common local terrain features and soil types, drainage conditions, and so on. This is what is known as **spatial autocorrelation.** One important problem with spatial autocorrelation is that the assumption of independent observations, required for parametric statistical procedures such as linear correlation and regression, may be violated. All locations are related to each other, but closer locations are more strongly related than those more distant. In other words, you should expect that for characteristics known to be related, data for two adjacent areal units will contain some redundancies. If test statistics support this contention, then statistical hypotheses based upon *independent data* will be flawed, often leading to incorrect statistical decisions. Research experiments involving correlation or regression of variables on adjacent or nearly adjacent areal units might display spatial autocorrelation. How one can tell statistically if there is autocorrelation in a particular situation is a procedure that is peripheral to our discussion. Interested readers are referred to Griffith and Amrhein (1991).

7. *An erroneous (nonlinear) functional form is specified.* In Figure 8.1, a moderately high linear relationship was evident except for the two largest values of distance, which are lower than expected for their given transportation costs. Viewing that scatter plot, we have cause to suspect that the relationship might be *nonlinear*—that is to say, over increasing distances transportation costs might increase at a *decreasing,* rather than a constant, rate. One or both of the variables have a distribution that is not normal. The consequence of this is that the t test is not appropriate. This non-normality could, of course, have occurred by chance in one sample distribution; sample correlations from a bivariate population are quite variable if n is small. Another sample, or a larger sample, might have produced a scatter plot more like that of Figure 8.2. However, if repeated samples produce the same nonlinear appearance, another recourse is to *transform* the raw data to a nonarithmetic base, which will be demonstrated later in this chapter. If the relationship is nonlinear, it will be necessary to employ a *nonparametric* statistical test, which will be demonstrated in Chapter 9.

REGRESSION ANALYSIS

The **correlation coefficient** estimates the degree of closeness of the linear relationship between two variables. Often, however, the most interesting questions about these variables are: How much does one variable change for a given change in the other? How accurately can the value of one variable be predicted from knowledge of the other? These questions may be answered with *regression analysis.* The term *regression* is an established and conventional concept based on the notion of allometry, or the change in proportion of various parts of an organism as a consequence of growth. For example, for any given height there is a range of observed weights. This variation will be partially due to measurement errors but primarily due to variation between individuals. Thus, no unique relationship between actual height and weight can be expected. But it can be observed that the average observed weight for a given observed height increases as height increases. This locus of average observed weights for given observed heights (as height varies) is called the *regression* curve of weight on height. It is thus an estimate of variation, based on average change.

The most elegant and simple way to express a relationship between two (or more) variables is by means of a mathematical equation. Physical laws such as $E = mc^2$ are examples of nearly perfect relationships because there are few exceptions to the rule. Often in geographic research, we deal with simple equations, including the equation for a straight line. In geography and related fields, some fairly convincing analyses involve simple linear relationships between two variables.

In regression analysis, there is a distinction between what is called the **dependent variable,** symbolically designated as *Y,* and the **independent variable,** which is designated as *X*. The purpose of regression is to suggest a possible explanation of the variation in the dependent variable, *Y*, by demonstrating a systematic covariation in some logically related variable, *X*. It is not necessary that the relationship be a causal one, nor should we imply that *X* is the only variable influencing *Y*. If *X* and *Y* are not related, that is, if they are independent, the values of *X* will not help us to predict values of *Y*, but the more *Y* is related to (dependent on) *X*, the more accurate will be our prediction of values of *Y*.

A functional relationship is **linear** when pairs of *X* and *Y* values fall into a pattern that is best described by the linear algebraic model,

$$\hat{Y}_i = a + bX_i \qquad (8.5)$$

where *a* is the *intercept* and *b* is the *slope* of the line, and $\hat{Y}_i$ is the expected value of *Y* for a given value of *X*. This function is usually called the **estimating equation.** The values *a* and *b* are estimates, and when values of *X* are substituted in the equation, the solution of the equation provides estimates of *Y* for given values of *X*. The proximity of the actual values of *Y* to a straight line derived through regression provides an indication of the strength of the relationship, as we observed in the section on correlation.

For example, Figure 8.6 indicates that there is a perfect linear relationship between the distance and the time that an object travels at a fixed speed. If the object's speed is 60 miles per hour, it will go 60 miles in one hour, 30 miles in 30 minutes, 15 miles in 15 minutes, and so on. Metric units may be substituted for miles, beginning with the assumption that the speed is 100 kilometers per hour. Equation 8.4 then expresses the relationship between the distance (*Y*) and the time (*X*) an object travels. The **regression coefficient,** *b,* is the **rate of change constant** and represents the **slope** of the linear function between the two variables. In the time-distance example, *b* is equal to 1.0. In other words, distance increases on the average by one mile per hour for each minute increase in time.

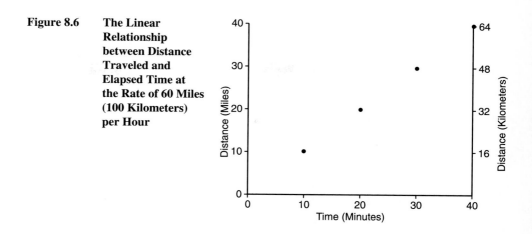

Figure 8.6 The Linear Relationship between Distance Traveled and Elapsed Time at the Rate of 60 Miles (100 Kilometers) per Hour

The constant *a* in Equation 8.4 is the value of *Y* when *X* is equal to 0.0. Referring again to the time-distance example, when time is equal to 0.0, distance also equals 0.0. In this case, *a* is equal to zero. This constant is called the **Y-intercept,** and represents the point at which the linear function passes through the *Y* axis. For this example, Equation 8.5 will be $Y = 0.0 + 1.0 X$, or simply $Y = X$.

Each of the four regression lines in Figure 8.7 has different values for *a* and *b*. The three different values of *a* are the result of the three different intersections of the lines with the *Y* axes. The different values of *b* reflect the steepness of their slopes. The higher the value of *b*, the steeper the slope. Remember that *b* indicates the rate of change in the

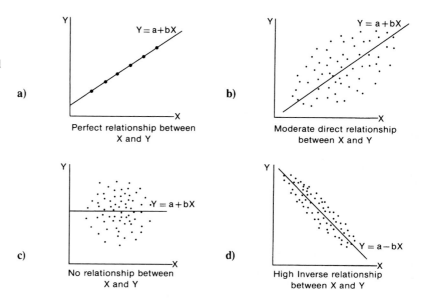

Figure 8.7 Scatter Diagrams of Two-Variable Relationships with Varying Slopes and Intercepts

dependent variable (Y) given a one unit change in X. Finally, the sign of b expresses the direction of the relationship between Y and X: when b is positive (Figure 8.7a and b), an increase in X is accompanied by an increase in Y; when b is negative (Figure 8.7d), Y decreases as X increases. In Figure 8.7c the relationship is neither positive nor negative, and the value of b is zero.

Computing Values of *a* and *b*

It would be unusual to find phenomena of research interest whose paired values all fall on a straight line, as did those of time and distance in this example. The object of regression is to find the estimates for values of a and b that best describe the relationship between any two logically related variables Y and X. Ordinarily, an analysis begins with a scatter plot. It is a good idea to develop the habit of drawing a scatter plot before proceeding with further analysis. Before the scatter plot is even completed, you may see that there is nothing to gain by proceeding further. For example, if dots appear to be randomly distributed on the diagram, there is clearly only a weak relationship at best between the two variables.

If the values representing mean annual runoff were exactly predictable by mean annual precipitation, the dots representing these sample years would form a straight line upward and to the right on the diagram. However, the pattern demonstrates that there is some variation from that *expected* relationship. You will recall that the sum of deviations between each element in a distribution and the mean for that distribution is equal to zero.

$$\sum_{i=1}^{n} (Y_i - \overline{Y}) = 0 \qquad (8.6)$$

Likewise, the sum of deviations between the actual and predicted values of Y is equal to zero.

$$\sum_{i=1}^{n} (Y_i - \hat{Y}) = 0 \qquad (8.7)$$

The squared values of ($Y_i - \hat{Y}_i$) may be summed to obtain the magnitude of the "error" around the regression line. The **linear model** obtained through regression is the line that produces the *minimum sum of squares of error*. This technique is known as the **least squares method**.

The computation of regression statistics is more complex than what is required for the univariate statistics described in preceding chapters. In practice the values necessary for regression-correlation analyses are seldom calculated by hand; we recommend using a computer to obtain these values if possible. To illustrate, however, a procedure follows for calculating a and b by the least squares criterion with a pocket calculator. Refer again to the runoff-precipitation example of Table 8.1.

The value of the regression coefficient b is obtained with the expression

$$b = \frac{\sum_{i=1}^{n}(X_i - \overline{X})(Y_i - \overline{Y})}{\sum_{i=1}^{n}(X_i - \overline{X})^2} \qquad (8.8)$$

However, this is more easily computed with the sums obtained from Table 8.1, employing Equation 8.9:

$$b = \frac{\Sigma XY - \overline{X}(\Sigma Y)}{\Sigma X^2 - \overline{X}(\Sigma X)} \qquad (8.9)$$

Substituting for the terms in this equation (where $\overline{X} = 50.34$), we obtain

$$b = \frac{12{,}092.2 - (50.34)(230.9)}{41{,}191.66 - (50.34)(805.4)} = \frac{468.69}{647.82} = .723$$

The value of the *intercept a* is then calculated in the expression

$$a = \overline{Y} - b\overline{X} \qquad (8.10)$$

Substituting for the terms on the right-hand side (where $\overline{Y} = 14.43$), we obtain

$$a = 14.43 - .723(50.34) = -21.97$$

The regression model,

$$\hat{Y} = -21.97 + .723(X)$$

for the runoff-precipitation sample data is the result. Note that when $X = 0$ (no annual precipitation), the estimated annual runoff is negative. This simply means that it takes a certain threshold amount of rainfall in a year before any runoff occurs, due to abstractions like evaporation and absorption. One may assume that, if the sample for the dependent variable is from a normally distributed population and the sampling is independent and random, this model represents a reliable approximation of the relationship between runoff and precipitation in this study area, that is, if all other features of the physical systems that produced these levels of precipitation and runoff are held constant. Figure 8.8 shows the scatter plot for the runoff-precipitation data, with the regression line fitted according to the least squares criterion.

One known point on the regression line is *a*, which is the value of $\hat{Y}$ when $X = 0$. The Y-intercept does not appear in this case because in order to display that intercept, the diagram would have to be approximately doubled in size. To avoid having a scatter plot with a large blank space between $X = 0$ and the lowest sample value of X, and between $Y = 0$ and the lowest sample value of Y, the lower extent of the ranges of both X and Y have been truncated.

Figure 8.8 The Relationship between Mean Annual Runoff and Mean Annual Precipitation, Showing Placement of the Regression Line

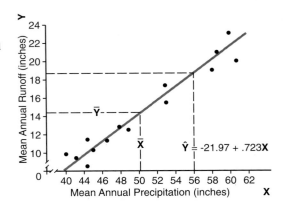

Another known point on the regression line is the intersection of the means of Y and X. We may connect the intercept with the intersection of the means to position the regression line within the scatter of points. In cases such as our runoff model, where it is difficult to plot a accurately, an additional point within the range of variable X must be computed with the predictive model (Equation 8.5). If we arbitrarily substitute the value of $X = 56$ in the predictive equation, we obtain $\hat{Y} = a + b(56) = -21.92 + .723(56) = 18.5$. It happens that one of the *observed* pairs of values happens to equal the *expected* pairs of values exactly. Thus, at the intersection of $X = 56$, and $Y = 18.5$, locate a point which can be used as a second reference point for the regression line of annual precipitation and annual runoff.

In cases where the slope of b is negative (the regression line slopes downward to the right), we can readily compute the point on the X-axis where the regression line will intersect. At that point, the value of $\hat{Y}$ may or may not be zero. If the minimum value of $Y = 0$, the X-intercept is computed simply as a/b. If the minimum value of Y on the graph is greater than 0, the value for the X-intercept is computed as $a - [Y_{MIN} \div b]$, where Y_{MIN} is the value of Y corresponding to the intercept of the X-axis.

The Goodness of Fit Around a Regression Line

In virtually all applications, you should expect at least some deviations about the regression line, and not all of them are necessarily due to sampling variation. Causality is frequently *multivariate*. That is, there are usually several, often interrelated, causes for any observable effect. For example, yields of grain in an agricultural area cannot be explained simply by amount of rainfall, but must include soil qualities, amount of capital applied in the forms of technology and agrochemicals, and other factors. You should therefore expect variations around a simple bivariate regression line similar to the deviations around the mean of a frequency distribution. Assuming that deviations are normally distributed around the regression line, the regression equivalent to Figure 6.1 may be diagrammed (Figure 8.9).

It is instructive to examine variations around the mean of the dependent variable (Y) and variations around the regression line. If every place were identical with respect to

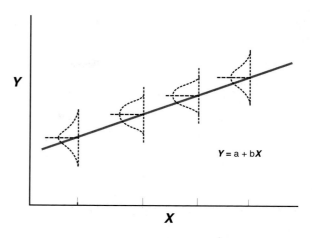

Figure 8.9 Representation of the Normal Distribution of Y about the Regression Line for Four Selected Values of X

some phenomenon, Y, regardless of their values of X, there would be no deviations, and every observation would equal the mean ($\overline{Y}$). On the other hand, if the correlation between Y and X is ± 1.00, all the paired values will fall on the regression line. Consequently, all of the variation in Y will have been explained by the variation in X, and there will be no residual variation. Figure 8.10 depicts these components of variation. Figure 8.10a shows the variation due to regression, which may be expressed as

$$\text{regression sum of squares} = \Sigma(\hat{Y}_i - \overline{Y})^2 \qquad (8.11)$$

The error, or *residual* sum of squares—variation that is not explained by regression—is expressed as

$$\text{residual sum of squares} = \Sigma(Y_i - \hat{Y})^2 \qquad (8.12)$$

Residual deviations are analogous to the deviations from the mean of a variable, as described in Chapter 4. If you divide the sum of squares of deviations about the regression line by $n - 2$, and take the square root of this result, you obtain the bivariate equivalent to the univariate standard deviation. In regression analysis, the standardized error term is known as the **standard error of estimate.** Symbolically, this is expressed as

$$S_{Y \cdot X} = \sqrt{\frac{\sum_{i=1}^{N}(Y_i - \hat{Y})^2}{n - 2}} \qquad (8.13)$$

Two degrees of freedom are lost, one each for the estimates of a and b. The standard error of estimate provides a measure of how well the regression equation predicts Y, given values of X. The smaller the standard error of estimate is, the less dispersed the paired X and Y values are around the regression line. If the residual variations above and below the regression line are normally distributed, then we expect approximately 68 per-

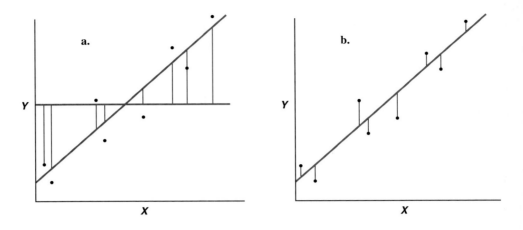
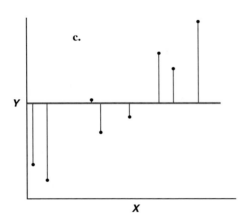

Figure 8.10 Partitioning the Variation of the Dependent Variable: (a) Due to Regression $(\hat{Y} - \overline{Y})$, (b) Due to Error $(Y - \hat{Y})$, and (c) Total Variation $(Y - \overline{Y})$

cent of the observations to fall within one standard error about the line and 95 percent of the observations to fall within 2 standard errors.[1]

In univariate statistics, we used the coefficient of variation to compare the variability of two or more variables with different means and standard deviations. We also have a

---

[1] If the entire population is represented, the standard error is computed as

$$S_{Y \cdot X} = \sqrt{\frac{\sum_{i=1}^{N} (Y_i - \hat{Y})^2}{N}}$$

form of the **coefficient of variation** that is applicable to regression, using the standard error of estimate. In this context, the formula is expressed as follows:

$$CV = \frac{S_{Y \cdot X}}{\overline{Y}} \times 100 \tag{8.14}$$

Multiplying by 100 enables us to express *CV* as a percentage. This statistic ranges from 0 to 100, corresponding to a range from a perfect correlation to zero correlation. The lower the value of *CV,* the better the relative fit of the regression line.

Finally, Equation 8.14 shows that the *total variation* consists of the sum of the two components just described:

$$\text{total sum of squares} = \Sigma(Y_i - \hat{Y})^2 + \Sigma(\hat{Y}_i - \overline{Y}) \tag{8.15}$$

The ratio of the explained variation to the total variation is referred to as the **coefficient of determination,** and is symbolized by r^2. This statistic, incidentally, may also be computed by squaring the correlation coefficient, *r*. The coefficient of determination is a measure of the proportion of the total variation in the data that is accounted for by the regression. When r^2 is unity, $S_{Y \cdot X}$ will be zero, since all points will be exactly on the regression line.

Returning to the unemployment–education experiment for the Baltimore metropolitan area, recall that the correlation coefficient was moderately high (.832) and the *t* test (−6.4) verified the significance of the relationship. Table 8.2 shows the twenty observations of tract unemployment rate (*Y*), percent high school graduates (*X*), predicted unemployment rate ($\hat{Y}$), and residuals from regression ($Y - \hat{Y}$). The residuals represent errors in prediction. Predicted values were computed from the estimating equation:

$$\hat{Y} = 26.27 - .289X$$

The first observation, which represents a lower middle-income suburban census tract, had only 54.6 percent of its population (age twenty-five and older) reported as high school graduates. Substituting this value into the estimating equation, we obtain $\hat{Y} = 26.27 - .289(54.6) = 10.47$. Since the *actual* unemployment rate for this tract was 6.7 percent, the equation *overpredicted* unemployment by 3.77 percent. Summing all twenty (*n*) of the squared residual variations yields a value of 409.12. Dividing by 18 ($n - 2$), the **residual variance** of 22.73 is obtained. The *standard error of estimate,* which is the square root of residual variance, is 4.77. In this analysis, the standard error provides a measure of how well the regression equation predicts unemployment for the Baltimore metropolitan area, given sample data on high school graduation rates. Figure 8.11 shows the scatter plot for this problem, with the regression line and its standard error bands.

An alternative formula for computing $S_{Y \cdot X}$ is

$$S_{Y \cdot X} = s_Y \sqrt{1 - r^2} \tag{8.16}$$

Given that $r^2 = .692$ and $s_Y = 8.59$, compute

Table 8.2 Predicted Percent Unemployment and Residuals from Regression, Baltimore Metropolitan Area Census Tracts, 1980

Y Unemployment Rate (Percent)	X High School Graduates (Percent)	$\hat{Y}$ Predicted Unemployment Rate	$Y - \hat{Y}$ Residuals from Regression
6.7	54.6	10.47	−3.77
19.8	32.2	16.95	2.85
4.5	78.1	3.67	.83
9.4	34.1	16.40	−7.00
29.9	29.9	17.62	12.28
3.1	81.1	2.80	.30
18.4	25.8	18.81	−.41
17.4	23.5	19.47	−2.07
21.5	30.6	17.42	4.08
6.1	56.5	9.92	−3.82
13.0	50.2	11.74	1.26
3.3	62.9	8.09	−4.77
12.9	33.8	16.49	−3.59
3.3	88.5	.66	2.64
2.0	91.1	−.09	2.09
4.9	94.1	−.96	5.86
3.2	61.6	8.45	−5.25
3.3	74.9	4.60	−1.30
22.7	27.1	18.43	4.27
4.6	59.4	9.08	−4.48

$$S_{Y \cdot X} = 8.59 \sqrt{.308} = 4.77$$

An r^2 of .692 indicates that a fairly large proportion of the total variation in the observed values of unemployment have been explained by high school graduation rates. One can use the sum of squares to compute variances that in turn can be used to test the null hypothesis of no linear relationship between X and Y. In particular, the regression

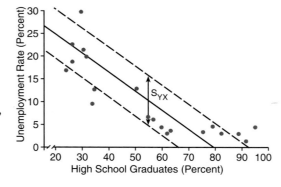

Figure 8.11 Scatter Plot Showing the Relationship between Unemployment Rate and High School Graduation Rate, Baltimore, 1980, with Regression Line and Standard Error Bands

sum of squares can be used to estimate variance about the regression line. Using the sums of squares, an analysis of variance can be constructed that will lead to rejection if either the observations are too variable or the postulated regression model is incorrect. The form of the ANOVA table for simple linear regression is:

Source of Variation	Sum of Squares	v	Estimated Variance	F Test
Linear Regression	RSS	1	MS_R	MS_R/MS_E
Residuals (Error)	ESS	$n-2$	MS_E	
Total	TSS	$n-1$		

where *RSS*, *ESS*, and *TSS* refer to regression sum of squares (Equation 8.11), residual sum of squares (Equation 8.12), and total sum of squares (Equation 8.15); MS_R and MS_E refer to mean square variances, which are computed by dividing *RSS* and *ESS* by their respective degrees of freedom.

As an example, consider the unemployment-education experiment. Sums of squares and degrees of freedom are substituted into the table (Table 8.3). In order to reject H_0 at $\alpha = .05$, the variance due to regression must exceed that due to error by an amount $\geq F_{\alpha = .05, df = 1, 18}$. Consulting Appendix Table 7, we find the critical value to be 4.41. The computed ratio of regression variance to residual variance is 40.42, so the probability that a difference that large could have occurred due to sampling variation is infinitesimal.

Finally, we may employ the coefficient of variation as an additional measure of goodness of fit. Recall that $S_{Y \cdot X} = 4.77$ and $\bar{Y} = 10.5$. Substituting for the terms of Equation 8.14:

$$CV = \frac{S_{Y \cdot X}}{\bar{Y}} \times 100 = \frac{4.77}{10.5} \times 100 = 45.43$$

Even though significance tests affirm that there is an areal association between these two variables, the scatter of dots suggests that the relationship might not be linear. Hence, *CV* is only moderately low. As you will see in the next section, the goodness of fit may be improved with a transformation of the original data base.

Table 8.3 ANOVA on Unemployment-Education Regression Equation

Source of Variation	Sum of Squares	v	Variance	F Ratio
Regression	918.60	1	918.60	40.42
Residual	409.12	18	22.73	
Total	1327.72	19		

NONLINEAR MODELS

There are numerous examples of nonlinear relationships in both human and physical geography. A regression model is linear, as we stated earlier, when the parameters present in the model are in the linear form, $Y = a + bX$. On the other hand,

$$Y = aX^b \tag{8.17}$$

is a **nonlinear model** because the parameter b does not enter the model in a linear fashion. Equation 8.17 is what is known as a *power function*. For example, the relationship between consumer attraction to shopping centers and distance from consumer origins is often an *inverse power function* as shown in Figure 8.12. Again, if a scatter diagram is prepared before proceeding to compute regression and correlation statistics, it may be evident that the distribution of dots is such that a linear equation cannot be fit to the data.

If the scatter plot of sample data should suggest a nonlinear relationship, as in the unemployment–education experiment (Figure 8.11), it is because the sample distribution of X, or Y, or both, is not normal. Recall that one of the standard assumptions of linear regression is that the variables under scrutiny are normally distributed. The most commonly violated assumptions are those concerning the linearity of the model and the constancy of the error variance.

Data Transformations

To satisfy the assumptions of the standard regression model, instead of working with the original variables, we sometimes work with variable **transformations.** The notion of altering sample data to make them fit a linear model understandably tends to make students uncomfortable. In statistics, however, there is no scientific necessity to adhere to the common linear or arithmetic scales to which most people are accustomed. It takes ex-

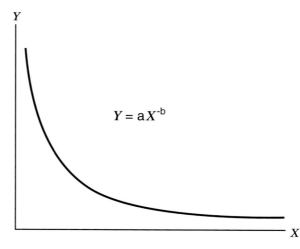

Figure 8.12 A Typical "Power Function" Relationship between Consumer Travel (Y) and Distance to Center of Attraction (X)

tensive experience in science and statistics to appreciate the fact that the linear scale occupies a similar position with relation to other scales of measurement as does the decimal system of numbers with respect to alternative numbering systems. If numbers can be multiplied on a linear scale, it might make more sense to think of adding them on a logarithmic scale.

Transformations of the scale of sample data to *base-10 logarithms* or *square roots* are frequently used. Depending on the source of the skew, logarithms, square roots, reciprocals, or even angular transformations may be employed. As the scale of measurement is arbitrary, you simply observe the distributions of transformed variables and decide which transformation most closely satisfies the assumptions of the analysis before carrying out a statistical test.

Transformation of data involves transferring the measurements from the original linear scale to a new one. The result is to change the *intervals* between the individual observations on the measurement scale. Transformation changes the value of an observation, but not the *rank* of a particular observation. The overall effect is to change the shape of the frequency distribution of the data. A few of the many techniques for transforming arithmetic data are shown in Figure 8.13. When a distribution is extremely peaked (leptokurtic), as in Figure 8.13c or d, these transformations will be of little help simply because changing the intervals between closely bunched values will not create enough separation to lower the peak relative to the observations in either tail.

Transformations may be necessary for several reasons:

1. The theory underlying the analysis may specify that the relationship between two variables is linear. An appropriate data transformation can make the relationship between the transformed variables linear. The regression equation resulting from the log transform of both the X and Y axes appears as follows:

$$\log Y = b \log (X) \qquad (8.18)$$

which is the logarithmic form of Equation 8.17.

2. The dependent variable, Y, may have a probability distribution whose variance is related to the mean. If the mean is related to the value of the independent variable, X, then the variance will not be constant, but will vary with X. The distribution of Y will usually also be nonnormal under these conditions, which invalidates the parametric tests of significance (t and F), especially with small samples ($n < 30$). The result is that the estimating equation will produce estimated values of Y that are imprecise. As a remedy, the data may be transformed to ensure normality and constancy of error variance.

3. There are neither *a priori* theoretical nor probabilistic reasons to suspect that a transformation is required. Instead, you may discover that the residuals produced by a linear regression equation are not randomly distributed about the mean of Y.

Before applying a logarithmic transformation, you must eliminate all zeros or negative numbers in your arithmetic data, since there are no logarithm equivalents for these values. This may be accomplished by first adding a constant to each observation. The constant can be any number that will ensure that the lowest value in the data set is greater than zero.

Figure 8.13 Some Data Transformations That Correct for Skew and Kurtosis

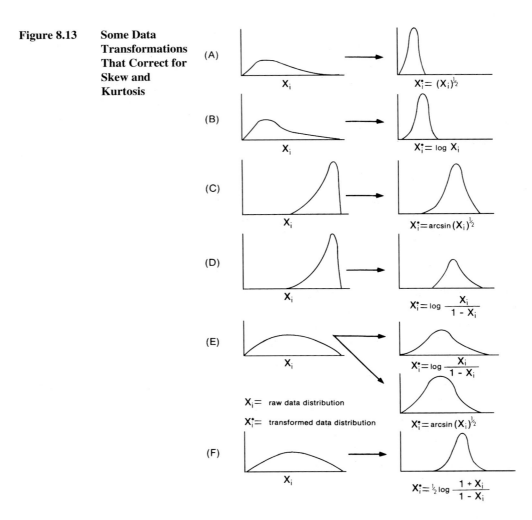

Frequently, transformation of both Y and X to logarithms prior to their use in a correlation analysis will yield a satisfactory fit to a linear model. For example, consider the hypothetical (nontransformed) data of Table 8.4. Plotting values of attraction versus distance (Figure 8.14a), we obtain a scatter which is quite well represented by the model of Figure 8.12. Substituting the values of log Y and log X improves the fit to a linear model (Figure 8.14b). Even when the raw data are forced into a linear model, the value of r is significantly high: $-.847$ ($t = 4.78$), with a standard error of 1.76. However, the transformed data yield a coefficient of correlation of $-.96$ ($t = 10.29$), and a standard error of only .102. Clearly, the transformation of the raw data has given us a stronger linear model. It is perfectly valid to perform correlation analysis on transformed data because r is a pure number without units or dimensions.

After fitting a linear model, you should examine a plot of the residuals. For example, briefly reconsider the linear model of unemployment and education in the Baltimore

Table 8.4 Original and Log Transformed Data for a Bivariate Curvilinear Relationship

Y	X	log Y	log X
10.0	1.5	1.00000	.17609
8.5	1.1	.92942	.04139
7.5	1.8	.87506	.25527
5.8	1.8	.76343	.25527
3.9	2.1	.59106	.32222
3.8	3.6	.57978	.55630
2.5	5.2	.39794	.71600
1.6	7.2	.20412	.85733
2.1	9.2	.32222	.96379
1.1	9.6	.04139	.98227
1.1	11.0	.04139	1.04139

metropolitan area. Figure 8.15a shows a plot of residuals from that model. Note on the right-hand side of the plot, the residuals tend to line up in one direction. This is a condition known as **heteroscedasticity**. Heteroscedasticity is revealed if the residuals tend to increase or decrease with the values of X. If heteroscedasticity is present, the least squares estimates will be less than precise. Heteroscedasticity can be removed by transforming

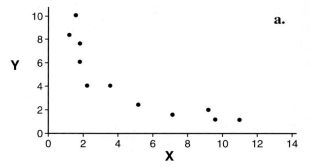
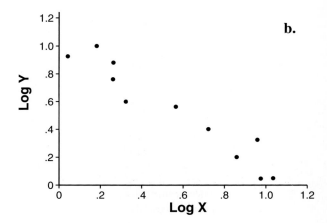

Figure 8.14 (a) Curvilinear Relationship between Y and X, and (b) Same Relationship with Log Y and Log X.

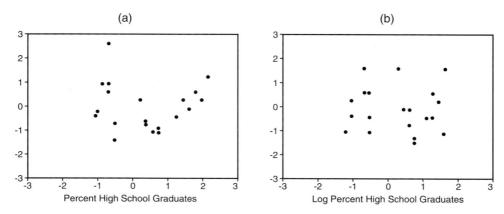

Figure 8.15 Standardized Residuals from Regression of Percent High School Graduates on Unemployment Rate, Baltimore, 1980: (a) Arithmetic Scales, and (b) Logarithmic Scales

these data to a logarithmic base. This yields what is called a *log-normal* or *log-linear* model. Figure 8.15b is a plot of the residuals from regression with the high school graduate and unemployment data transformed to logarithms. The residuals from the transformed model now show no evidence of heteroscedasticity. The distribution of residuals shows no distinct pattern, and we may conclude that the transformed model is adequate. Furthermore, r has increased from .83 to .91, $S_{Y \cdot X}$ has fallen from 4.77 to about 1.42, and the coefficient of variation decreased from 45.43 to 13.52.

The ability to transform data before you analyze them or after preliminary analysis is often as important as the analysis itself. Having a computer and statistical software that performs these transformations makes the task far easier than it is when only calculators are available. Appendix A contains SAS and SPSS-X commands necessary to perform correlation and regression analyses, including data transformations.

KEY TERMS

areal association   199
causation   205
coefficient of determination   215
coefficient of variation   204
correlation coefficient   199
covariance   199
dependent variable   208
estimating equation   208
heteroscedasticity   221
homoscedasticity   200
independent variable   205
inverse relationship   198
least squares method   210
linear model   210
negative relationship   198
nonlinear models   218
Pearson product-moment correlation (r)   199
positive relationship   198
rate of change constant   208
regression coefficient   208
residual variance   215
scatter plot   197
slope   208
spatial autocorrelation   207
standard error of estimate   208
transformations   207
Y-intercept   209

REFERENCES

Griffith, Daniel A., and Carl G. Amrhein. (1991). *Statistical Analysis for Geographers*. Englewood Cliffs, NJ: Prentice Hall.

Haan, Charles T. (1977). *Statistical Methods in Hydrology*. Ames, IA: Iowa State University Press.

Mauldin, W. Parker, and Sheldon J. Segal. (1988). *Studies in Family Planning*. 19, No. 6.

Meade, Melinda S., John W. Florin, and Wilbert M. Gesler. (1988). *Medical Geography*. New York: Guilford Press.

Thouez, J-P., P. Ghadirian, and A. Rannou. (1991). "L'incidence des cancers de l'estomac / pancreas et du colon / rectum au Québec: Interprétation géographique." Unpublished manuscript.

U.S. Bureau of the Census. (1986). *State and Metropolitan Area Data Book, 1983*. Washington, DC: U.S. Government Printing Office.

U.S. Bureau of the Census. (1988). *County and City Data Book, 1988*. Washington, DC: U.S. Government Printing Office.

U.S. Bureau of the Census. (1992). *Census of Population and Housing, 1990: Summary Tape File 3 on CD-ROM (Maryland)*. Washington, DC: U.S. Bureau of the Census.

EXERCISES

1. Appendix B-3 lists seven variables for a sample of U.S. metropolitan areas. Using the methods outlined for calculator solution of regression and correlation problems, compute the parameters a, b, r, r^2, and $S_{Y \cdot X}$ for a random sample of twenty metropolitan areas for *one of the pairs of variables, below*. Also construct a scatter plot on graph paper, fit the regression line, and label that line with the estimating equation. Finally, compute and draw parallel lines representing the standard error of estimate, $S_{Y \cdot X}$, on each side of the regression line.

   a. *Y:* Percent families below poverty
   X: Percent women in labor force

   b. *Y:* Percent non-English speakers
   X: Percent foreign born

   c. *Y:* Median family income
   X: Percent women in labor force

2. The basic inferential test of r states a null hypothesis that there is no correlation between variable X_1 and variable X_2. Which statement below expresses this null hypothesis?
   a. $H_0: r < 0$ **b.** $H_0: r_{X_1} \neq r_{X_2}$ **c.** $H_0: r = 0$ **d.** $H_0: r \neq 0$

3. Does the null hypothesis call for a one-tailed or a two-tailed test of significance?

4. Referring to the correlation computed in Problem 1, compute a value of t and determine the significance level (P) for r for a two-tailed test.

5. A typical alternative hypothesis for a one-tailed test of r is $H_1: r < 0$. Which of the following alternative hypotheses below would call for $r < 0$?

   a. The literacy rate among women in countries is positively related to their rate of contraceptive use.

   b. The fertility rate in countries is inversely related to the percent of families using contraceptives.

   c. There is a relationship between energy use per capita in countries and the GNP per capita in those countries.

6. In the general equation for the straight line, $Y = a + bX$, what is the value of Y when $X = 0$?
   a. Unknown **b.** a **c.** Zero **d.** Infinity

7. Which equation below describes a linear relationship with a positive slope of 1.5, and an intercept of 12?
   a. $Y = 12 - 1.5X$ b. $Y = 12 + 1.5X$ c. $Y = 1.5 + 12X$

8. Suppose that the linear equation, $Y = 19.08 - .21X$ describes the relationship in American cities between (Y) percent of families below poverty and (X) percent of women in the labor force. If an American city has a rate of .85 (85 percent) for women in its labor force, what percent of families below poverty would you predict for that city?

9. Which variable is the dependent variable in the following problem? "Researchers wish to use latitude values in North America to predict length of growing season for small grains."
   a. length of growing season
   b. latitude values
   c. size of grains

10. Using the data in Table 8.1, compute a, b and $S_{Y \cdot X}$. Construct a scatter diagram on graph paper, fit a regression line and label it with the estimating equation. Finally, construct $S_{Y \cdot X}$ lines above and below the regression line as demonstrated in Figure 8.11.

Nonparametric Statistics   9

In earlier chapters we introduced inferential statistics that are based on the assumptions that sample data are drawn from metric-scaled, normally distributed populations. Chapter 6 pointed out that the object of a statistical test was to minimize the chances of committing of a Type I error. The *power* of a test against a specified alternative hypothesis is the probability of (correctly) rejecting the null hypothesis. Tests such as z, t, and F, which are based on these assumptions, have also been described by statisticians as *robust*. The term robust refers to the efficiency of a statistical test in terms of its performance in the face of data limitations. The t test was robust, for example, with reasonably large samples even when it was known that the population was not normal. As with all aspects of statistics, robustness is a matter of degree and interpretation.

Unfortunately, many of the variables that geographers wish to analyze simply do not have normally distributed populations, or they are not measured on either the interval or ratio scale. Often only frequency or ranked data are available. What is needed is a set of robust tests whose power relative to competing tests remains high when the data are not suited to a parametric test. Such tests have been developed, and some of them are presented in this chapter.

It should be emphasized that the so-called *nonparametric* tests are not necessarily inferior statistics. For instance, comparing a parametric t test and a nonparametric rank test, statisticians George Snedecor and William Cochran (1989) state:

In large normal samples, the rank tests have an efficiency of about 95 percent relative to the *t* test; and in small normal samples, the signed-rank test has been shown to have an efficiency slightly higher than this. With non-normal data from a continuous distribution, the efficiency of the rank tests relative to *t* never falls below 85 percent in large samples and may be much greater than 100 percent for distributions that have long tails. Since they are relatively quickly made, the rank tests are highly useful for the investigator who is doubtful that the data can be regarded as normal. (p. 144)

Strictly speaking, only procedures that test hypotheses that are not statements about population parameters are classified as *nonparametric,* while those that make no assumption about the sampled population are termed **distribution-free** procedures. The terms *nonparametric* and *distribution-free* are customarily used interchangeably.

Nonparametric statistics have certain advantages and disadvantages. Among the former are:

1. They allow for the testing of hypotheses other than those concerned with population parameter values.
2. They may be applied when the sample data are measured on the ordinal or nominal scale.
3. They may be used when the form of the sampled population is known to be extremely non-normal, or is unknown.
4. They are easier to compute by hand or calculator, and are consequently more quickly applied than parametric tests on small samples (however, this should not be used as the sole *justification* for the choice of a nonparametric test over a parametric one). Some nonparametric techniques become laborious when large samples are involved.
5. They are a suitable alternative to a parametric test when the sample n is small (less than about fifteen), when violations of the assumptions of parametric tests are particularly difficult to detect. Based on the central limit theorem, the larger the sample, the more advantageous a parametric test becomes. As we pointed out above, however, nonparametric tests may be less efficient than parametric statistics.

The nonparametric tests in this chapter have been designed to apply to ordinal or nominal scale measurements. In some tests, the assumption is made that the sample median is no different from the population median, and the ranks of the sample observations are thus posited to be symmetrically distributed about their median. There are procedures for testing hypotheses regarding the frequency of occurrences of observations in various nominal categories; still other methods test hypotheses about cumulative probabilities derived from frequencies of occurrences of observations in nominal categories. We will examine procedures that are based on nonmetric measurements in one, two (paired and unpaired), and three or more samples, tests that are based on the "goodness of fit" to a cumulative probability distribution, tests that are based on the "goodness of fit" between frequency distributions, and a test that is based on the comparison of the ranks of two variables.

THE SIGN TEST FOR THE MEDIAN

We introduced our first parametric test in Chapter 6 with a comparison of a sample mean with a known population mean. Although the *t* test has been touted as rather insensitive to violations of the normality assumption, there are times when a nonparametric test is desirable, particularly if the available measurements are ordinal. The **sign test for the median** provides a simple, distribution-free test of hypotheses about symmetry of sample data around their median. The median and mean are of course equal in symmetric distributions. The hypothesis to be tested is that the value of the sample median is not significantly different from an expected population median.

The sign test gets its name from the fact that pluses and minuses, rather than numerical values, provide the data used in the computations. As we will demonstrate, a sign test is available for both single samples and paired samples. Essentially, the sign test counts the number of positive and negative signs among the differences between observations and the median. As an illustration (Box 9.1), an economic geographer consulting for a federal government agency is presented with what is explained to be a random sample of $n = 20$ observations of median family income for renter-occupied households in an eastern U.S. metropolis. The statistical question is, do the sample data support the null hypothesis that the true median family income for this metropolis is $12,217, at the .05 level of significance?

As a first step in testing the hypothesis, the geographer determines which tract incomes lie above and which ones lie below the hypothesized median of $12,217. A plus sign is attached to those tract incomes that are above the hypothesized median and a minus sign to those that fall below the median. Observed values that equal the hypothesized median are counted neither as plus or minus, and the sample size is reduced accordingly.

From the listing of the sample of household incomes in Box 9.1, it may be easily seen that eight of the tract values are below the hypothesized median of $12,217, and twelve of the values are above the median. The second step in this test is to calculate the probability of obtaining as few or fewer of the less frequently occurring sign, or the probability of obtaining as many or more of the more frequently occurring sign. The critical upper and lower bounds are found with Equations 9.1 and 9.2:

$$\text{Lower Bound} = .5 - z_{\alpha/2} \sqrt{\frac{(.5)(.5)}{n}} \qquad (9.1)$$

$$\text{Upper Bound} = .5 + z_{\alpha/2} \sqrt{\frac{(.5)(.5)}{n}} \qquad (9.2)$$

It does not matter which proportion—the most frequently occurring sign or the less frequently occurring sign—is computed, since the test assumption is that half the observations will fall above and half below the median. For a two-tailed test, the area under a symmetrical distribution between the critical upper and lower bounds constitutes an acceptance region, or $1 - \alpha$. The area in each tail $= \alpha/2$ (Figure 9.1). The decision follows

BOX 9.1 — The Sign Test for the Median Household Income Experiment

Observations of median household incomes in renter-occupied households are drawn at random from a hypothetical metropolitan area. The $n = 20$ income values (in dollars) are shown below. For convenience, they have been arranged in ascending order of value.

4,417   4,459   6,186   7,977   10,122   11,535   11,786   12,193   12,312   12,435   13,358
14,324   15,144   15,735   16,818   20,455   21,068   21,319   27,708   42,765

The null hypothesis to be tested, at $\alpha = .05$, is that the median household income of the hypothetical population from which this sample is drawn is $12,217. The alternative is that the median is not $12,217. This statement implies a two-tailed test.

The result of tract income differences from the median is as follows:

– – – – – – – – + + + + + + + + + + + +

The null hypothesis may be stated as follows: The probability of a plus is equal to the probability of a minus, and these probabilities are each equal to 0.5. Symbolically, this is stated as

$$H_0: p(+) = p(-) = .5$$

The test statistic is the proportion of observations in the sample with values above or below the value of the population median specified above. Assuming that we use the proportion (out of twenty) of negative signs, the null hypothesis may be rewritten as

$$H_0: p = f_{NEG}/n = .5$$

where p is the proportion of observations below the median. In this case $p = 8/20 = 0.4$.

The critical values ($p = .05$, two-tailed test) for p are computed as the normal distribution of the binomial:

$$\text{Lower Bound} = .5 - z_{\alpha/2}\sqrt{\frac{(.5)(.5)}{n}}$$

$$= .5 - 1.96\sqrt{\frac{.25}{20}} = .28$$

$$\text{Upper Bound} = .5 + z_{\alpha/2}\sqrt{\frac{(.5)(.5)}{n}}$$

$$= .5 + 1.96\sqrt{\frac{.25}{20}} = .72$$

We conclude that, since p falls within the bounds dictated by sample size $n = 20$, we accept the null hypothesis and conclude that the proportion of negative scores is not significantly different from 0.5, or that the median = $12,217.

Figure 9.1 The Acceptance and Rejection Regions for The Sign Test and The Wilcoxon Test

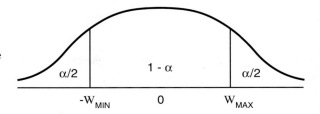

the convention that if the proportion, *p*, falls within the critical bounds, the null hypothesis is accepted. As the example in Box 9.1 illustrates, $p = 0.4$ when negative signs are employed. If the proportion of positive signs were counted instead, *p* would have been equal to 0.6. Since the area of acceptance is symmetrical, the decision is the same in either case.

PAIRED-SAMPLE TEST WITH ORDINAL DATA

The second technique, the **Wilcoxon signed-ranks test,** is the nonparametric equivalent to the matched-pairs *t* test presented in Chapter 6. It is, however, less robust than the matched-pairs *t* test because it is more likely to lead to a Type I error in marginal cases. The Wilcoxon test requires a level of measurement higher than the ordinal scale. As you will see, not only will pairs of interval values for a sample of places be used, but the *differences* between observations on each place will be ranked. The Wilcoxon test is *distribution-free,* as it does not assume a normal population. Also, the observations are not independent. The robustness of this test is greater than that of the sign test. Compared with a *t* test, the robustness of the Wilcoxon test is about 95 percent for both small and large samples. It is thus useful for experiments involving sample sizes that are too small to justify the normality assumption.

The Wilcoxon test applies essentially the same null hypothesis as the *t* test for paired samples—that there are no differences between the values of the two populations represented by the pairs of sample values. As an example, we draw a random sample of Third World countries and list their urban population proportions for the years 1982 and 1990 (Table 9.1). The signed ranks test will ascertain whether or not a significant change in these proportions occurred in the eight-year interim.

The first step is to obtain the magnitude of the difference between values for each pair. These differences are then ranked, *without regard to the sign of the difference.* Thus, a difference of −1 would receive a higher rank than a difference of +3. In the case of **tied ranks** in these absolute differences, *use average ranks for the tied observations.* For instance, differences for Guatemala, Iran, and Nicaragua are ranked, not 9, 10, and 11, but 10, 10, and 10, and for Indonesia and Honduras the ranks of 12 and 13 must be averaged. *After* ranking the absolute values of the differences, the signs of the observed differences are attached to the ranks, hence the term *signed ranks.* If the null hypothesis is correct, we expect that the sum of the ranks of the positive differences will cancel out the sum of the ranks for negative differences, leaving a balance of zero. As the difference varies, however, at some point the difference reaches an unacceptable magnitude, which increases the risk of committing a Type I error.

Table 9.1 Change in Proportion of Urban Population in Random Sample of Third-World Countries

Country	Urban as Percent of Total Population	
	1982	1990
Ethiopia	14	11
Nigeria	13	31
Western Samoa	20	21
Indonesia	20	26
Guatemala	36	40
Honduras	36	42
Mauritius	43	41
El Salvador	41	43
Costa Rica	43	45
Egypt	45	45
Iran	50	54
Panama	51	52
Nicaragua	53	57
Mexico	67	66

Source: Population Reference Bureau, Inc., *World Population Data Sheet, 1982* and *1990,* Washington, DC.

Next, the signed rank column is summed. The resulting sum is the test statistic ΣR. The variance of the distribution of ΣR is given by the expression,

$$\sigma_W^2 = \frac{n(n+1)(2n+1)}{6} \quad (9.3)$$

To test the significance of ΣR, the decision must be based on its value with respect to the critical value of W_α. Tables of the sample distribution of W_α may be found in many statistics textbooks or a book of statistical tables.[1] However, these tables are rarely needed, since the normal approximation is applicable for sample sizes of twenty-five or more. For a two-tailed test, where H_0 is applicable, the acceptance range is defined as:

$$W = \pm z_{\alpha/2} \sqrt{\frac{n(n+1)(2n+1)}{6}} \quad (9.4)$$

and, for a one-tailed test, H_1 is accepted if ΣR is either less than or greater than the critical values of W:

$$W = \pm z_\alpha \sqrt{\frac{n(n+1)(2n+1)}{6}} \quad (9.5)$$

---

[1]This and other nonparametric tables may be found in Beyer's, *Basic Statistical Tables*. See References for a complete citation.

BOX 9.2 — Application of the Wilcoxon Signed-Ranks Test: Significance of Change in the Urban Proportions of Third World Countries

Urban proportions of the populations of $n = 14$ randomly selected Third World countries are collected for two years, 1982 and 1990 (Table 9.1).

The null hypothesis to be tested, at $\alpha = .05$, is H_0:
There is no difference in urban population proportions of Third World countries between 1982 and 1990.

1. Compute the differences between 1982 values and 1990 values. Ignoring the signs, rank the differences. Note how ties are allocated the same rank. After ranking, carry over the signs from the difference column to the signed rank column.

Nation	% Urban 1982	% Urban 1990	Difference	Signed Rank
Ethiopia	14	11	3	8
Nigeria	13	31	−18	−14
Western Samoa	20	21	−1	−3
Indonesia	20	26	−6	−12.5
Guatemala	36	40	−4	−10
Honduras	36	42	−6	−12.5
Mauritius	43	41	2	6
El Salvador	41	43	−2	−6
Costa Rica	43	45	−2	−6
Egypt	45	45	0	1
Iran	50	54	−4	−10
Panama	51	52	−1	−3
Nicaragua	53	57	−4	−10
Mexico	67	66	1	3

2. Sum the signed ranks column:

$$\Sigma R = 8 + 6 + 1 + 3 - 14 - 3 - 12.5 - 10 - 12.5 - 6 - 6 - 10 - 3 - 10 = -69$$

3. Compute test statistic W. For a two-tail test, employ Equation 9.3:

$$W = \pm 1.96 \sqrt{\frac{14(15)(29)}{6}} = \pm 62.44$$

4. Decision: Since ΣR falls into the negative tail, reject H_0; conclude that there has been a significant positive change in urbanization in Third World countries.

In the example of change in percent urbanization of Third World countries (Box 9.2), the computed value of ΣR exceeds $W_{\alpha/2}$. The null hypothesis must be rejected. From the data we can conclude that the percent urban population in Third World countries increased over the period 1982–1990, and that this increase was significant.

TWO INDEPENDENT SAMPLES TEST

In Chapter 6 we employed the parametric t test to compare two populations on the basis of independent random samples. There is a comparable nonparametric test that can be used in situations where the parametric test on two independent random samples is inappropriate. This test is commonly referred to as the **Mann-Whitney U test,** although it was also developed by Wilcoxon. The U test is based on the assumption that populations differ with respect to their medians. Another way of stating it is that if the two independent samples are drawn from the same population, then the mean ranks of the two samples should be approximately equal. The two samples need not be equal in size, and although the exact form of their distributions need not be specified, they should have the same shapes. The U test is a robust alternative to the t test for independent samples.

The null hypothesis is that the two populations from which samples have been drawn have equal medians. The alternatives are (1) the populations do not have equal medians, (2) the median of population X_1 is larger than the median of population X_2, or (3) the median of population X_1 is smaller than the median of population X_2. If the two populations are symmetrical, so that within each population the mean and median are not significantly different, the conclusions reached regarding the two population medians will also apply to the population means.

As an example, suppose two biogeographers, working independently, collect samples ($n_1 = 10$ and $n_2 = 10$) of soil infiltration rates in liters per hour from two different study areas (Table 9.2). They state a null hypothesis (H_0) that there is no difference between median infiltration rates in these soils and assign a significance level (α) of .05. The values of the infiltration rates for both samples are combined and arrayed in order from least to greatest. The object is to keep track of the elements of each sample. As Box

Table 9.2 Independent Samples of Soil Infiltration Rates in Two Soil Types

Soil Type A (L/hr)	Soil Type B (L/hr)
25.4	17.8
21.4	19.3
18.9	13.7
25.6	16.1
24.1	14.5
18.5	13.8
27.3	9.9
29.4	11.3
28.6	10.4
24.3	11.8

9.3 illustrates, this may be done by underlining the infiltration rates of one sample. Each value is then replaced with its rank in the combined sample. The procedure for ties that was explained in the Wilcoxon test applies here as well. In this experiment there are no ties among infiltration rates, so the ranks are twenty discrete values. The ranks that belong to the underlined values are themselves identified by underlining. Now, the sums of the ranks of each sample are calculated.

A test statistic, U, is computed based on sample sizes and the sums of the ranks of the two samples. Equations 9.6 and 9.7 are used with the appropriate sums and sample sizes:

BOX 9.3

Procedure for Mann-Whitney U Test on Two Independent Samples of Soil Infiltration Rates

H_0: There is no significant difference between median infiltration rates (L/hr) for the two soil types (data from Table 9.2), or $Md_{X_1} = Md_{X_2}$. $\alpha = .05$.

H_1: $Md_{X_1} \neq Md_{X_2}$

Step 1. Combine the two samples for purposes of ranking them in ascending order of infiltration rate. Keep track of the ranks of the two soil sample sets, as shown below (second sample underlined).

<u>9.9</u> 10.4 <u>11.3</u> 11.8 <u>13.7</u> 13.8 14.5 <u>16.1</u> <u>17.8</u> 18.5 18.9 <u>19.3</u> 21.4 24.1 24.3 25.4 25.6 27.3 28.6 29.4

<u>1</u> 2 <u>3</u> 4 <u>5</u> 6 <u>7</u> <u>8</u> <u>9</u> 10 11 <u>12</u> 13 14 15 16 17 18 19 20

Step 2. Compute values of U for the ranks not underlined (n_1) and for the ranks that are underlined (n_2), using Equations 9.6 and 9.7:

$$U_1 = n_1 n_2 + \frac{n_1(n_1 + 1)}{2} - \Sigma R_1$$

$$= 10(10) + \frac{10(11)}{2} - 153 = 2$$

$$U_2 = n_1 n_2 + \frac{n_2(n_2 + 1)}{2} - \Sigma R_2$$

$$= 10(10) + \frac{10(11)}{2} - 57 = 98$$

Step 3. Evaluation and decision. The smaller of the two computed values of U is 2, which is far below the critical value of 23 found in Appendix Table 8 for $n_1 = n_2 = 10$. We therefore reject H_0, and conclude that the ranks of the infiltration rates differ significantly. Based on these rates, we cannot conclude that the two samples are from the same population of soils.

$$U_1 = n_1 n_2 + \frac{n_1(n_1 + 1)}{2} - \Sigma R_1 \tag{9.6}$$

$$U_2 = n_1 n_2 + \frac{n_2(n_2 + 1)}{2} - \Sigma R_2 \tag{9.7}$$

Summing the ranks of the sample infiltration rates, we obtain (for nonunderlined ranks) $\Sigma R_1 = 153$, and (underlined) $\Sigma R_2 = 57$. Substituting for the terms of Equations 9.6 and 9.7, we obtain $U_1 = 2$ and $U_2 = 98$. As a check, we should find

$$R_1 + R_2 = \frac{(n_1 + n_2)(n_1 + n_2 + 1)}{2} \tag{9.8}$$

and for the infiltration experiment,

$$153 + 57 = \frac{20(21)}{2} = 210$$

Appendix Table 8 contains critical values of Mann-Whitney U for sample sizes ≤ 20. In that table, bracketed values on the second line refer to two-tailed tests with $\alpha = .05$. The appendix table has been established in terms of the *smallest* of the values provided by Equations 9.6 and 9.7. With the Mann-Whitney U test, the computed value *must be less than the critical value in order to reject H_0*. For the soil infiltration experiment, $n_1 = n_2 = 10$, and $U = 23$. Since $U_{MIN} = 2$, we easily reject the null hypothesis, and can safely conclude that the two soil samples come from different populations, based on their infiltration rates.

The appendix table is designed for a maximum sample size of twenty. For larger sample experiments, we may test H_0 with a z statistic using Equation 9.9:

$$z = \frac{R_1 - R_2 - \frac{(n_1 - n_2)(n_1 + n_2 + 1)}{2}}{\sqrt{\frac{n_1 n_2 (n_1 + n_2 + 1)}{3}}} \tag{9.9}$$

MORE THAN TWO (INDEPENDENT) SAMPLES WITH ORDINAL DATA

In Chapter 7, a one-way analysis of variance test was demonstrated. The null hypothesis to be tested was that more than two subsample means are equal. The nonparametric equivalent of that parametric one-way ANOVA test is the **Kruskal-Wallis test.** When the assumptions underlying ANOVA are not met, that is, when the populations from which the samples are drawn are not normally distributed with equal variances, or when the data do not meet the restrictions imposed for parametric tests, the Kruskal-Wallis procedure is employed. This test again involves combining the values for the subsamples and ranking them, but keeping track of the ranks for each subsample as you proceed. This test extends the Mann-Whitney U test of medians to accommodate the situation where there are more than two independent samples. Again ordinal data are involved, either because they are the only figures available, or because the metric data from which the ordinal numbers have been derived are themselves unreliable in their prior form. Ranks obtained from ties in raw data are managed as previously explained.

As an example, refer to the data of Table 9.3. An index of national development has been formed for developing nations from two common measures, per capita caloric intake and per capita gross domestic product. Being unsure of the distribution characteristics of the populations of these two variables, the indices can probably be converted to ranks without substantial loss of information. Sampling these data for Third World regions in 1988–1990 yielded $N = 15$ values, which were combined and converted to ranks. The assumption in this test is that the sum of mean squared ranks of the three regions are not significantly different. Box 9.4 illustrates the application of the Kruskal-Wallis test statistic, H, to these data.

To apply this test, first combine the $n_1, n_2, \ldots, n_k$ elements of the raw data of the k subsamples into a single series of size N. Then rearrange the raw data in order of magnitude from smallest to largest. The observations are then replaced by ranks from 1, which is assigned in this case to the smallest index, to N, which is assigned to the largest index.

There are no ties in this example. If there were, it would be necessary to adjust the computed H value by a correction factor,

$$1 - \frac{\Sigma T}{n^3 - n} \qquad (9.10)$$

where $T = t^3 - t$. The letter t is the number of observations tied for a given rank. The value of H is adjusted by dividing it by Equation 9.10. Some statisticians believe that if there are several groups of tied values, a value of T should be computed for each group.

Table 9.3 Index of Development, Based on Caloric Intake and Gross Domestic Product, Fifteen Developing Countries, 1988–1990

Africa		Asia		Latin America	
Nation	Index	Nation	Index	Nation	Index
Côte d'Ivoire	.717	Bhutan	.504	Colombia	.891
Lesotho	.832	China	.744	Ecuador	.757
Malawi	.589	Hong Kong	1.160	Nicaragua	.826
Sierra Leone	.491	Sri Lanka	.770	Peru	.740
Zambia	.714	Syria	.787	Venezuela	.931

Source: Population Reference Bureau (1991a)

BOX 9.4 — **Procedure for Kruskal-Wallis Test of Three Independent Samples of National Development Indices**

H_0: The populations represented by the three regional samples do not differ with respect to the development index. $\alpha = .05$.

H_1: The populations represented by the three regional samples are significantly different with respect to the development index.

Step 1. The $n_1, n_2, \ldots, n_k$ observations from the k groups are combined into a single series of size N and arranged in ascending order. The observations are then replaced by ranks from 1 to N (with ties receiving the mean of the ranks for which they are tied):

Africa		*Asia*		*Latin America*	
Nation	Rank	Nation	Rank	Nation	Rank
Côte d'Ivoire	11	Bhutan	14	Colombia	3
Lesotho	4	China	9	Ecuador	8
Malawi	13	Hong Kong	1	Nicaragua	5
Sierra Leone	15	Sri Lanka	7	Peru	10
Zambia	12	Syria	6	Venezuela	2

Step 2. The ranks assigned to each class (region) are summed.

$$\text{SUM}_{\text{AFRICA}} = 55; \text{SUM}_{\text{ASIA}} = 37; \text{SUM}_{\text{LATIN AMERICA}} = 28$$

Step 3. The test statistic H (Equation 9.9) is computed:

$$H = \left[\frac{12}{N(N+1)} \left(\frac{\sum_{i=1}^{n_1} R_1^2}{n_1} + \frac{\sum_{i=1}^{n_2} R_2^2}{n_2} + \ldots + \frac{\sum_{i=1}^{n_k} R_k^2}{n_k} \right) \right] - 3(N+1)$$

Substituting the values from the national development experiment, we obtain

$$H = \left[\frac{12}{15(16)} \left(\frac{55^2}{5} + \frac{37^2}{5} + \frac{28^2}{5} \right) \right] - 3(16) = 3.78$$

Step 4. Consult Appendix Table 9 for the critical value of $H_{\alpha=.05}$, for $n_1 = 5$, $n_2 = 5$, and $n_3 = 5$. Find 5.78.

Step 5. Evaluation and decision. Since the computed value of H does not exceed the critical value, we can safely accept H_0: There is no significant difference in the ranks of national development indices in the three regions.

The ranks assigned to observations in each of the k groups are now summed separately to yield k rank sums. Then, the test statistic, H, is computed as follows:

$$H = \left[\frac{12}{N(N+1)} \left(\frac{\sum_{i=1}^{n_1} R_1^2}{n_1} + \frac{\sum_{i=1}^{n_2} R_2^2}{n_2} + \ldots + \frac{\sum_{i=1}^{n_k} R_k^2}{n_k} \right) \right] - 3(N+1) \quad (9.11)$$

where k = the number of subsamples,
n_j = the number of observations in the jth subsample,
N = the number of observations in the entire sample, and
ΣR_j = the sum of the ranks in the jth subsample.

When there are three subsamples and five or fewer observations in each subsample, the critical value of the computed value of H is determined by consulting Appendix Table 9. When there are more than five observations in one or more of the subsamples, or more than three subsamples, H is compared with values of χ^2 (chi-square) with $k-1$ degrees of freedom (Appendix Table 6). In any case, the null hypothesis is rejected if the computed value of H is so large that the probability of obtaining a value that large or larger when H_0 is true is equal to or less than the chosen significance level, α.

As shown in Box 9.4, the value of $H = 3.78$, which is below the critical value (5.78) established for sample sizes $n_1 = 5$, $n_2 = 5$, and $n_3 = 5$, at $\alpha = .05$. The null hypothesis cannot be rejected. Based on this sample, there is no reason to believe that these three world regions differ with respect to this index of economic development.

NONPARAMETRIC GOODNESS-OF-FIT TESTS

When one wishes to compare an observed distribution of frequencies or probabilities to some preconceived or theoretical distribution, it is possible to employ one of several **goodness-of-fit** significance tests. For example, you may wish to determine whether or not a sample of observed frequencies is compatible with the hypothesis that it was drawn from a population with a normal frequency distribution. A distribution may appear to have been the result of a Poisson process, one that yields rare and random events. A test for this situation will be demonstrated later.

It is common for policy makers to establish a criterion that the proportion of some phenomenon in several geographic regions should be identical, or nearly so. Federal spending of tax dollars might be such a case. Each region is a mutually exclusive nominal category containing frequencies of things like post office branches, military installations, or recreation sites. The problem is reduced to one of comparing the observed frequencies by state or region with a set of theoretical frequencies in order to test the hypothesis that the two frequency distributions are not significantly different.

As an example, the data of Table 9.4 represent a single sample of survey respondents, aggregated by region within the United States, that reported having some knowl-

Table 9.4 Number of Survey Sample Respondents, and Proportion of the Population Who Have Heard the AIDS Virus Called HIV, by United States Region, 1990

U.S. Region	Number of Respondents	Percent Aware
West	19,174	85.5
South	29,031	86.0
Midwest	20,647	81.1
Northeast	12,704	83.5

Data sources: Centers for Disease Control (1991), p. 795, and Population Reference Bureau (1991b), p. 10.

edge and awareness of HIV/AIDS in 1990. The numbers of respondents are *frequencies,* and percent aware refers to the *proportion* of all respondents who have heard the AIDS virus called HIV. We might expect all regions' respondents to have been equally informed that the AIDS virus was called HIV. The median proportion of respondents nationally who had heard the AIDS virus called HIV was 83 percent. The hypothesis to be tested is that there is no difference between the percent aware by region and the median percent for the nation.

Table 9.5 was derived from the data of Table 9.4. The column titled *Observed* is the percent of the total sample for each region who were aware that the AIDS virus is called HIV; the column entitled *Expected* is the frequency of each region's respondents who would have been aware that the AIDS virus was called HIV if 83 percent of that region's respondents were aware of that fact. There is obviously some discrepancy between the frequencies observed by the survey and the frequencies one would expect. If we believe that knowledge about AIDS is uniform nationally, are these differences too great to be attributed to chance? If so, we might wish to explore the issue further. If the differences were *not* significant, however, we would have a *good fit* of the expected to the observed data.

The χ^2 Test

The χ^2 **(chi-square) test** provides a standard for deciding whether two sets of frequencies are statistically independent. If the differences between the observed and ex-

Table 9.5 Observed Frequencies of Survey Respondents Who Have Heard the AIDS Virus Called HIV, and Expected Frequencies of Aware Respondents, by United States Region, 1990

U.S. Region	Observed Aware Respondents	Expected Aware Respondents
West	16,393	15,914
South	24,967	24,096
Midwest	16,745	17,137
Northeast	10,608	10,544

pected frequencies are small, the conclusion is that the differences could have arisen by chance. On the other hand, if the discrepancies between what is expected and what is observed are large, one might decide to reject the null hypothesis in favor of an alternative.

Chi-square is computed as

$$\chi^2 = \sum_1^k \frac{(f_O - f_E)^2}{f_E} \qquad (9.12)$$

where k is the number of class intervals in the nominal classification. The chi-square distribution is actually a family of distributions which vary in shape with their sample size. Figure 9.2 shows four members of the family, with $n = 2, 3, 4,$ and 7. Degrees of freedom associated with the single-sample test are simply $k - r$, where r is the number of restrictions or constraints imposed on a given comparison. For instance, a restriction is imposed if we force the sum of the expected frequencies to equal the sum of the observed frequencies, and another restriction might involve the computing of any parameter needed to produce a distribution. If the basis is the Poisson process, we lose only one degree of freedom, that associated with estimation of λ. If the basis is the normal distribution, two degrees are lost with the estimation of μ and σ.

Table 9.6 is derived from Table 9.5. The difference between each region's observed and expected frequencies is squared and divided by expected frequencies. Obviously, the sum of the last column, $(f_O - f_E)^2 \div f_E$, will be small if the observed and expected frequencies are a close fit, and will become larger as one or more of the differences diverge. The

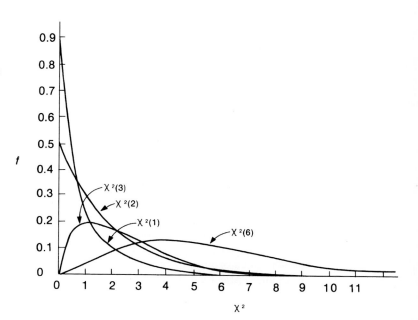

Figure 9.2 Frequency Curves of the χ^2 Distribution for 1, 2, 3, and 6 Degrees of Freedom

Table 9.6 Observed and Expected Frequencies and Computation for χ^2 for the AIDS Survey Experiment

U.S. Region	Observed Frequencies f_O	Expected Frequencies f_E	$(f_O - f_E)^2 / f_E$
West	16,393	15,914	14.42
South	24,967	24,096	31.48
Midwest	16,745	17,137	8.97
Northeast	10,608	10,544	.39
Totals	68,713	67,691	55.26

computed value of χ^2, in this case 55.26, is compared with a critical value of χ^2 with $k - r$ degrees of freedom. In this case, there are no restrictions imposed. Recall from Chapter 6, however, that one degree of freedom is lost because, given the total of 68,713 respondents, we can arbitrarily fill in all but one region's frequencies, and that last region's value is then determined, so degrees of freedom $(\nu) = 3$.

Appendix Table 6 is a table of the chi-square distribution based on a one-tailed distribution. You will note that, although we are testing a null hypothesis, the possible values of the sum of the last column range from zero (if f_O and f_E for each region matched exactly) to ∞. All such values are positive due to the squaring procedure. The table of chi-square distribution is based on computed values within that range. Thus, we may read directly from the table for the number of degrees of freedom in the problem at hand.

For $\nu = 3$, we can interpret the outcome of the AIDS survey experiment as follows: The probability that a computed value of $\chi^2 = 55.26$ could occur by sampling variation (chance) is less than .001. At the .05 level of significance we can reject the null hypothesis that the sample observed frequencies represent similar proportions (83 percent) of knowledgeable respondents in all regions. In other words, when H_0 is true, a computed χ^2 of 55.26 is rarely going to occur due to chance alone for only four categories, so we must assume that the null hypothesis is false.

In the analysis of frequencies there is one important caveat that is stressed in the literature. Frequently, you will encounter situations where the *expected frequencies* for one or more of the categories will be small—fewer than five. If a category f_E is small, with fewer than 30 degrees of freedom, the comparison of computed χ^2 to the theoretical distribution may be invalid. There is a lack of agreement over how small these so called *cell frequencies* may be without invalidating the test. Estimates of the lower limits range from 10 to 1! If your computations are based on less than 30 degrees of freedom, it is probably unwise to proceed with any category with expected cell frequencies of two or fewer. If that situation should arise, the category with small f_E count should be combined with an adjacent category to achieve the suggested minimum expected frequency. Reducing the number of categories, of course, reduces the number of degrees of freedom, which results in smaller critical values of χ^2. This, in turn, gives less latitude for accepting the null hypothesis at any given significance level.

Nonparametric Statistics

The Contingency Table. A common problem in geography is a situation in which two nominal variables are cross-classified in a **contingency table.** Statistical procedures essentially test for *independence* of the criteria of classification. We say that the categories are independent if the distribution of one category is the same no matter what the distribution of the other category. For instance, if average annual family income and area of residence of the inhabitants of a certain metropolitan area are independent, we would expect to find the same proportion of families in the low-, middle-, and high-income groups in all neighborhoods of the metropolis.

The classification of a set of frequencies, such as river flow rates at specified levels, can be shown by a table in which the r rows represent the various levels of one criterion of classification and the c columns represent the various levels of the second criterion. The data of Table 9.7 provide a simple example of a 2×2 contingency table to which the χ^2 test is applicable. This table shows the number of years between 1902 and 1956 that mean annual river flow was below and above mean annual average discharge (MAD) in two rivers. Suppose we were interested in comparing these flow patterns to test for independence.

Under the null hypothesis, we assume that there are no differences between the rivers in their annual flow rates above and below their mean annual flow; that is, we would expect that in the population of river flow data, the two criteria of classification—rivers and flows—are independent. If the hypothesis is rejected, we would conclude that the two criteria of classification are not independent. A sample of size n is selected from the population of annual measurements, and the frequency of occurrence of annual measurements in the samples below and above the mean annual discharge is displayed. Under the hypothesis of independence, we will compute expected frequencies and compare them with the observed frequencies. If the differences between f_O and f_E are too large to occur by chance, we can reject the null hypothesis and conclude that the two criteria of classification are not independent.

To compute the expected frequencies for the cells in a contingency table, we use the following equality:

$$f_E = \frac{(\Sigma f_r)(\Sigma f_c)}{n} \qquad (9.13)$$

Table 9.7 Frequencies of Mean Annual Discharge (MAD) Rates of Two Rivers (P and S) Above and Below Mean Flow, 1902–1956

	Below MAD	Above MAD	Total
River "P"	28	26	54
River "S"	19	26	45
Total	47	52	99

Data source: Hahn (1977), pp. 324–325.

where Σf_r is the row sum, Σf_c is the column sum, and n is the sample size. These "margin sums" and "grand sum" are also provided in Table 9.7. For example, the expected frequency for the upper left-hand (River "P," below MAD) cell can be obtained by multiplying the first row marginal sum by the first column marginal sum and dividing by n:

$$f_{E(1,1)} = \frac{(47 \cdot 54)}{99} = 25.64$$

Table 9.8 contains all of the computations necessary to arrive at χ^2 for this experiment. Note that $\Sigma (f_O - f_E)^2 \div f_E$ would have been zero if there had been no differences between the observed frequencies and the expected frequencies.

To evaluate the computed χ^2 statistic, we compare it to the critical value in Appendix Table 6, at the predetermined α and degrees of freedom (ν). Degrees of freedom for contingency table analyses are computed as $\nu = (r - 1)(c - 1)$, where r = number of rows, and c = number of columns. Note that ν is equal to the number of cell frequencies in Table 9.7, which could be filled in arbitrarily while maintaining the observed marginal totals. For example, fifty-four observations were taken from River "P," of which twenty-eight were below mean average discharge. We then have no choice but to enter "26" as the frequency above mean average discharge. There was but one degree of freedom in that row. When we attempt to fill in the cells of the second row, we find that these frequencies are already determined by the frequencies of the first row, and thus in this experiment, where $r = 2$ and $c = 2$, and the total degrees of freedom is 1. In most problems involving contingency tables, there are more class intervals in the two criteria of classification.

Numerous human geographic applications of the χ^2 test come to mind: a comparison where one criterion is subdivisions of a rural study area and the other criterion is classes of soil or vegetation types; or an analysis where one criterion is subdivisions of an urban study area and the other criterion contains classes of demographic or economic characteristics.

For a slightly more complex contingency table problem, consider Table 9.9. A random sample of 171 metropolitan areas was selected, and the 1980–1986 rate of change in population growth and region were noted. The regions are standard U.S. census regions,

Table 9.8 Computations for χ^2 for the River Flow Experiment

Cell	f_O	f_E	$\frac{(f_O - f_E)^2}{f_E}$
"P" Below MAD	28	25.64	.22
"P" Above MAD	26	28.36	.20
"S" Below MAD	19	21.36	.26
"S" Above MAD	26	23.64	.24
Total	99	99.00	.92

Table 9.9 Frequency Distribution of Metropolitan Area Percent Population Change, United States, 1980–1987, by Region and by Rate of Change

Region	Loss	Population Change (Percent)				Total
		0.0 – 4.0	4.1 – 8.3	8.4 – 14.9	15.0 +	
Pacific	1	1	4	7	8	21
Mountain	1	1	0	4	1	7
W. N. Cent.	2	2	6	1	0	11
W. S. Cent.	1	3	5	11	6	26
E. N. Cent.	16	7	4	0	0	27
E. S. Cent.	1	5	6	1	0	13
Mid-Atlantic	11	7	4	3	0	25
So. Atlantic	2	3	7	3	9	24
New England	4	10	1	1	1	17
Total	39	39	37	31	25	171

and the population change rate class intervals were obtained from a map (Figure 4) in the census publication, *Patterns of Metropolitan Area and County Population Growth: 1980 to 1987*. The map suggests that the population growth rate between 1980 and 1987 was higher in the South and West than in the rest of the nation. The question is, were growth rates independent of region in the United States during that time period?

The null hypothesis is that the frequencies in Table 9.9, classified by region and by rate category, are independent. Box 9.5 shows the expansion of Table 9.9 into a listing of observed and expected frequencies for the cells. For example, the expected frequency for cell 1,1 is obtained by multiplying the marginal totals, 39×21, and dividing by the total frequency, N (Equation 9.13), to yield 4.79. This value is subtracted from the observed frequency and squared, then divided by f_E.

The result of the analysis is that the null hypothesis of independence must be rejected, since the probability that a computed value of $\chi^2 = 125.14$ could arise by chance is less than .001 with $\nu = 32$. We conclude, then, that the map's portrayal of rapid southern and western growth in the United States between 1980 and 1987 is borne out. These data, because of the divergence of observed and expected frequencies, do not provide sufficient evidence to indicate independence between the two criteria of classification. The combined contribution to χ^2 of cells 5,1, 9,2, 4,4, 4,5, and 8,5 is enough to be significant.

The problem of small expected frequencies discussed above may also be encountered in contingency tables. Again, there is a lack of consensus on how to handle this problem. One suggestion is that, for contingency tables with more than 1 degree of freedom, minimum expected frequencies of 1 are allowable if no more than 20 percent of the cells have expected frequencies of less than 5. Combining adjacent rows or columns might be necessary to meet this constraint. Since more than 20 percent of the cells in the population change experiment had expected frequencies under 5, two separate analyses

BOX 9.5 Computations for χ^2 for the U. S. Population Change Experiment

Cell	f_O	f_E	$(f_O - f_E)^2 / f_E$	Cell	f_O	f_E	$(f_O - f_E)^2 / f_E$
1,1	1	4.79	2.99	1,4	7	3.81	2.67
2,1	1	1.60	0.22	2,4	4	1.27	6.01
3,1	2	2.51	0.10	3,4	1	1.99	0.50
4,1	1	5.93	4.10	4,4	11	4.71	8.39
5,1	16	6.16	15.73	5,4	0	4.89	4.90
6,1	1	2.96	1.30	6,4	1	2.36	0.78
7,1	11	5.70	4.92	7,4	3	4.53	0.52
8,1	2	5.47	2.21	8,4	3	4.35	0.42
9,1	4	3.88	0.00	9,4	1	3.08	1.41
1,2	1	4.80	2.99	1,5	8	3.07	7.92
2,2	1	1.60	0.22	2,5	1	1.02	0.00
3,2	2	2.51	0.10	3,5	0	1.61	1.61
4,2	3	5.93	1.45	4,5	6	1.90	8.84
5,2	7	6.16	0.11	5,5	0	3.95	3.94
6,2	4	2.96	0.37	6,5	0	1.90	1.90
7,2	7	5.70	0.30	7,5	0	3.65	3.65
8,2	3	5.47	1.12	8,5	9	3.51	8.59
9,2	10	3.88	9.65	9,5	1	2.49	0.89
1,3	4	4.54	0.07				
2,3	0	1.51	1.52	Total			125.14
3,3	6	2.38	5.51				
4,3	5	5.63	0.07				
5,3	4	5.84	0.58	$v = (r-1)(c-1) = 8 \times 4 = 32$			
6,3	6	2.81	3.62				
7,3	4	5.41	0.37	Critical value of $\chi^2 \approx 46.14$ ($P = .05$)			
8,3	7	5.19	0.63				
9,3	1	3.68	1.95				

were conducted—one that reduced nine regions to four (Northeast, Midwest, South, and West), and one that retained the nine regions but reduced the growth categories to two ("above average" and "below average"). Both supplementary analyses eliminated the problem of small expected frequencies, but in both cases, computed values of χ^2 exceeded the critical values found in the table. Thus, only in situations where the outcome of the test is close to the critical value will the combining of rows and/or columns likely make a difference to the decision on the hypothesis.

Assessing the Strength of χ^2. Chi-square only enables us to decide whether frequencies are independent or related to theoretical distributions. It does not tell us how strongly they are related. Several statistics measure this strength, and we will mention two. The first is known as the **phi (ϕ) statistic.** Phi makes a correction for the fact that the value of

χ^2 is directly proportional to sample size by adjusting the computed χ^2 value. ϕ is computed with the expression,

$$\phi = \sqrt{\frac{\chi^2}{N}} \tag{9.14}$$

Phi takes on the value of 0 when no relationship exists, and the value of ± 1 when the frequencies are perfectly related. It should only be used to compare tables having the same number of rows and columns.

When rows and columns are not equal, ϕ is inappropriate, and we may employ a statistic called **Cramer's V.** V adjusts either the number of rows or number of columns in the table, depending on which of the two is smaller. Cramer's V also has a range of 0 to 1, and the closer its value is to 1, the greater the degree of relationship between f_O and f_E. V is computed with the following expression:

$$V = \sqrt{\frac{\phi^2}{\text{Min }(r-1), (c-1)}} \tag{9.15}$$

where the denominator is the smaller of the two values $(r - 1)$, $(c - 1)$. It is, of course, necessary to compute the value of ϕ prior to computing V. Substituting into Equations 9.14 and 9.15, we may evaluate the computed χ^2 for the river flow problem:

$$\phi^2 = \frac{.92}{99} = .009$$

and

$$V = \sqrt{\frac{.009}{1}} = .096$$

Clearly, the relationship of the observed to expected frequencies is weak in this case, and the hypothesis of independence cannot be rejected.

Point Pattern Analysis

One particularly geographical application for the chi-square test is to evaluate a distribution of points in space, where each point is regarded as being located by a *random selection* from an underlying fixed probability distribution. The principle can be demonstrated readily by repeated generation of dot maps, using a table of random numbers to generate X and Y coordinates for the points.

Note that the word *random* has been used to describe the *method* by which the dots are located, *not the patterns that might result*. This does not mean that geographic patterns from the real world occur only by chance. Every office building, tree, and statue has a plausible reason for its location. If you observe these phenomena in aggregate, there are

Figure 9.3 Spatial Distribution of Thirty Locations Chosen at Random

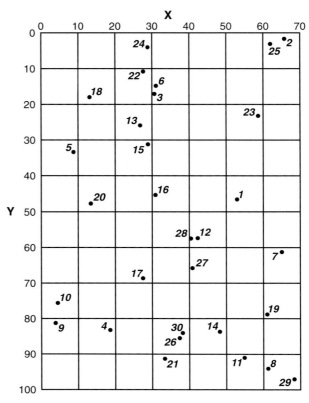

many processes that can generate their geographic patterns. By far the simplest is called the **independent stochastic process.** If it were not for the influences of distance, topography, climate, human preference, and so on, a pattern of independent random locations might well result for all phenomena.

As an example, suppose we wish to take a random sample of thirty soil pH readings over a study area of 100 feet by 70 feet, or 7,000 square feet. Suppose the unit areas, or **quadrats,** are arbitrarily set as 10 × 10 foot cells. The point locations for the sample are chosen by a random numbers table, such as that in Appendix Table 1. Each point has an *equal probability* of being in any place in the study area, and the positioning of any point is *independent* of that of any other point. Figure 9.3 shows the location of the thirty randomly selected points in the study area. The first two digits taken from the table may represent the X-coordinate, and two additional digits may represent the Y coordinate. For example, say the first 2-digit number drawn is 53, and the second is 47. Therefore, the first sample point (labeled "1") will be located 53 feet to the right of the origin along the X axis, and 47 feet down the Y axis. The same procedure is used to position the remaining twenty-nine points.[2]

---

[2] Any random number exceeding 70 for X must be ignored, since there are no X coordinates that exceed that value.

Nonparametric Statistics

Recall the discussion in Chapter 5 on the nature of the Poisson distribution. A **Poisson process** is one that involves events that occur rarely and at random in both space and time. Since the location of each point in Figure 9.3 was chosen at random, it follows that the process generating the point pattern was a random one. The Poisson distribution, therefore, should provide an excellent approximation of the number of occurrences in each grid cell. When unit areas within the study area are not defined by specific boundaries, *optimal quadrat size* for outlining cells can be determined by using the formula, $Q = 2A/n$, where A = size of study area, and n = the total number of points.

Table 9.10 lists the frequency of occurrences (*X*) per grid cell shown in Figure 9.3. Compiling this list is simply termed *quadrat counting*. Using Equation 5.1 to compute **Poisson probabilities** for the number of points per unit of space, we must find the density of points per grid cell (λ). There are seventy grid cells and thirty points. The average points per quadrat is given by:

$$\lambda = \frac{n}{k} \tag{9.16}$$

where k = the number of quadrats. For this experiment, $v = 30 \div 70 = .429$. Then, using Equation 5.1, the Poisson probability for two points per cell is:

$$P(2) = \frac{2.7183^{-.429} \times .429^2}{2!} = .0599$$

Since there are seventy cells, it follows that four cells (.0599 × 70) should contain two points each.

The Poisson probability for three points per cell is:

$$P(3) = \frac{2.7183^{-.429} \times .429^3}{3!} = .0086$$

With seventy cells, the Poisson probability indicates that there will be less than one cell (0.6) with three points. It is usually more convenient to consult a table of Poisson probabilities for a specific value of λ. Appendix Table 5 provides such probabilities for λ ranging from 0.1 to 7.0. Table 9.11 shows the Poisson probabilities, and observed and expected frequencies for each category.

Table 9.10	Quadrat Counts for an Experimental Random Spatial Distribution	(X)	Number of Quadrats
		0	45
		1	20
		2	5
		3	0
			$k = 70$

Table 9.11 Observed and Expected Quadrat Counts in Random Spatial Distribution

x	f_O	$x \cdot f_O$	$P(x)$	f_E
0	45	0	.6429	45
1	20	20	.2793	20
2	5	10	.0599	4
3	0	0	.0086	<1

Since the Poisson function describes a set of probabilities that obey the rules of Poisson variables, we might state a hypothesis that our *observed* frequency counts correspond to frequencies *expected* according to the Poisson function. A visual comparison of Columns 2 and 5 in Table 9.11 shows a reasonable degree of correspondence. However, we need a rigorous statistical test for goodness-of-fit comparable to those of earlier chapters for testing this distribution, and the χ^2 test is suited to this task.

As Figure 9.2 demonstrates, the chi-square distribution is skewed to the right, since it was derived from experiments that produced higher probabilities of smaller frequencies of occurrences (X), such as those predicted by the Poisson process. Note that as quadrat counts are performed for each cell, they form a discrete random variable. To compare the statistical similarity, or goodness-of-fit, between the observed frequencies and those generated by Poisson probabilities, we can employ Equation 9.12. The procedure for computing χ^2 in quadrat analysis is shown in Box 9.6, using the previously described random spatial distribution.

The degrees of freedom associated with this chi-square test are equal to $k - 1$, where k is equal to the number of Poisson variables, which in this case is three. The previously mentioned sensitivity of the statistic to low expected frequencies has been avoided by combining all frequencies of two or more points to give expected frequencies all greater than two. Referring to Appendix Table 6, at the 95 percent level of confidence the value of χ^2 at 2 degrees of freedom is equal to 5.99. Our calculated value of χ^2 is .021 and falls easily within the region of acceptance. Since the dots were purposefully located at random, we did not expect the decision to be otherwise.

An alternative test to χ^2 that avoids the problem of low frequencies and does not forfeit as much information is the **Kolmogorov-Smirnov D test,** which uses the maximum difference between observed and predicted cumulative frequency distributions instead of the cumulative differences between observed and expected frequencies. This test will be demonstrated later, and the random spatial distribution experiment will be reexamined at that time.

Nearest Neighbor Analysis. The Poisson probability function may not be able to approximate point patterns over space. If, in fact, spatially discrete data are not independent and the underlying process is not a random one, then maps may produce clustered patterns. Maps may also display regular point patterns with each point nearly equidistant from its nearest neighbor point.

Distance, as discussed in Chapter 2, is one of the fundamental spatial concepts. It is useful to be able to describe distribution patterns using a spacing approach based on dis-

BOX 9.6 Procedure for Computing χ^2 for a Spatial Distribution (Quadrat Analysis)

H_0: The two sets of frequencies f_O and f_E are not statistically different.

$\alpha = .05$.

1. Compute differences between observed and expected frequencies for each state $(x) = f_O - f_E$.
2. Square the values obtained in Step 1: $(f_O - f_E)^2$.
3. Divide the values just obtained by the expected frequencies for that state: $(f_O - f_E)^2 \div f_E$.
4. Sum the values obtained in Step 3, which yields test statistic χ^2.

The data for the random spatial distribution experiment yield the following:

(x)	f_O	f_E	$(f_O - f_E)^2$	$(f_O - f_E)^2/f_E$
0	45	45.581	.338	.007
1	20	19.554	.199	.010
≥2	5	4.865	.018	.004
Totals	70	70		.021

5. Consult table of χ^2, Appendix Table 6, under $p = .05$, $df = 2$. Find 5.991.
6. Decision: Accept H_0, since computed value of χ^2 falls within body of curve. The points are randomly distributed within the study area.

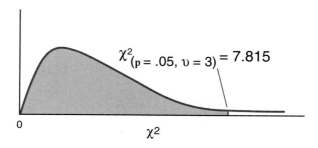

tances between point symbols. When points are clustered, the average spacing between points is at a minimum. When points display regular patterns the distance between points is at a maximum (Figure 9.4). A random distribution of spatially discrete items (such as one relating to the Poisson) lies between the limits of maximum and minimum spacing imposed by regular and clustered patterns.

It is in fact possible to classify point patterns on a continuum. A technique called **nearest neighbor analysis** was formulated by plant ecologists to describe the pattern of

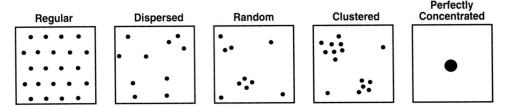

Figure 9.4 Dispersion of Points in Space from Perfectly Regular to Perfectly Concentrated

plant species.[3] The computational process involves the measurement of distance between mapped points. Distances between points can be computed by using Pythagoras's equation (Chapter 2), or, if only a limited number of points are involved such that automation is not required, one can simply use a ruler to measure the distance between each point and its nearest neighbor.

Once the actual distances are measured and recorded, the mean *observed* distance can be compared to a *hypothesized,* or theoretical distance, as developed in the work by Clark and Evans. For example, if all points of a discrete spatial distribution are *clustered* at one location, then the mean hypothesized distance ($\bar{d}_E$) between each point and its nearest neighbor will be zero. If the points are dispersed in a *regular* fashion so as to form a perfect square lattice, the mean hypothesized distance is given by

$$\bar{d}_E = 1.00 \div \sqrt{n/A} \qquad (9.17)$$

where n/A is equal to the density of the distribution. If the points are dispersed at **random,** the mean hypothesized distance between each point and its nearest neighbor will lie between 0.0 and $1.0\sqrt{n/A}$, and

$$\bar{d}_E = 0.50 \div \sqrt{n/A} \qquad (9.18)$$

If the points are dispersed in a regular manner so as to form a *hexagonal lattice,* then the mean hypothesized distance is given by

$$\bar{d}_E = 1.0746 \div \sqrt{n/A} \qquad (9.19)$$

Thus we have expected mean distance equations that describe the patterns shown in Figure 9.4. These can be compared with the *observed mean* distance measured on a map.

To illustrate the technique, the thirty-point example described in Figure 9.3 may be

---

[3]This original work was done by P. J. Clark and F. C. Evans (1954).

used to calculate a mean nearest-neighbor distance. The distances along the solid lines connecting each point to its nearest neighbor are measured with a ruler to the nearest 100th of an inch (Figure 9.5). The mean nearest-neighbor distance is equal to 0.86 inches.

Recall that the locations of the points in Figure 9.3 were determined from a table of random numbers. It has already been demonstrated that the point pattern was generated by a Poisson distribution. Therefore, it should come as no surprise if an analysis of this pattern is best described by Equation 9.18. Thus, the expected near neighbor distance would be

$$\bar{d}_E \text{(random)} = .5 \div \sqrt{n/A}$$
$$= 0.5 \div \sqrt{30/70}$$
$$= 0.76''$$

The difference between the observed mean and the hypothesized (expected) mean is:

$$(.86 - .76) = 0.10''$$

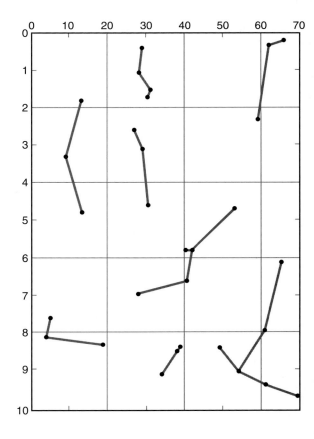

Figure 9.5 Nearest Neighbor Distances Between Thirty Points of a Random Spatial Distribution

Table 9.12 shows the absolute differences between the hypothesized mean distances and the observed distances, under differing assumptions about the dispersion of points. The observed distribution indeed approximates the hypothesized value for *a random* distribution.

The significance of the departure of the actual distribution from the theoretical random distribution may be tested with a goodness-of-fit test. The outcome of this test will allow us to decide how probable it is that the observed arrangement of points occurred by chance. Following on the procedure established in earlier chapters, we state a null hypothesis that $\bar{d}_o$ is not significantly different from $\bar{d}_E$, that is, the distribution of points is random. The appropriate test statistic is similar in form to that of Student's t (Chapter 6):

$$c = \frac{\bar{d}_o - \bar{d}_E}{SE_{\bar{d}}} \quad (9.20)$$

where c is the test statistic,
$\bar{d}_o$ is the mean nearest-neighbor distance,
$\bar{d}_E$ is the expected mean nearest-neighbor distance for a random arrangement, and
$SE_{\bar{d}}$ is the standard error of the mean nearest-neighbor distance.

This standard error is exactly analogous to the ordinary standard error of the mean identified in Chapter 6. For this application, it is computed as:

$$SE_{\bar{d}} = \frac{0.26136}{\sqrt{\frac{n^2}{A}}} \quad (9.21)$$

For the present experiment, compute:

$$SE_{\bar{d}} = \frac{0.26136}{\sqrt{\frac{30^2}{70}}} = \frac{0.26136}{3.59} = .07$$

substituting into Equation 9.20, compute:

$$c = \frac{.857 - .763}{.073} = 1.33$$

Table 9.12 Expected and Observed Mean Nearest-Neighbor Distances, under Differing Distribution Assumptions

	$\bar{d}_E$	$\bar{d}_o$	$|\bar{d}_E - \bar{d}_o|$
Clustered	0.0	0.857	0.857
Random	0.763	0.857	0.094
Regular			
Square	1.537	0.857	0.680
Hexagonal	1.641	0.857	0.784

Nonparametric Statistics

Since c is a *standard normal deviate*, like Z, its significance is determined by referring to the cumulative normal frequency distribution (Appendix Table 2). With an α level of .05, the critical value for this two-tailed test is 1.96, which is greater than the observed value of c. Our computed c value lies in the acceptance region. The distribution of Figure 9.3 is not significantly different from random.

If the calculated value of c had been negative ($\bar{d}_o < \bar{d}_E$), it would have implied that, on average, the individual points were closer than would be expected in a spatially random pattern, indicating a clustered pattern. A positive value of c would indicate a tendency toward a regular pattern. In practice, the geographers usually conclude, for instance, that a pattern is "more regular than random" or "more clustered than random."

THE NUMBER-OF-RUNS TEST FOR RANDOMNESS

Randomness is a key assumption in statistics. Do events occur in a random sequence? Does a sample characteristic exhibit a truly random dispersion on the landscape? Is any given event a function of an event in an adjacent time or place (recall the discussion on spatial autocorrelation in Chapter 8)? Are residuals randomly distributed? The following procedure, called the **number-of-runs test,** is a nonparametric test for randomness of the data of a single sample.

A simple number-of-runs test is based on an analysis of the number of unbroken sequences in binary samples with n_1 observations of one type and n_2 observations of the other type. If there are too many runs, the possibility exists that there is a repeating or alternating pattern that departs from random. For example consider a sequence of n observations of precipitation within a study area. The state of precipitation is observed in each place at the same time. An experiment yields n_1 dry observations and n_2 wet observations (where "wet" is any amount of rainfall greater than a trace), where $n_1 + n_2 = n$. Suppose there are eight observations of "wet" and seven observations of "dry" arranged across the study area in a sequence

where ■ represents a wet observation and ❏ represents a dry observation. This arrangement of observations yields the maximum number of runs ($r = 15$). A run is counted each time there is a change in the status of the phenomenon. If there are too few runs, that is, unbroken sequences, the series becomes less random and more clustered. An extreme example of clustering would appear as follows:

This sequence of wet-dry observations represents a minimum number of runs ($r = 2$).

Statisticians have established the expected mean and distribution of the number of runs for every sample size. The critical numbers representing minimum and maximum runs for $\alpha = .05$ have been computed and tabulated for sample sizes between 2 and 20 for n_1 and n_2 (Appendix Table 14). Any number of runs that is less than or equal to the minimum value shown in the table at the intersection of n_1, n_2 or is greater than or equal to the maximum value shown at that intersection, results in a rejection of the hypothesis of random arrangement.

For larger n's, the sampling distribution of r is approximately normal with

$$\text{mean} = \mu_r = \frac{2n_1 n_2}{n_1 + n_2} + 1 \tag{9.22}$$

and

$$\text{standard error} = \sigma_r = \sqrt{\frac{2n_1 n_2 (2n_1 n_2 - n_1 - n_2)}{(n_1 + n_2)^2 (n_1 + n_2 - 1)}} \tag{9.23}$$

Then,

$$z = (r - \mu_r) \div \sigma_r \tag{9.24}$$

Recall from several earlier examples in this book that at $\alpha = .05$ for a two-tailed test, the critical values of z are ± 1.96.

Problems that would benefit from this simple test may be drawn from both physical and human geography. From the former, a geographer might examine sequences of wet and dry days in one or more places; sequences of acidic and alkaline soil samples from a transect across a glaciated area; or a sequence of rock samples—some with fossils and some without—arranged in order of occurrence. To illustrate the test, we return to the example of the precipitation events. The null hypothesis in this test is that the series of wet and dry observations is generated by a random process. Box 9.7 illustrates a runs test for these data.

It is obvious how a sequence of some characteristic through census tracts across a city might lack independence. In a typical city, a number of census tracts typically contain predominantly low-income residents, followed by several tracts with middle-income residents, and so forth, in almost a checkerboard pattern. The runs test provides a measure of the randomness of these patterns.

Other applications for this procedure include a test for dichotomized data above and below a median. Such a test was illustrated at the beginning of the chapter. Another application is a test for heteroscedasticity, as displayed in residuals from regression, which was presented in Chapter 8. The runs test could be applied to the scatter of these residuals. Moving left to right along the X axis, the pattern of runs of negative and positive residuals should be random if the regression model is to be considered linear.

THE KOLMOGOROV-SMIRNOV ONE-SAMPLE GOODNESS-OF-FIT TEST

A class of nonparametric statistics has been formulated to test hypotheses that the probability distribution based on a set of observed frequencies is comparable to a specified theoretical probability distribution. One familiar case involves the hypothesis

> **BOX 9.7** **A Runs Test for a Sequence of Precipitation Observations on a Transect Through a Hypothetical Study Area**
>
> Observations of precipitation status (wet–dry) along a transect in a hypothetical study area taken from a weather record for a specified day are listed below, where W symbolizes any reading above a trace of precipitation and D indicates no more than a trace. There are $n_1 = 11$ wet observations and $n_2 = 7$ dry observations. The number of runs, $r = 8$ (sequences of one or more like states preceded and/or followed by unlike states). The runs have been underlined.
>
> <u>D D D</u> <u>W W</u> <u>D</u> <u>W</u> <u>D</u> <u>W W W W W W W</u> <u>D D</u> <u>W</u>
>
> The critical values ($P = .05$) for r obtained from Appendix Table 15 is:
>
> Lower bound = $r = 5$
> Upper bound = $r = 14$
>
> We conclude that the wet and dry observations appear in a random sequence. A sample sequence of 5 or less would have indicated a clustered precipitation pattern, and a sequence of 14 or more would have indicated an alternating sequence of wet and dry conditions.

that a sample has been drawn from a population of values that is normally distributed. Recall the discussion in Chapter 5 (Figures 5.6 and 5.8) where it was demonstrated that the cumulative normal distribution takes on the form of the S-shaped *logistic* curve.

The *Kolmogorov-Smirnov goodness-of-fit test* provides an alternative to the chi-square goodness-of-fit test. The premise of this test is that the *cumulative probability distribution* of a random sample matches a theoretical cumulative probability distribution. This is termed a *one-sample* test. The Kolmogorov-Smirnov (K-S) test may also be used to compare two independent-sample cumulative probability distributions, as we will see. These tests are named after two Russian mathematicians, A. Kolmogorov and N. V. Smirnov, who introduced two closely related statistical tests in the 1930s.

In the **Kolmogorov-Smirnov one-sample test,** the null hypothesis states that the population distribution from which an independent, random sample is drawn is not significantly different from some theoretical, or *expected,* cumulative probability distribution. If the maximum absolute difference between the theoretical and observed cumulative distribution functions at any point along those distributions is too great to be attributed to sampling error, the null hypothesis is rejected.

The test value, D_{max}, is the maximum vertical distance between the sample cumulative distribution function and the theoretical cumulative distribution function. As we will see, usually the cumulative probability for one sample observation will exceed the cumulative probability expected for that observation more than for any other, and that becomes the test value. The null hypothesis is rejected at a specified α level of significance if the value of D_{max} exceeds the value shown in Appendix Table 10 at sample size n. The assumptions underlying the K-S test include the following:

1. The samples are random and mutually exclusive; and
2. The samples come from the same continuous population distribution.

As an example of the K-S test, suppose we have sample data on children's blood chemistry from schools throughout a hypothetical metropolitan area. One measure of health risk for small children is lead in their environment, which is a product of either lead-based paint or leaded gasoline, the particles of which become part of the ambient dust in residential neighborhoods. Clearly, inner-city residential areas, with their older housing, industry, and dense vehicular traffic, might be *expected* to have more lead in their environment.

Table 9.13 shows the frequency of fifty sample children from a hypothetical metropolitan area with blood lead content ranging from 17 to 45 micrograms (mg) per deciliter (dl) of blood. In this hypothetical case, most of the sampled children with low blood lead live in suburban areas while virtually all those with high lead content live in inner-city residential areas. The null hypothesis states that the sample cumulative probability distribution of observed frequencies of varying blood lead levels is normally distributed. The alternative hypothesis states that at least one level of observed frequencies causes the cumulative probability distribution to deviate significantly from the normal state. The first step is to compute cumulative probabilities for our observed frequencies of lead content. This is accomplished by dividing the observed frequency of each lead level by n and accumulating these probabilities through to the highest lead level, as shown in Table 9.13.

Expected probabilities are now obtained by first converting each f_o to a value of the standard normal variable, z (Equation 5.2), substituting the expected mean and standard deviation for this posited normal distribution. Then, from Appendix Table 2, find the area between ∞ and z. Accumulate these probabilities from one observation to the next. This is demonstrated in Table 9.14.

The test statistic D_{max} may be computed by subtracting the observed cumulative probability of each level from its expected cumulative probability. Although this can be

Table 9.13 Observed Frequency of Children with Blood Lead between 17 and 45 mg/dl in Hypothetical Metropolitan Area, Cumulative Frequencies and Cumulative Distribution

Blood Lead (mg/dl)	f_o	Cumulative f_o	Cumulative Probability
17	4	4	.08
19	4	8	.16
22	5	13	.26
23	7	20	.40
25	7	27	.54
29	7	34	.68
34	7	41	.82
36	4	45	.90
39	3	48	.96
45	2	50	1.00

Table 9.14 Steps in Computation of Expected Cumulative Probabilities for Blood Lead Problem

Blood Lead (mg/dl)	z $(X_i - 29)/6.5$	Cumulative Probability
17	−1.85	.0322
19	−1.54	.0618
22	−1.08	.1401
23	−.92	.1788
25	−.62	.2676
29	0	.5000
34	.77	.7794
36	1.08	.8599
39	1.54	.9382
45	2.46	.9931

a tedious procedure if one has a large number of values, it is usually possible to narrow the possibilities down quickly by visual observation. In this problem, the difference between the two cumulative probabilities is clearly that associated with 29 mg/dl, with a value of $D = .2724$. A graph of the differences between the two distributions is shown in Figure 9.6.

The critical values of D are given in Appendix Table 10 for sample sizes up to 35. For a larger sample size, it is necessary to compute this value. For a sample size of 50, if the confidence level (α) is .05, D is computed as $1.36/\sqrt{n} = .192$. To accept H_0, D_{max} must be less than the critical value. In this example, however, D_{max} far exceeds the critical value even at the $\alpha = .01$ level, so we are unable to accept the null hypothesis. In terms of the problem at hand, since a value as large as D_{max} could occur so rarely by chance alone, it is at least statistically possible that inner-city children have significantly different levels of blood lead than suburban children.

As stated earlier, the Kolmogorov-Smirnov test will now be applied as an alternative test to χ^2 to the quadrat analysis of Box 9.6. Recall the Poisson probabilities associat-

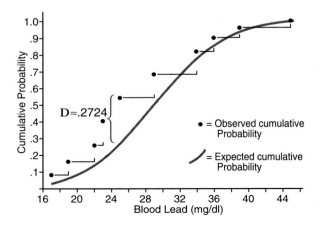

Figure 9.6 Graph of Observed versus Cumulative Distributions of Blood Lead Content, Tables 9.13 and 9.14

Table 9.15 Kolmogorov-Smirnov Analysis for the Quadrat Count Data for the Random Experiment (Figure 9.3)

X	f_o	Observed Cumulative Proportion	Expected Cumulative Proportion	Difference
0	45	.6428	.6429	−.0001
1	20	.9286	.9222	.0064
2	5	1.000	.9821	.0179
3	0	1.000	.9907	.0093

ed with $\lambda = .429$ (Table 9.11). These probabilities represent the *expected* form of the distribution. They may then be converted to cumulative probabilities. A set of *observed* cumulative probabilities may be computed from the column of observed frequencies as shown in Table 9.15.

For $v = 3$, the critical value for D_{max} is .708. The maximum difference for the experiment was .0179, which means that we cannot reject the null hypothesis of an independent random generating process at the .05 significance level. This test is attractive in that it avoids much of the computational labor associated with the chi-square test. According to Davis (1986),

> Ordinarily, the Kolmogorov-Smirnov test is used when the hypothetical model can be completely specified. That is, the parameters of the distribution are known (or assumed) from information other than that contained within the sample itself. . . . however, . . . [we may] use the kolmogorov-smirnov procedure for testing the fit of a sample to a normal distribution with an unspecified mean and variance. This . . . permits use of the procedure in a manner exactly analogous to the χ^2 procedure. (p. 101)

THE KOLMOGOROV-SMIRNOV TWO-SAMPLE GOODNESS-OF-FIT TEST

The K-S test may also be used to compare two independent sample distributions in a **Kolmogorov-Smirnov two-sample test.** In the two-sample test, the null hypothesis to be tested is that two cumulative frequency distributions come from the same population. The basic procedure is identical to that of the one-sample test except that instead of comparing one sample with a theoretical distribution, one sample is matched against another.

As an example, suppose a geographer has data on tree diameters from a subalpine coniferous forest (Table 9.16). Diameters are given as a continuous ratio value ranging from less than 4 cm to 63 cm. Two samples are drawn, one of lodgepole pine and another of subalpine fir. To compare two-sample distributions, we must first convert the frequency distributions to probability distributions, then accumulate them from the smallest diameter class to the largest (Table 9.17).

The probability for each pine diameter class is its frequency divided by the sum of the sample of pine trees, and the probability for each fir diameter class is its frequency di-

Table 9.16 Frequencies of Two Tree Species Samples by Tree Diameter Class

Diameter Class (cm)	Lodgepole Pine (f)	Subalpine Fir (f)
<4	1	81
4–7	3	34
8–11	9	18
12–15	9	7
16–19	14	2
20–23	22	4
24–27	15	1
28–31	6	0
32–35	4	1
36–39	1	0
40–43	0	0
44–47	1	0
48–51	1	0
52–55	0	0
56–59	0	0
60–63	1	0
Total	87	148

vided by the sum of the sample of firs. Since there are no firs larger than 35 cm, the maximum cumulative probability occurs there. In both species, zero frequencies in any class result in a continuation of the previous cumulative probability for that class.

It quickly becomes evident that the maximum D value will occur close to the smallest diameter class. For the < 4 cm class, $D = -.5358$; for the 4–7 cm class, $D = -.731$; and

Table 9.17 Probability and Cumulative Probability Distributions for Two Tree Species Samples by Tree Diameter Class

Diameter Class (cm)	Pine (P)	Fir (P)	Pine (Cum. P)	Fir (Cum. P)
<4	.0115	.5473	.0115	.5473
4–7	.0345	.2297	.0460	.7770
8–11	.1034	.1216	.1494	.8987
12–15	.1034	.0473	.2528	.9459
16–19	.1609	.0135	.4137	.9595
20–23	.2529	.0270	.6666	.9865
24–27	.1724	.0068	.8390	.9933
28–31	.0690	0	.9080	.9933
32–35	.0460	.0068	.9540	1.000
36–39	.0115		.9655	
40–43	0		.9655	
44–47	.0115		.9770	
48–51	.0115		.9885	
52–55	0		.9885	
56–59	0		.9885	
60–63	.0115		1.000	

for the 8–11 cm class, $D = -.7493$. Beyond that class, the distributions converge. Hence, $D_{max} = -.7493$. This is a two-tailed test, as either sample cumulative distribution can be the greater of the two, and the sign of D_{max} can be either negative or positive. Consult Appendix Table 11, which contains critical values for the K-S two-sample test. Note that the experiment at hand has large and unequal sample sizes, and we must compute the critical value from the expression:

$$D = 1.36 \sqrt{\frac{n_1 + n_2}{n_1 n_2}} = .1837$$

In order to reject H_0, D_{max} must exceed the critical value at $\alpha = .05$. In this case, it exceeds even the critical value for $\alpha = .001$, so the probability of obtaining a computed difference as large as .7493 by chance alone is less than 1 in 1,000. It is likely that either one species has inherently larger diameter trunks, or if firs are younger trees than the pines, firs may be displacing pines as pines are cut or as they die.

The advantages and disadvantages of the K-S test versus χ^2 are:

1. The K-S test does not require that the observations be grouped as is the case with χ^2 test. Thus, the K-S test makes use of all the information present in a set of data.
2. The K-S test may be used with any size sample. The χ^2 test requires a cell minimum expected frequency of at least 2.
3. The K-S test is not applicable when parameters have to be estimated from the sample. The χ^2 test may be used in these situations, reducing the degrees of freedom by 1 for each parameter estimated.

A NONPARAMETRIC METHOD OF CORRELATION

Often the population of one or both variables in a bivariate analysis is far from normally distributed. As we previously stated, in some cases a transformation of one or both of the variables will bring their joint distribution close to the bivariate normal, making it possible to estimate r. Failing this, a method of expressing the amount of correlation in nonnormal metric or ratio data is needed. Often only a small sample of raw data are available, or perhaps the only data available are ordinal.

It is, however, still possible to examine whether two variables are independent or whether they vary in the same or in opposite directions if we convert the raw data to rankings. The use of a nonparametric test is indicated, as we are not concerned with estimating or testing a hypothesis about a parameter for a population that is assumed to be normally distributed. As an example, suppose a political geographer wished to test the hypothesis that voter turnout was a function of the proportion of below-poverty population in the United States. Table 9.18 contains these data for the twenty-four largest states, along with rankings of the raw data. A casual glance at these rankings suggests that the relationship may be an inverse one, as the two top ranked states in percent voting in 1980

Table 9.18 Voting Turnout and Poverty Populations, Ranks and Differences, in the Twenty-Four Largest United States, 1980

State	Percent Reported Voting, 1980 (Y)	Percent Persons Below Poverty Level, 1979 (X)	Rankings (Y)	Rankings (X)	D (Y − X)	D^2
Alabama	49.0	18.9	18	1	17	289
California	49.5	11.4	17	12	5	25
Connecticut	61.2	8.0	3	24	−21	441
Florida	49.6	13.5	16	8	8	64
Georgia	41.7	16.6	24	4	20	400
Illinois	57.8	11.0	8	13	−5	25
Indiana	57.7	9.7	9	20	−11	121
Iowa	62.9	10.1	2	17	−15	225
Kentucky	50.0	17.6	15	3	12	144
Louisiana	53.7	18.6	12	2	10	100
Maryland	50.2	9.8	14	18.5	−4.5	20.25
Massachusetts	59.3	9.6	5	21	−16	256
Michigan	59.8	10.4	4	15	−11	121
Minnesota	70.4	9.5	1	22.5	−21.5	462.25
Missouri	58.9	12.2	6	10	−4	16
New Jersey	55.1	9.5	11	22.5	−11.5	132.25
New York	48.0	13.4	20.5	9	11.5	132.25
North Carolina	43.9	14.8	23	6	17	289
Ohio	55.4	10.3	10	16	−6	36
Pennsylvania	52.0	10.5	13	14	−1	1
Tennessee	48.9	16.5	19	5	14	196
Texas	45.6	14.7	22	7	15	225
Virginia	48.0	11.8	20.5	11	8.5	72.25
Washington	58.0	9.8	7	18.5	−11.5	132.25
Total					0	3925.50

Data sources: Percent voting—U.S. Bureau of the Census (1981), p. 498; and Poverty—U.S. Bureau of the Census (1986).

rank as 22.5 and 17, respectively, in percent minority population. Ties in either Y or X must be accorded the same ranks.

A frequently used correlation technique (invented by Spearman) is the **rank correlation coefficient,** symbolized as r_S. It is attractive because of the simplicity of the calculations involved. This statistic provides us with a measure of the strength of the relationship between paired rankings. Spearman's procedure is recommended for sample n's between 4 and 30. The technique involves first ranking the values of the raw data (if this has not already been done), then subtracting the ranks of X from the ranks of Y, squaring each, and summing the squared values. These differences are symbolized as D and D^2 respectively. This step has been performed in Table 9.18.

Spearman's rank correlation coefficient, which is interpreted identically to its parametric counterpart, the Pearson's product-moment coefficient, is obtained with the expression,

$$r_S = 1 - \frac{6\Sigma D^2}{n(n^2 - 1)} \qquad (9.25)$$

where ΣD^2 is the sum of the squared differences between each pair of ranks, and where n is the number of observations in the sample. Substitute for ΣD^2 and n in Equation 9.25:

$$r_S = 1 - \frac{6(3{,}925.5)}{24(24^2 - 1)} = -.71$$

The resulting coefficient is negative, which verifies our original supposition: As state poverty population percentages increase, the proportions who vote decrease. Is r_S significant at the $\alpha = .05$ level? To answer this question the t test introduced in Chapter 8 to verify the significance of the product-moment correlation coefficient may be employed. The hypothesis to be tested is that there is no correlation in the population. The critical value of t is found in Appendix Table 4 at the appropriate number of degrees of freedom $(n - 2)$. For the experiment at hand, $\nu = 22$, and for the two-tail test at $\alpha = .05$, we find 2.074. Substituting values of n, r, and r^2 into Equation 8.4, we obtain

$$t = r\sqrt{\frac{n-2}{1-r^2}}$$

$$= -.71 \times \sqrt{\frac{24-2}{1-.5}} = 4.71$$

In the table of t distribution for $\nu = 23$, we find that a value of 4.71 could have occurred due to sampling error less than 1 time in 1,000.

Thus, we might conclude that poverty played *at least a statistical role* in reducing voter turnout in the 1980 presidential election in the United States. Two caveats are in order here. This generalization might not hold up at a different scale, such as within state or metropolitan-area election districts. Furthermore, it might not be the poverty condition that was the root cause of voter turnout. One can never be too cautious about interpreting simple statistical cause-and-effect relationships.

The t test is inappropriate if n is greater than 30, because such a large sample is likely to be distributed normally. In situations with large samples, substitute

$$z = r_S \sqrt{n-1} \qquad (9.26)$$

and use Appendix Table 2 to find the probability that a computed z value could have arisen due to sampling error. The t test is reliable only if none of the ranks in either variable are tied. If there are ties, the Kendall coefficient of rank correlation, τ (tau), may be used. Because of its complexity, we will not introduce the technique here. Sokal and Rohlf (1981) contains a demonstration.

The nonparametric techniques discussed here do not even approach exhausting the menu of such tests. In their book, *Nonparametric Statistical Methods,* Hollander and Wolfe (1973) demonstrate close to fifty techniques. However, the techniques demonstrated here will cover most relatively simple test situations found in geographic research.

KEY TERMS

χ^2 (chi-square) test 238
contingency table 240
Cramer's *V* statistic 245
distribution free 226
goodness-of-fit 237
independent stochastic process 246
Kolmogorov-Smirnov *D* test 248
Kolmogorov-Smirnov one-sample test 255
Kolmogorov-Smirnov two-sample test 258
Kruskal-Wallis test 234
Mann-Whitney U test 232
nearest neighbor analysis 248
number-of-runs test 253
phi φ statistic 244
point pattern analysis 245
Poisson probability 247
Poisson process 245
quadrat 246
random 245
rank correlation coefficient 260
sign test for the median 227
tied ranks 229, 235
Wilcoxon signed-ranks test 229

REFERENCES

Beyer, W. H. (ed.). (1971). *Basic Statistical Tables.* Cleveland, OH: The Chemical Rubber Co.
Centers for Disease Control. (1991). *Morbidity and Mortality Weekly Report,* November 22.
Clark, P. J., and F. C. Evans. (1954). "Distance to nearest neighbor as a measure of spatial relationships in populations," *Ecology* 35: 445–453.
Daniel, W. W. (1983). *Biostatistics: A Foundation for Analysis in the Health Sciences,* 3d ed. New York: John Wiley & Sons.
Davis, John C. (1986). *Statistics and Data Analysis in Geology.* New York: John Wiley and Sons.
Hahn, Charles T. (1977). *Statistical Methods in Hydrology.* Ames, IA: Iowa State University Press.
Hollander, M., and D. A. Wolfe. (1973). *Nonparametric Statistical Methods.* New York: John Wiley and Sons.
Population Reference Bureau, Inc. (1982). *World Population Data Sheet, 1982.* Washington, DC: Population Reference Bureau.
Population Reference Bureau, Inc. (1987). *World Population Data Sheet, 1987.* Washington, DC: Population Reference Bureau.
Population Reference Bureau, Inc. (1990). *World Population Data Sheet, 1990.* Washington, DC: Population Reference Bureau.
Population Reference Bureau, Inc. (1991a). *Connections: Linking Population and the Environment.* Washington, DC: Population Reference Bureau.
Population Reference Bureau, Inc. (1991b). *Population Today.* February. Washington, DC: Population Reference Bureau.

Snedecor, George W., and William G. Cochran. (1989). *Statistical Methods,* 8th ed. Ames, IA: Iowa State University Press.

Sokal, Robert R. and F. James Rohlf. (1981). *Biometry,* Second Edition. San Francisco, CA: W. H. Freeman and Company.

U.S. Bureau of the Census. (1981). *Statistical Abstract of the United States, 102nd ed.* Washington, DC: U.S. Government Printing Office.

U.S. Bureau of the Census. (1986). *State and Metropolitan Area Data Book, 1986.* Washington, DC: U.S. Government Printing Office.

U.S. Bureau of the Census. (1988a) *Statistical Abstract of the United States,* 108th ed. Washington, DC: U.S. Government Printing Office.

U.S. Bureau of the Census. (1988b). *County Business Patterns, 1988.* Washington, DC: U.S. Government Printing Office.

U.S. Bureau of the Census. (1989). *Patterns of Metropolitan Area and County Population Growth: 1980 to 1987.* Current Population Reports, Series P-25, No. 1039. Washington, DC: U.S. Government Printing Office.

U.S. Bureau of the Census. (1990). *Statistical Abstract of the United States,* 1990. 110th ed. Washington, DC: U.S. Government Printing Office.

World Resources Institute. (1990). *World Resources, 1990–91.* New York: Oxford University Press.

EXERCISES

1. A survey of city and neighborhood quality is carried out in hundreds of cities throughout the developed world. One value that is reported for each city in the population is a neighborhood quality score. The scores range from 0 to 1.00. The median score for this variable in the population is .34. A sample of ($n = 24$) scores are drawn at random from the subset of responses from the United Kingdom. The question to be answered is, do the sample scores vary symmetrically around the median of the population? Write a null hypothesis, select a significance level, and carry out a sign test on the hypothesis. State a probability level for rejection.

    Sample scores of average overall neighborhood quality for twenty-four hypothetical cities:
    .61 .56 .53 .51 .44 .40 .35 .35 .33 .30 .30 .29 .28 .26 .24
    .25 .22 .22 .19 .18 .18 .18 .18 .11

2. From the table of death rates from lung disease in the exercises in Chapter 5, draw a random sample of fifteen states and record their death rates. Are the ranks symmetrically distributed about the United States median death rate? Write a null hypothesis, select a significance level, and apply a sign test to the hypothesis.

3. Monthly stream runoff has been measured for years at the same site on a hypothetical stream. Samples of stream runoff, in cubic feet per second (cfs), are obtained for February and March runoff for seven years, selected at random from the population of this variable. Do the February–March differences vary significantly from what would be expected by chance? Write a null hypothesis, select a significance level, and apply the Wilcoxon test to the hypothesis. State a probability level for rejection.

Hypothetical Monthly Stream Runoff (cfs) for February and March in Seven Sample Years

Year	February	March
1958	2.21	1.17
1959	1.95	1.12
1963	1.46	5.48
1966	2.81	.79
1967	1.61	4.66
1968	.98	4.25
1970	3.89	2.91

4. Two independent, random samples ($n_1 = 10$, $n_2 = 11$) of peak stream flow data (in cubic feet per second) have been taken from the records for two hypothetical streams, "Cave Creek" and "Hilton Branch." Are the two samples drawn from the same population? Write a null hypothesis, choose a significance level, and apply a Mann-Whitney test to the hypothesis. State a probability level for rejection.

Sample Peak Flows (cfs) for Two Hypothetical Streams

Cave Creek	Hilton Branch
98	76
198	136
154	54
30	18
71	65
44	32
184	182
127	130
27	21
54	46
	69

5. A major international development agency classified countries into three categories in 1987, based on relative levels of national income: low-income economies, middle-income economies, and high-income economies. Three small samples ($n_1 = n_2 = n_3 = 5$) of countries were drawn at random from among countries in each income category, and the infant mortality rates (IMR) per 1,000 live births for these sample countries for the year 1987 were recorded. Are the three samples drawn from the same population of infant mortality rates? Write a null hypothesis, select a level of significance, and apply a Kruskal-Wallis test to the hypothesis. State a probability level for rejection.

Infant Mortality Rates in Sample Countries Classified by Income Group, 1987

Sample I		Sample II		Sample III	
Nation	IMR	Nation	IMR	Nation	IMR
Zaire	103	Philippines	50	Ireland	9
Niger	141	Guatemala	71	Italy	11
Kenya	76	Peru	94	France	8
Ghana	94	Brazil	63	Sweden	7
Burma	103	Libya	90	United States	11

6. A hypothetical national park has been divided into seven "wildlife management districts." A bird flyway across the park is situated such that district B is 2 kilometers farther from the southern entry point than district A, district C is two kilometers farther than B, and so forth. It is observed over the years that, on the average, as birds enter the park from the south, some 40 percent of each group tends to stop in district A. If the attrition rate is about 40 percent in each successive district, out of each 100 birds entering the park, the distribution over the seven districts *should* be about

$$40, 24, 14, 9, 5, 4, 4$$

A group of biogeographers decide to test this "systematic distance decay" hypothesis, by establishing a visual and voice communication line along the flyway and counting birds according to their initial district of choice. During one week, 1,000 birds are observed arriving from the south, and their distribution along the seven-district line is as follows (from district A through G):

$$401, 254, 132, 83, 52, 49, 29$$

Are the observed frequencies a "good fit" to the expected "distance decay" frequencies? State a null hypothesis, choose a significance level, use the chi-square procedure to test this hypothesis, and state the probability level for rejection.

7. A sample survey of 782 home buyers in a hypothetical city are distributed according to their family incomes and four neighborhoods of varying social class (selected at random) within which all bought homes. The distribution of purchase matrix is as follows:

Are the frequencies in this matrix independent of their two criteria of classification? Are buyers in their three income classes indifferent to the social status of the four neighborhoods in which they bought homes? State a null hypothesis, choose a level of significance, and use a chi-square test on this contingency table to assess the independence of the neighborhoods and buyer income levels. What is the probability that the computed χ^2 could have arisen by chance? Apply Cramer's *V* to test the strength of the χ^2 statistic.

Neighborhood	Low	Medium	High	Total
A	78	71	100	249
B	90	42	53	185
C	41	96	57	194
D	53	41	60	154
Total	262	250	270	782

8. The map shows the locations of air sampling stations in Wayne County, Michigan, as of 1980. Does the distribution of air sampling stations within this study area appear to be the result of an independent stochastic process? Write a null hypothesis and establish an α level. Compute Poisson probabilities for quadrat frequencies of 0, 1, 2, 3, and 4. Then, employ a χ^2 test for randomness of this distribution.

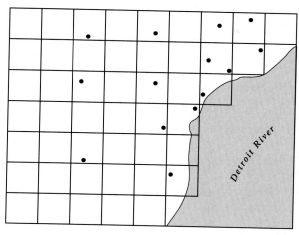

Wayne County, Michigan, Showing Locations of Air Sampling Stations within Quadrats

9. Using a base map of your city or county and a classified telephone directory, plot the locations of convenience retail stores within that area. Construct a uniform grid on a scale of one kilometer (or mile) for each vertical and horizontal line. Follow the procedures of point pattern analysis and Box 9.6 to test for randomness in the distribution of your convenience stores.

10. Perform a nearest neighbor analysis on the Wayne County air pollution station distribution to test the null hypothesis that this distribution is more random than either regular or concentrated.

11. Perform a nearest neighbor analysis on the convenience store distribution of Problem 9 to test the null hypothesis that this distribution is more random than either regular or concentrated.

12. The diagram shows residuals from regression for the relationship between unemployment and income in the Baltimore metropolitan area (from Chapter 8). Working from left to right, count

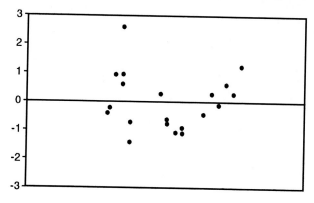

the number of runs of positive and negative residuals. Are these residuals randomly distributed above and below zero, or is there homoscedasticity in this regression analysis? Write a null hypothesis, select an α level, and perform a runs test. Box 9.7 illustrates this procedure.

13. Recalling Problem 7, a second set of four neighborhoods is selected at random from the same city, and the home purchase patterns of 650 low-, medium-, and high-income buyers were recorded. The table at the right shows the results. Do these two independent sample distributions come from the same population? Write a null hypothesis, establish an α level, and use the Kolmogorov-Smirnov two-sample test to assess the hypothesis.

Neighborhood	Low	Medium	High	Total
E	66	43	34	143
F	41	87	52	180
G	58	23	85	166
H	32	67	62	161
Total	197	220	233	650

14. A political plum in American politics is federal employment. Each state benefits from federal agencies within their borders, because of the jobs they provide. But are federal jobs located within U.S. regions in proportion to the available civilian labor force in those regions? The table at the right shows the ranks of (1) federal employment rates (per 10,000 population), and (2) the proportion of the total civilian labor force, by region, in 1986. Use Spearman's rank correlation to test the null hypothesis that there is no correlation between the ranks of these two variables ($\alpha = .05$).

Paid Civilian Employment in the Federal Government, and Proportion of Civilian Labor Force, Ranked by U.S. Region, 1986

Region	Employment Rate	Proportion of Civilian Labor Force
New England	8	8
Mid-Atlantic	7	3
E. N. Central	9	1
W. N. Central	6	6
S. Atlantic	1	2
E. S. Central	4	7
W. S. Central	5	5
Mountain	2	9
Pacific	3	4

Data source: U.S. Bureau of the Census (1988b), pp. 308, 367.

Appendix Tables

1. Table of Random Numbers
2. Cumulative Normal Frequency Distribution
3. Confidence Limits for Fractile Diagrams
4. t Distribution
5. Poisson Probabilities
6. Chi-Square (χ^2) Distribution
7. 5% and 1% Points F Distribution
8. Critical Values of U in the Mann-Whitney Test
9. Critical Values of H in the Kruskal-Wallis Test
10. Critical Values of D in the Kolmolgorov-Smirnov One-Sample Test
11. Critical Values of D in the Kolmolgorov-Smirnov Two-Sample Test
12. Critical Values of the Standard Normal Deviate z
13. Sample Sizes Required for 50% of Population, for 95% Certainty, and Specified Percent Accuracy
14. Run Statistic Critical Values for Sample Sizes n_1 and n_2 (Two-Tail Test) $\alpha = .05$

appendix table **1**

Table of Random Numbers

26687	74223	43546	45699	94469	82125	37370	23966	68926	37664
60675	75169	24510	15100	02011	14375	65187	10630	64421	66745
45418	98635	83123	98558	09953	60255	42071	40930	97992	93085
69872	48026	89755	28470	44130	59979	91063	28766	85962	77173
03765	86366	99539	44183	23886	89977	11964	51581	18033	56239
84686	57636	32326	19867	71345	42002	96997	84379	27991	21459
91512	49670	32556	85189	28023	88151	62896	95498	29423	38138
10737	49307	18307	22246	22461	10003	93157	66984	44919	30467
54870	19676	58367	20905	38324	00026	98440	37427	22896	37637
48967	49579	65369	74305	62085	39297	10309	23173	74212	32272
91430	79112	03685	05411	23027	54735	91550	06250	18705	18909
92564	29567	47476	62804	73428	04535	86395	12162	59647	97726
41734	12199	77441	92415	63542	42115	84972	12454	33133	48467
25251	78110	54178	78241	09226	87529	35376	90690	54178	08561
91657	11563	66036	28523	83705	09956	76610	88116	78351	50877
00149	84745	63222	50533	50159	60433	04822	49577	89049	16162
53250	73200	84066	59620	61009	38542	05758	06178	80193	26466
25587	17481	56716	49749	70733	32733	60365	14108	52573	39391
01176	12182	06882	27562	75456	54261	38564	89054	96911	88906
83531	15544	40834	20296	88576	47815	96540	79462	78666	25353
19902	98866	32805	61091	91587	30340	84909	64047	67750	87638
96516	78705	25556	35181	29064	49005	29843	68949	50506	45862
99417	56171	19848	24352	51844	03791	72127	57958	08366	43190
77699	57853	93213	27342	28906	31052	65815	21637	49385	75406
32245	83794	99528	05150	27246	48263	62156	62469	97048	16511
12874	72753	66469	13782	64330	00056	73324	03920	13193	19466
63899	41910	45484	55461	66518	82486	74694	07865	09724	76490
16255	43271	26540	41298	35095	32170	70625	66407	01050	44225
75553	30207	41814	74985	40223	91223	64238	73012	83100	92041
41772	18441	34685	13892	38843	69007	10362	84125	08814	66785
09270	01245	81765	06809	10561	10080	17482	05471	82273	06902
85058	17815	71551	36356	97519	54144	51132	83169	27373	68609
80222	87572	62758	14858	36350	23304	70453	21065	63812	29860
83901	88028	56743	25598	79349	47880	77912	52020	84305	02897
36303	57833	77622	02238	53285	77316	40106	38456	92214	54278
91543	63886	60539	96334	20804	72692	08944	02870	74892	22598
14415	33816	78231	87674	96473	44451	25098	29296	50679	07798
82465	07781	09938	66874	72128	99685	84329	14530	08410	45953
27306	39843	05634	96368	72022	01278	92830	40094	31776	41822
91960	82766	02331	08797	33858	21847	17391	53755	58079	48498
59284	96108	91610	07483	37943	96832	15444	12091	36690	58317
10428	96003	71223	21352	78685	55964	35510	94805	23422	04492
65527	41039	79574	05105	59588	02115	33446	56780	18402	36279
59688	43078	93275	31978	08768	84805	50661	18523	83235	50602
44452	10188	43565	46531	93023	07618	12910	60934	53403	18401
87275	82013	59804	78595	60553	14038	12096	95472	42736	08573
94155	93110	49964	27753	85090	77677	69303	66323	77811	22791
26488	76394	91282	03419	68758	89575	66469	97835	66681	03171
37073	34547	88296	68638	12976	50896	10023	27220	05785	77538
83835	89575	55956	93957	30361	47679	83001	35056	07103	63072

appendix table 2

Cumulative Normal Frequency Distribution

(area under standard normal curve from μ to Z)

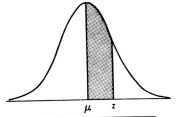

Z	0.00	0.01	0.02	0.03	0.04	0.05	0.06	0.07	0.08	0.09
0.0	0.0000	0.0040	0.0080	0.0120	0.0160	0.0199	0.0239	0.0279	0.0319	0.0359
0.1	.0398	.0438	.0478	.0517	.0557	.0596	.0636	.0675	.0714	.0753
0.2	.0793	.0832	.0871	.0910	.0948	.0987	.1026	.1064	.1103	.1141
0.3	.1179	.1217	.1255	.1293	.1331	.1368	.1406	.1443	.1480	.1517
0.4	.1554	.1591	.1628	.1664	.1700	.1736	.1772	.1808	.1844	.1879
0.5	.1915	.1950	.1985	.2019	.2054	.2088	.2123	.2157	.2190	.2224
0.6	.2257	.2291	.2324	.2357	.2389	.2422	.2454	.2486	.2517	.2549
0.7	.2580	.2611	.2642	.2673	.2704	.2734	.2764	.2794	.2823	.2852
0.8	.2881	.2910	.2939	.2967	.2995	.3023	.3051	.3078	.3106	.3133
0.9	.3159	.3186	.3212	.3238	.3264	.3289	.3315	.3340	.3365	.3389
1.0	.3413	.3438	.3461	.3485	.3508	.3531	.3554	.3577	.3599	.3621
1.1	.3643	.3665	.3686	.3708	.3729	.3749	.3770	.3790	.3810	.3830
1.2	.3849	.3869	.3888	.3907	.3925	.3944	.3962	.3980	.3997	.4015
1.3	.4032	.4049	.4066	.4082	.4099	.4115	.4131	.4147	.4162	.4177
1.4	.4192	.4207	.4222	.4236	.4251	.4265	.4279	.4292	.4306	.4319
1.5	.4332	.4345	.4357	.4370	.4382	.4394	.4406	.4418	.4429	.4441
1.6	.4452	.4463	.4474	.4484	.4495	.4505	.4515	.4525	.4535	.4545
1.7	.4554	.4564	.4573	.4582	.4591	.4599	.4608	.4616	.4625	.4633
1.8	.4641	.4649	.4656	.4664	.4671	.4678	.4686	.4693	.4699	.4706
1.9	.4713	.4719	.4726	.4732	.4738	.4744	.4750	.4756	.4761	.4767
2.0	.4772	.4778	.4783	.4788	.4793	.4798	.4803	.4808	.4812	.4817
2.1	.4821	.4826	.4830	.4834	.4838	.4842	.4846	.4850	.4854	.4857
2.2	.4861	.4864	.4868	.4871	.4875	.4878	.4881	.4884	.4887	.4890
2.3	.4893	.4896	.4898	.4901	.4904	.4906	.4909	.4911	.4913	.4916
2.4	.4918	.4920	.4922	.4925	.4927	.4929	.4931	.4932	.4934	.4936
2.5	.4938	.4940	.4941	.4943	.4945	.4946	.4948	.4949	.4951	.4952
2.6	.4953	.4955	.4956	.4957	.4959	.4960	.4961	.4962	.4963	.4964
2.7	.4965	.4966	.4967	.4968	.4969	.4970	.4971	.4972	.4973	.4974
2.8	.4974	.4975	.4976	.4977	.4977	.4978	.4979	.4979	.4980	.4981
2.9	.4981	.4982	.4982	.4983	.4984	.4984	.4985	.4985	.4986	.4986
3.0	.4987	.4987	.4987	.4988	.4988	.4989	.4989	.4989	.4990	.4990
3.1	.4990	.4991	.4991	.4991	.4992	.4992	.4992	.4992	.4993	.4993
3.2	.4993	.4993	.4994	.4994	.4994	.4994	.4994	.4995	.4995	.4995
3.3	.4995	.4995	.4995	.4996	.4996	.4996	.4996	.4996	.4996	.4997
3.4	.4997	.4997	.4997	.4997	.4997	.4997	.4997	.4997	.4997	.4998
3.6	.4998	.4998	.4999	.4999	.4999	.4999	.4999	.4999	.4999	.4999
3.9	.5000									

appendix table 3

Confidence Limits for Fractile Diagrams (Level of 95% and 99%)*

Width of CONFIDENCE BAND = $\pm \sqrt{PQ/N}$ (Z) (100)							
P = .50 Q = .50 PQ = .25	.60 .40 .24	.70 .30 .21	.80 .20 .16	.90 .10 .09	.95 .05 .0475	.99 .01 .0099	.999 .001 .000999

N								
25	19.6 25.8	19.2 25.3	18.0 23.6	15.7 20.6	11.8 15.5	8.5 11.2	3.9 5.1	1.23 1.62
30k	17.9 23.6	17.5 23.1	16.4 21.6	14.3 18.8	10.7 14.1	7.8 10.3	3.6 4.7	1.13 1.48
35	16.6 21.8	16.2 21.4	15.2 20.0	13.3 17.4	9.9 13.1	7.2 9.5	3.3 4.3	1.04 1.37
40	15.5 20.4	15.2 20.0	14.2 18.6	12.4 16.3	9.3 12.2	6.8 8.9	3.1 4.1	.98 1.28
45	14.6 19.2	14.3 18.8	13.4 17.6	11.7 15.4	8.8 11.5	6.4 8.4	2.9 3.8	.92 1.21
50	13.9 18.2	13.6 17.9	12.7 16.7	11.1 14.6	8.3 10.9	6.0 8.0	2.8 3.6	.87 1.15
55	13.2 17.4	12.9 17.0	12.1 15.9	10.6 13.9	7.9 10.4	5.8 7.6	2.6 3.5	.83 1.09
60	12.7 16.7	12.4 16.3	11.6 15.3	10.1 13.3	7.6 10.0	5.5 7.3	2.5 3.3	.80 1.05
65	12.2 16.0	11.9 15.7	11.1 14.7	9.7 12.8	7.3 9.6	5.3 7.0	2.4 3.2	.76 1.01
70	11.7 15.4	11.5 15.1	10.7 14.1	9.4 12.3	7.0 9.3	5.1 6.7	2.3 3.1	.74 .97
75	11.3 14.9	11.1 14.6	10.4 13.7	9.1 11.9	6.8 8.9	4.9 6.5	2.3 3.0	.71 .94
80	11.0 14.4	10.7 14.1	10.0 13.2	8.8 11.5	6.6 8.7	4.8 6.3	2.2 2.9	.69 .91
85	10.6 14.0	10.4 13.7	9.7 12.8	8.5 11.2	6.4 8.4	4.6 6.1	2.1 2.8	.67 .88
90	10.3 13.5	10.1 13.3	9.5 12.5	8.3 10.9	6.2 8.2	4.5 5.9	2.1 2.7	.65 .86
95	10.1 13.2	9.9 13.0	9.2 12.1	8.0 10.6	6.0 7.9	4.4 5.8	2.0 2.6	.63 .83
100	9.8 12.9	9.6 12.6	9.0 11.8	7.8 10.3	5.9 7.7	4.3 5.6	2.0 2.6	.62 .81
110	9.3 12.3	9.2 12.1	8.6 11.3	7.5 9.8	5.6 7.4	4.1 5.4	1.9 2.4	.59 .77
120	8.9 11.8	8.8 11.5	8.2 10.8	7.2 9.4	5.4 7.1	3.9 5.1	1.8 2.3	.56 .74
130	8.6 11.3	8.4 11.1	7.9 10.4	6.9 9.1	5.2 6.8	3.7 4.9	1.7 2.3	.59 .71
140	8.3 10.9	8.1 10.7	7.6 10.0	6.6 8.7	5.0 6.5	3.6 4.8	1.6 2.2	.52 .69
150	8.0 10.5	7.8 10.3	7.3 9.7	6.4 8.4	4.8 6.3	3.5 4.6	1.6 2.1	.50 .66
160	7.7 10.2	7.6 10.0	7.1 9.3	6.2 8.2	4.6 6.1	3.4 4.4	1.5 2.0	.49 .64
170	7.5 9.9	7.4 9.7	6.9 9.1	6.0 7.9	4.5 5.9	3.3 4.3	1.5 2.0	.47 .62
180	7.3 9.6	7.2 9.4	6.7 8.8	5.8 7.7	4.4 5.8	3.2 4.2	1.5 1.9	.46 .61
190	7.1 9.4	7.0 9.2	6.5 8.6	5.7 7.5	4.3 5.6	3.1 4.1	1.4 1.9	.45 .50
200	6.9 9.1	6.8 8.9	6.4 8.4	5.5 7.3	4.2 5.5	3.0 4.0	1.4 1.8	.44 .57
220	6.6 8.7	6.5 8.5	6.1 8.0	5.3 7.0	4.0 5.2	2.9 3.8	1.3 1.7	.42 .55
240	6.3 8.3	6.2 8.2	5.8 7.6	5.1 6.7	3.8 5.0	2.8 3.6	1.3 1.7	.40 .52
260	6.1 8.0	6.0 7.8	5.6 7.3	4.9 6.4	3.6 4.8	2.6 3.5	1.2 1.6	.38 .50
280	5.9 7.7	5.7 7.6	5.4 7.1	4.7 6.2	3.5 4.6	2.6 3.4	1.2 1.5	.37 .49
300	5.7 7.4	5.5 7.3	5.2 6.8	4.5 6.0	3.4 4.5	2.5 3.2	1.1 1.5	.36 .47
350	5.2 6.9	5.1 6.8	4.8 6.3	4.2 5.5	3.1 4.1	2.3 3.0	1.0 1.4	.33 .43
400	4.9 6.5	4.8 6.3	4.5 5.9	3.9 5.2	2.9 3.9	2.1 2.8	1.0 1.3	.31 .41
450	4.6 6.1	4.5 6.0	4.2 5.6	3.7 4.9	2.8 3.6	2.0 2.7	.9 1.2	.29 .38
500	4.4 5.8	4.3 5.7	4.0 5.3	3.5 4.6	2.6 3.5	1.9 2.5	.9 1.1	.29 .36

*Use the upper row in any N to obtain the .95 fractile for any value of P and the lower row for the .99 fractile.

appendix table 4

t Distribution

[Note: This table refers to critical values for both tails of the curve. Therefore, values in the body of the table indicate probabilities for the null hypothesis]

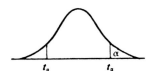

ν	Level of significance (P)										Level of significance (P)						ν
	.90	.80	.70	.60	.50	.40	.30	.25	.20	.10	.05	.025	.02	.01	.005	.001	
1	.158	.325	.510	.727	1.000	1.376	1.963	2.414	3.078	6.314	12.706	25.452	31.821	63.657	127.32	636.619	1
2	.142	.289	.445	.617	.816	1.061	1.386	1.604	1.886	2.920	4.303	6.205	6.965	9.925	14.089	31.598	2
3	.137	.277	.424	.584	.765	.978	1.250	1.423	1.638	2.353	3.182	4.176	4.541	5.841	7.453	12.941	3
4	.134	.271	.414	.569	.741	.941	1.190	1.344	1.533	2.132	2.776	3.495	3.747	4.604	5.598	8.610	4
5	.132	.267	.408	.559	.727	.920	1.156	1.301	1.476	2.015	2.571	3.163	3.365	4.032	4.773	6.859	5
6	.131	.265	.404	.553	.718	.906	1.134	1.273	1.440	1.943	2.447	2.969	3.143	3.707	4.317	5.959	6
7	.130	.263	.402	.549	.711	.896	1.119	1.254	1.415	1.895	2.365	2.841	2.998	3.499	4.029	5.405	7
8	.130	.262	.399	.546	.706	.889	1.108	1.240	1.397	1.860	2.306	2.752	2.896	3.355	3.832	5.041	8
9	.129	.261	.398	.543	.703	.883	1.100	1.230	1.383	1.833	2.262	2.685	2.821	3.250	3.690	4.781	9
10	.129	.260	.397	.542	.700	.879	1.093	1.221	1.372	1.812	2.228	2.634	2.764	3.169	3.581	4.587	10
11	.129	.260	.396	.540	.697	.876	1.088	1.214	1.363	1.796	2.201	2.593	2.718	3.106	3.497	4.437	11
12	.128	.259	.395	.539	.695	.873	1.083	1.209	1.356	1.782	2.179	2.560	2.681	3.055	3.428	4.318	12
13	.128	.259	.394	.538	.694	.870	1.079	1.204	1.350	1.771	2.160	2.533	2.650	3.012	3.372	4.221	13
14	.128	.258	.393	.537	.692	.868	1.076	1.200	1.345	1.761	2.145	2.510	2.624	2.977	3.326	4.140	14
15	.128	.258	.393	.536	.691	.866	1.074	1.197	1.341	1.753	2.131	2.490	2.602	2.947	3.286	4.073	15
16	.128	.258	.392	.535	.690	.865	1.071	1.194	1.337	1.746	2.120	2.473	2.583	2.921	3.252	4.015	16
17	.128	.257	.392	.534	.689	.863	1.069	1.191	1.333	1.740	2.110	2.458	2.567	2.898	3.222	3.965	17
18	.127	.257	.392	.534	.688	.862	1.067	1.189	1.330	1.734	2.101	2.445	2.552	2.878	3.197	3.922	18
19	.127	.257	.391	.533	.688	.861	1.066	1.187	1.328	1.729	2.093	2.433	2.539	2.861	3.174	3.883	19
20	.127	.257	.391	.533	.687	.860	1.064	1.185	1.325	1.725	2.086	2.423	2.528	2.845	3.153	3.850	20
21	.127	.257	.391	.532	.686	.859	1.063	1.183	1.323	1.721	2.080	2.414	2.518	2.831	3.135	3.819	21
22	.127	.256	.390	.532	.686	.858	1.061	1.182	1.321	1.717	2.074	2.406	2.508	2.819	3.119	3.792	22
23	.127	.256	.390	.532	.685	.858	1.060	1.180	1.319	1.714	2.069	2.398	2.500	2.807	3.104	3.767	23
24	.127	.256	.390	.531	.685	.857	1.059	1.179	1.318	1.711	2.064	2.391	2.492	2.797	3.090	3.745	24
25	.127	.256	.390	.531	.684	.856	1.058	1.178	1.316	1.708	2.060	2.385	2.485	2.787	3.078	3.725	25
26	.127	.256	.390	.531	.684	.856	1.058	1.177	1.315	1.706	2.056	2.379	2.479	2.779	3.067	3.707	26
27	.127	.256	.389	.531	.684	.855	1.057	1.176	1.314	1.703	2.052	2.373	2.473	2.771	3.056	3.690	27
28	.127	.256	.389	.530	.683	.855	1.056	1.175	1.313	1.701	2.048	2.368	2.467	2.763	3.047	3.674	28
29	.127	.256	.389	.530	.683	.854	1.055	1.174	1.311	1.699	2.045	2.364	2.462	2.756	3.038	3.659	29
30	.127	.256	.389	.530	.683	.854	1.055	1.173	1.310	1.697	2.042	2.360	2.457	2.750	3.030	3.646	30
40	.126	.255	.388	.529	.681	.851	1.050	1.167	1.303	1.684	2.021	2.329	2.423	2.704	2.971	3.551	40
60	.126	.254	.387	.527	.679	.848	1.046	1.162	1.296	1.671	2.000	2.299	2.390	2.660	2.915	3.460	60
120	.126	.254	.386	.526	.677	.845	1.041	1.156	1.289	1.658	1.980	2.270	2.358	2.617	2.860	3.373	120
∞	.126	.253	.385	.524	.674	.842	1.036	1.150	1.282	1.645	1.960	2.241	2.326	2.576	2.807	3.291	∞

appendix table 5

Poisson Probabilities ($\lambda = 0.1$ to $\lambda = 7.0$)

Poisson probabilities

x	0.1	0.2	0.3	0.4	0.5	0.6	0.7	0.8	0.9	1.0
0	.9048	.8187	.7408	.6703	.6065	.5488	.4966	.4493	.4066	.3679
1	.0905	.1637	.2222	.2681	.3033	.3293	.3476	.3595	.3659	.3679
2	.0045	.0164	.0333	.0536	.0758	.0988	.1217	.1438	.1647	.1839
3	.0002	.0011	.0033	.0072	.0126	.0198	.0284	.0383	.0494	.0613
4		.0001	.0002	.0007	.0016	.0030	.0050	.0077	.0111	.0153
5				.0001	.0002	.0004	.0007	.0012	.0020	.0031
6							.0001	.0002	.0003	.0005
7										.0001

x	1.1	1.2	1.3	1.4	1.5	1.6	1.7	1.8	1.9	2.0
0	.3329	.3012	.2725	.2466	.2231	.2019	.1827	.1653	.1496	.1353
1	.3662	.3614	.3543	.3452	.3347	.3230	.3106	.2975	.2842	.2707
2	.2014	.2169	.2303	.2417	.2510	.2584	.2640	.2678	.2700	.2707
3	.0738	.0867	.0998	.1128	.1255	.1378	.1496	.1607	.1710	.1804
4	.0203	.0260	.0324	.0395	.0471	.0551	.0636	.0723	.0812	.0902
5	.0045	.0062	.0084	.0111	.0141	.0176	.0216	.0260	.0309	.0361
6	.0008	.0012	.0018	.0026	.0035	.0047	.0061	.0078	.0098	.0120
7	.0001	.0002	.0003	.0005	.0008	.0011	.0015	.0020	.0027	.0034
8	.0000	.0000	.0001	.0001	.0001	.0002	.0003	.0005	.0006	.0009
9	.0000	.0000	.0000	.0000	.0000	.0000	.0001	.0001	.0001	.0002

x	2.1	2.2	2.3	2.4	2.5	2.6	2.7	2.8	2.9	3.0
0	.1225	.1108	.1003	.0907	.0821	.0743	.0672	.0608	.0550	.0498
1	.2572	.2438	.2306	.2177	.2052	.1931	.1815	.1703	.1596	.1494
2	.2700	.2681	.2652	.2613	.2565	.2510	.2450	.2384	.2314	.2240
3	.1890	.1966	.2033	.2090	.2138	.2176	.2205	.2225	.2237	.2240
4	.0992	.1082	.1169	.1254	.1336	.1414	.1488	.1557	.1622	.1680
5	.0417	.0476	.0538	.0602	.0668	.0735	.0804	.0872	.0940	.1008
6	.0146	.0174	.0206	.0241	.0278	.0319	.0362	.0407	.0455	.0504
7	.0044	.0055	.0068	.0083	.0099	.0118	.0139	.0163	.0188	.0216
8	.0011	.0015	.0019	.0025	.0031	.0038	.0047	.0057	.0068	.0081
9	.0003	.0004	.0005	.0007	.0009	.0011	.0014	.0018	.0022	.0027
10	.0001	.0001	.0001	.0002	.0002	.0003	.0004	.0005	.0006	.0008
11	.0000	.0000	.0000	.0000	.0000	.0001	.0001	.0001	.0002	.0002
12	.0000	.0000	.0000	.0000	.0000	.0000	.0000	.0000	.0000	.0001

x	1.5	2.0	2.5	3.0	3.5	4.0	4.5	5.0	6.0	7.0
0	.2231	.1353	.0821	.0498	.0302	.0183	.0111	.0067	.0025	.0009
1	.3347	.2707	.2052	.1494	.1057	.0733	.0500	.0337	.0149	.0064
2	.2510	.2707	.2565	.2240	.1850	.1465	.1125	.0842	.0446	.0223
3	.1255	.1804	.2138	.2240	.2158	.1954	.1687	.1404	.0892	.0521
4	.0471	.0902	.1336	.1680	.1888	.1954	.1898	.1755	.1339	.0912
5	.0141	.0361	.0668	.1008	.1322	.1563	.1708	.1755	.1606	.1277
6	.0035	.0120	.0278	.0504	.0771	.1042	.1281	.1462	.1606	.1490
7	.0008	.0034	.0099	.0216	.0385	.0595	.0824	.1044	.1377	.1490
8	.0001	.0009	.0031	.0081	.0169	.0298	.0463	.0653	.1033	.1304
9		.0002	.0009	.0027	.0066	.0132	.0232	.0363	.0688	.1014
10			.0002	.0008	.0023	.0053	.0104	.0181	.0413	.0710
11				.0002	.0007	.0019	.0043	.0082	.0225	.0452
12				.0001	.0002	.0006	.0016	.0034	.0113	.0264
13					.0001	.0002	.0006	.0013	.0052	.0142
14						.0001	.0002	.0005	.0022	.0071
15							.0001	.0002	.0009	.0033
16									.0003	.0014
17									.0001	.0006
18										.0002
19										.0001

appendix table 6

Chi-Square (χ^2) Distribution

For given degrees of freedom and for specified values of α.

$P(\chi^2_{20} \leq 31.410) = \alpha_{.05}$

df	Value of P																					df
	.999	.995	.99	.98	.975	.95	.90	.80	.75	.70	.50	.30	.25	.20	.10	.05	.025	.02	.01	.005	.001	
1	.0⁵157	.0⁴393	.0³157	.0³628	.0³982	.00393	.0158	.0642	.102	.148	.455	1.074	1.323	1.642	2.706	3.841	5.024	5.412	6.635	7.879	10.827	1
2	.00200	.0100	.0201	.0404	.0506	.103	.211	.446	.575	.713	1.386	2.408	2.773	3.219	4.605	5.991	7.378	7.824	9.210	10.597	13.815	2
3	.0243	.0717	.115	.185	.216	.352	.584	1.005	1.213	1.424	2.366	3.665	4.108	4.642	6.251	7.815	9.348	9.837	11.345	12.838	16.268	3
4	.0908	.207	.297	.429	.484	.711	1.064	1.649	1.923	2.195	3.357	4.878	5.385	5.989	7.779	9.488	11.143	11.668	13.277	14.860	18.465	4
5	.210	.412	.554	.752	.831	1.145	1.610	2.343	2.675	3.000	4.351	6.064	6.626	7.289	9.236	11.070	12.832	13.388	15.086	16.750	20.517	5
6	.381	.676	.872	1.134	1.237	1.635	2.204	3.070	3.455	3.828	5.348	7.231	7.841	8.558	10.645	12.592	14.449	15.033	16.812	18.548	22.457	6
7	.598	.989	1.239	1.564	1.690	2.167	2.833	3.822	4.255	4.671	6.346	8.383	9.037	9.803	12.017	14.067	16.013	16.622	18.475	20.278	24.322	7
8	.857	1.344	1.646	2.032	2.180	2.733	3.490	4.594	5.071	5.527	7.344	9.524	10.219	11.030	13.362	15.507	17.535	18.168	20.090	21.955	26.125	8
9	1.152	1.735	2.088	2.532	2.700	3.325	4.168	5.380	5.899	6.393	8.343	10.656	11.389	12.242	14.684	16.919	19.023	19.679	21.666	23.589	27.877	9
10	1.479	2.156	2.558	3.059	3.247	3.940	4.865	6.179	6.737	7.267	9.342	11.781	12.549	13.442	15.987	18.307	20.483	21.161	23.209	25.188	29.588	10
11	1.834	2.603	3.053	3.609	3.816	4.575	5.578	6.989	7.584	8.148	10.341	12.899	13.701	14.631	17.275	19.675	21.920	22.618	24.725	26.757	31.264	11
12	2.214	3.074	3.571	4.178	4.404	5.226	6.304	7.807	8.438	9.034	11.340	14.011	14.845	15.812	18.549	21.026	23.337	24.054	26.217	28.300	32.909	12
13	2.617	3.565	4.107	4.765	5.009	5.892	7.042	8.634	9.299	9.926	12.340	15.119	15.984	16.985	19.812	22.362	24.736	25.472	27.688	29.819	34.528	13
14	3.041	4.075	4.660	5.368	5.629	6.571	7.790	9.467	10.165	10.821	13.339	16.222	17.117	18.151	21.064	23.685	26.119	26.873	29.141	31.319	36.123	14
15	3.483	4.601	5.229	5.985	6.262	7.261	8.547	10.307	11.036	11.721	14.339	17.322	18.245	19.311	22.307	24.996	27.488	28.259	30.578	32.801	37.697	15
16	3.942	5.124	5.812	6.614	6.908	7.962	9.312	11.152	11.912	12.624	15.338	18.418	19.369	20.465	23.542	26.296	28.845	29.633	32.000	34.267	39.252	16
17	4.416	5.697	6.408	7.255	7.564	8.672	10.085	12.002	12.792	13.531	16.338	19.511	20.489	21.615	24.769	27.587	30.191	30.995	33.409	35.718	40.790	17
18	4.905	6.265	7.015	7.906	8.231	9.390	10.865	12.857	13.675	14.440	17.338	20.601	21.605	22.760	25.989	28.869	31.526	32.346	34.805	37.156	42.312	18
19	5.407	6.844	7.633	8.567	8.907	10.117	11.651	13.716	14.562	15.352	18.338	21.689	22.718	23.900	27.204	30.144	32.852	33.687	36.191	38.582	43.820	19
20	5.921	7.434	8.260	9.237	9.591	10.851	12.443	14.578	15.452	16.266	19.337	22.775	23.828	25.038	28.412	31.410	34.170	35.020	37.566	39.997	45.315	20
21	6.447	8.034	8.897	9.915	10.283	11.591	13.240	15.445	16.344	17.182	20.337	23.858	24.935	26.171	29.615	32.671	35.479	36.343	38.932	41.401	46.797	21
22	6.983	8.643	9.542	10.600	10.982	12.338	14.041	16.314	17.240	18.101	21.337	24.939	26.039	27.301	30.813	33.924	36.781	37.659	40.289	42.796	48.268	22
23	7.529	9.260	10.196	11.293	11.688	13.091	14.848	17.187	18.137	19.021	22.337	26.018	27.141	28.429	32.007	35.172	38.076	38.968	41.638	44.181	49.728	23
24	8.085	9.886	10.856	11.992	12.401	13.848	15.659	18.062	19.037	19.943	23.337	27.096	28.241	29.553	33.196	36.415	39.364	40.270	42.980	45.558	51.179	24
25	8.649	10.520	11.524	12.697	13.120	14.611	16.473	18.940	19.939	20.867	24.337	28.172	29.339	30.675	34.382	37.652	40.646	41.566	44.314	46.926	52.620	25
26	9.222	11.160	12.198	13.409	13.844	15.379	17.292	19.820	20.843	21.792	25.336	29.246	30.434	31.795	35.563	38.885	41.923	42.856	45.642	48.290	54.052	26
27	9.803	11.808	12.879	14.125	14.573	16.151	18.114	20.703	21.749	22.719	26.336	30.319	31.528	32.912	36.741	40.113	43.194	44.140	46.963	49.645	55.476	27
28	10.391	12.461	13.565	14.847	15.308	16.928	18.939	21.588	22.657	23.647	27.336	31.391	32.620	34.027	37.916	41.337	44.461	45.419	48.278	50.993	56.893	28
29	10.986	13.121	14.256	15.574	16.047	17.708	19.768	22.475	23.567	24.577	28.336	32.461	33.711	35.139	39.087	42.557	45.722	46.693	49.588	52.336	58.302	29
30	11.588	13.787	14.953	16.306	16.791	18.493	20.599	23.364	24.478	25.508	29.336	33.530	34.800	36.250	40.256	43.773	46.979	47.962	50.892	53.672	59.703	30

For values of $df > 30$, approximate values for X^2 may be obtained from the expression

$$df\left[1 - \frac{2}{9\,df} \pm \frac{x}{\sigma}\sqrt{\frac{2}{9\,df}}\right]^3,$$

where x/σ is the normal deviate cutting off the corresponding tails of a normal distribution. If x/σ is taken at the 0.02 level, so that 0.01 of the normal distribution is in each tail, the expression yields X^2 at the 0.99 and 0.01 points. For very large values of df, it is sufficiently accurate to compute $\sqrt{2X^2}$, the distribution of which is approximately normal around a mean of $\sqrt{2\,df-1}$ and with a standard deviation of 1. Source: Croxton, Cowden and Klein (1967:672).

appendix table 7

5% (Roman Type) and 1% (Boldface Type) *F* Distribution

n_1 df in Numerator

n_2	1	2	3	4	5	6	7	8	9	10	11	12	14	16	20	24	30	40	50	75	100	200	500	∞	n_2
1	161 **4,052**	200 **4,999**	216 **5,403**	225 **5,625**	230 **5,764**	234 **5,859**	237 **5,928**	239 **5,981**	241 **6,022**	242 **6,056**	243 **6,082**	244 **6,106**	245 **6,142**	246 **6,169**	248 **6,208**	249 **6,234**	250 **6,261**	251 **6,286**	252 **6,302**	253 **6,323**	253 **6,334**	254 **6,352**	254 **6,361**	254 **6,366**	1
2	18.51 **98.49**	19.00 **99.00**	19.16 **99.17**	19.25 **99.25**	19.30 **99.30**	19.33 **99.33**	19.36 **99.36**	19.37 **99.37**	19.38 **99.39**	19.39 **99.40**	19.40 **99.41**	19.41 **99.42**	19.42 **99.43**	19.43 **99.44**	19.44 **99.45**	19.45 **99.46**	19.46 **99.47**	19.47 **99.48**	19.47 **99.48**	19.48 **99.49**	19.49 **99.49**	19.49 **99.49**	19.50 **99.50**	19.50 **99.50**	2
3	10.13 **34.12**	9.55 **30.82**	9.28 **29.46**	9.12 **28.71**	9.01 **28.24**	8.94 **27.91**	8.88 **27.67**	8.84 **27.49**	8.81 **27.34**	8.78 **27.23**	8.76 **27.13**	8.74 **27.05**	8.71 **26.92**	8.69 **26.83**	8.66 **26.69**	8.64 **26.60**	8.62 **26.50**	8.60 **26.41**	8.58 **26.35**	8.57 **26.27**	8.56 **26.23**	8.54 **26.18**	8.54 **26.14**	8.53 **26.12**	3
4	7.71 **21.20**	6.94 **18.00**	6.59 **16.69**	6.39 **15.98**	6.26 **15.52**	6.16 **15.21**	6.09 **14.98**	6.04 **14.80**	6.00 **14.66**	5.96 **14.54**	5.93 **14.45**	5.91 **14.37**	5.87 **14.24**	5.84 **14.15**	5.80 **14.02**	5.77 **13.93**	5.74 **13.83**	5.71 **13.74**	5.70 **13.69**	5.68 **13.61**	5.66 **13.57**	5.65 **13.52**	5.64 **13.48**	5.63 **13.46**	4
5	6.61 **16.26**	5.79 **13.27**	5.41 **12.06**	5.19 **11.39**	5.05 **10.97**	4.95 **10.67**	4.88 **10.45**	4.82 **10.29**	4.78 **10.15**	4.74 **10.05**	4.70 **9.96**	4.68 **9.89**	4.64 **9.77**	4.60 **9.68**	4.56 **9.55**	4.53 **9.47**	4.50 **9.38**	4.46 **9.29**	4.44 **9.24**	4.42 **9.17**	4.40 **9.13**	4.38 **9.07**	4.37 **9.04**	4.36 **9.02**	5
6	5.99 **13.74**	5.14 **10.92**	4.76 **9.78**	4.53 **9.15**	4.39 **8.75**	4.28 **8.47**	4.21 **8.26**	4.15 **8.10**	4.10 **7.98**	4.06 **7.87**	4.03 **7.79**	4.00 **7.72**	3.96 **7.60**	3.92 **7.52**	3.87 **7.39**	3.84 **7.31**	3.81 **7.23**	3.77 **7.14**	3.75 **7.09**	3.72 **7.02**	3.71 **6.99**	3.69 **6.94**	3.68 **6.90**	3.67 **6.88**	6
7	5.59 **12.25**	4.74 **9.55**	4.35 **8.45**	4.12 **7.85**	3.97 **7.46**	3.87 **7.19**	3.79 **7.00**	3.73 **6.84**	3.68 **6.71**	3.63 **6.62**	3.60 **6.54**	3.57 **6.47**	3.52 **6.35**	3.49 **6.27**	3.44 **6.15**	3.41 **6.07**	3.38 **5.98**	3.34 **5.90**	3.32 **5.85**	3.29 **5.78**	3.28 **5.75**	3.25 **5.70**	3.24 **5.67**	3.23 **5.65**	7
8	5.32 **11.26**	4.46 **8.65**	4.07 **7.59**	3.84 **7.01**	3.69 **6.63**	3.58 **6.37**	3.50 **6.19**	3.44 **6.03**	3.39 **5.91**	3.34 **5.82**	3.31 **5.74**	3.28 **5.67**	3.23 **5.56**	3.20 **5.48**	3.15 **5.36**	3.12 **5.28**	3.08 **5.20**	3.05 **5.11**	3.03 **5.06**	3.00 **5.00**	2.98 **4.96**	2.96 **4.91**	2.94 **4.88**	2.93 **4.86**	8
9	5.12 **10.56**	4.26 **8.02**	3.86 **6.99**	3.63 **6.42**	3.48 **6.06**	3.37 **5.80**	3.29 **5.62**	3.23 **5.47**	3.18 **5.35**	3.13 **5.26**	3.10 **5.18**	3.07 **5.11**	3.02 **5.00**	2.98 **4.92**	2.93 **4.80**	2.90 **4.73**	2.86 **4.64**	2.82 **4.56**	2.80 **4.51**	2.77 **4.45**	2.76 **4.41**	2.73 **4.36**	2.72 **4.33**	2.71 **4.31**	9
10	4.96 **10.04**	4.10 **7.56**	3.71 **6.55**	3.48 **5.99**	3.33 **5.64**	3.22 **5.39**	3.14 **5.21**	3.07 **5.06**	3.02 **4.95**	2.97 **4.85**	2.94 **4.78**	2.91 **4.71**	2.86 **4.60**	2.82 **4.52**	2.77 **4.41**	2.74 **4.33**	2.70 **4.25**	2.67 **4.17**	2.64 **4.12**	2.61 **4.05**	2.59 **4.01**	2.56 **3.96**	2.55 **3.93**	2.54 **3.91**	10
11	4.84 **9.65**	3.98 **7.20**	3.59 **6.22**	3.36 **5.67**	3.20 **5.32**	3.09 **5.07**	3.01 **4.88**	2.95 **4.74**	2.90 **4.63**	2.86 **4.54**	2.82 **4.46**	2.79 **4.40**	2.74 **4.29**	2.70 **4.21**	2.65 **4.10**	2.61 **4.02**	2.57 **3.94**	2.53 **3.86**	2.50 **3.80**	2.47 **3.74**	2.45 **3.70**	2.42 **3.66**	2.41 **3.62**	2.40 **3.60**	11
12	4.75 **9.33**	3.88 **6.93**	3.49 **5.95**	3.26 **5.41**	3.11 **5.06**	3.00 **4.82**	2.92 **4.65**	2.85 **4.50**	2.80 **4.39**	2.76 **4.30**	2.72 **4.22**	2.69 **4.16**	2.64 **4.05**	2.60 **3.98**	2.54 **3.86**	2.50 **3.78**	2.46 **3.70**	2.42 **3.61**	2.40 **3.56**	2.36 **3.49**	2.35 **3.46**	2.32 **3.41**	2.31 **3.38**	2.30 **3.36**	12
13	4.67 **9.07**	3.80 **6.70**	3.41 **5.74**	3.18 **5.20**	3.02 **4.86**	2.92 **4.62**	2.84 **4.44**	2.77 **4.30**	2.72 **4.19**	2.67 **4.10**	2.63 **4.02**	2.60 **3.96**	2.55 **3.85**	2.51 **3.78**	2.46 **3.67**	2.42 **3.59**	2.38 **3.51**	2.34 **3.42**	2.32 **3.37**	2.28 **3.30**	2.26 **3.27**	2.24 **3.21**	2.22 **3.18**	2.21 **3.16**	13
14	4.60 **8.86**	3.74 **6.51**	3.34 **5.56**	3.11 **5.03**	2.96 **4.69**	2.85 **4.46**	2.77 **4.28**	2.70 **4.14**	2.65 **4.03**	2.60 **3.94**	2.56 **3.86**	2.53 **3.80**	2.48 **3.70**	2.44 **3.62**	2.39 **3.51**	2.35 **3.43**	2.31 **3.34**	2.27 **3.26**	2.24 **3.21**	2.21 **3.14**	2.19 **3.11**	2.16 **3.06**	2.14 **3.02**	2.13 **3.00**	14
15	4.54 **8.68**	3.68 **6.36**	3.29 **5.42**	3.06 **4.89**	2.90 **4.56**	2.79 **4.32**	2.70 **4.14**	2.64 **4.00**	2.59 **3.89**	2.55 **3.80**	2.51 **3.73**	2.48 **3.67**	2.43 **3.56**	2.39 **3.48**	2.33 **3.36**	2.29 **3.29**	2.25 **3.20**	2.21 **3.12**	2.18 **3.07**	2.15 **3.00**	2.12 **2.97**	2.10 **2.92**	2.08 **2.89**	2.07 **2.87**	15
16	4.49 **8.53**	3.63 **6.23**	3.24 **5.29**	3.01 **4.77**	2.85 **4.44**	2.74 **4.20**	2.66 **4.03**	2.59 **3.89**	2.54 **3.78**	2.49 **3.69**	2.45 **3.61**	2.42 **3.55**	2.37 **3.45**	2.33 **3.37**	2.28 **3.25**	2.24 **3.18**	2.20 **3.10**	2.16 **3.01**	2.13 **2.96**	2.09 **2.98**	2.07 **2.86**	2.04 **2.80**	2.02 **2.77**	2.01 **2.75**	16
17	4.45 **8.40**	3.59 **6.11**	3.20 **5.18**	2.96 **4.67**	2.81 **4.34**	2.70 **4.10**	2.62 **3.93**	2.55 **3.79**	2.50 **3.68**	2.45 **3.59**	2.41 **3.52**	2.38 **3.45**	2.33 **3.35**	2.29 **3.27**	2.23 **3.16**	2.19 **3.08**	2.15 **3.00**	2.11 **2.92**	2.08 **2.86**	2.04 **2.79**	2.02 **2.76**	1.99 **2.70**	1.97 **2.67**	1.96 **2.65**	17
18	4.41 **8.28**	3.55 **6.01**	3.16 **5.09**	2.93 **4.58**	2.77 **4.25**	2.66 **4.01**	2.58 **3.85**	2.51 **3.71**	2.46 **3.60**	2.41 **3.51**	2.37 **3.44**	2.34 **3.37**	2.29 **3.27**	2.25 **3.19**	2.19 **3.07**	2.15 **3.00**	2.11 **2.91**	2.07 **2.83**	2.04 **2.78**	2.00 **2.71**	1.98 **2.68**	1.95 **2.62**	1.93 **2.59**	1.92 **2.57**	18

r_2	\multicolumn{20}{c	}{r_1, df in Numerator}	r_2
	1	2	3	4	5	6	7	8	9	10	11	12	14	16	20	24	30	40	50	75	100	200	500	∞	
19	4.38 8.18	3.52 5.93	3.13 5.01	2.90 4.50	2.74 4.17	2.63 3.94	2.55 3.77	2.48 3.63	2.43 3.52	2.38 3.43	2.34 3.36	2.31 3.30	2.26 3.19	2.21 3.12	2.15 3.00	2.11 2.92	2.07 2.84	2.02 2.76	2.00 2.70	1.96 2.63	1.94 2.60	1.91 2.54	1.90 2.51	1.88 2.49	19
20	4.35 8.10	3.49 5.85	3.10 4.94	2.87 4.43	2.71 4.10	2.60 3.87	2.52 3.71	2.45 3.56	2.40 3.45	2.35 3.37	2.31 3.30	2.28 3.23	2.23 3.13	2.18 3.05	2.12 2.94	2.08 2.86	2.04 2.77	1.99 2.69	1.96 2.63	1.92 2.56	1.90 2.53	1.87 2.47	1.85 2.44	1.84 2.42	20
21	4.32 8.02	3.47 5.78	3.07 4.87	2.84 4.37	2.68 4.04	2.57 3.81	2.49 3.65	2.42 3.51	2.37 3.40	2.32 3.31	2.28 3.24	2.25 3.17	2.20 3.07	2.15 2.99	2.09 2.88	2.05 2.80	2.00 2.72	1.96 2.63	1.93 2.58	1.89 2.51	1.87 2.47	1.84 2.42	1.82 2.38	1.81 2.36	21
22	4.30 7.94	3.44 5.72	3.05 4.82	2.82 4.31	2.66 3.99	2.55 3.76	2.47 3.59	2.40 3.45	2.35 3.35	2.30 3.26	2.26 3.18	2.23 3.12	2.18 3.02	2.13 2.94	2.07 2.83	2.03 2.75	1.98 2.67	1.93 2.58	1.91 2.53	1.87 2.46	1.84 2.42	1.81 2.37	1.80 2.33	1.78 2.31	22
23	4.28 7.88	3.42 5.66	3.03 4.76	2.80 4.26	2.64 3.94	2.53 3.71	2.45 3.54	2.38 3.41	2.32 3.30	2.28 3.21	2.24 3.14	2.20 3.07	2.14 2.97	2.10 2.89	2.04 2.78	2.00 2.70	1.96 2.62	1.91 2.53	1.88 2.48	1.84 2.41	1.82 2.37	1.79 2.32	1.77 2.28	1.76 2.26	23
24	4.26 7.82	3.40 5.61	3.01 4.72	2.78 4.22	2.62 3.90	2.51 3.67	2.43 3.50	2.36 3.36	2.30 3.25	2.26 3.17	2.22 3.09	2.18 3.03	2.13 2.93	2.09 2.85	2.02 2.74	1.98 2.66	1.94 2.58	1.89 2.49	1.86 2.44	1.82 2.36	1.80 2.33	1.76 2.27	1.74 2.23	1.73 2.21	24
25	4.24 7.77	3.38 5.57	2.99 4.68	2.76 4.18	2.60 3.86	2.49 3.63	2.41 3.46	2.34 3.32	2.28 3.21	2.24 3.13	2.20 3.05	2.16 2.99	2.11 2.89	2.06 2.81	2.00 2.70	1.96 2.62	1.92 2.54	1.87 2.45	1.84 2.40	1.80 2.32	1.77 2.29	1.74 2.23	1.72 2.19	1.71 2.17	25
26	4.22 7.72	3.37 5.53	2.98 4.64	2.74 4.14	2.59 3.82	2.47 3.59	2.39 3.42	2.32 3.29	2.27 3.17	2.22 3.09	2.18 3.02	2.15 2.96	2.10 2.86	2.05 2.77	1.99 2.66	1.95 2.58	1.90 2.50	1.85 2.41	1.82 2.36	1.78 2.28	1.76 2.25	1.72 2.19	1.70 2.15	1.69 2.13	26
27	4.21 7.68	3.35 5.49	2.96 4.60	2.73 4.11	2.57 3.79	2.46 3.56	2.37 3.39	2.30 3.26	2.25 3.14	2.20 3.06	2.16 2.98	2.13 2.93	2.08 2.83	2.03 2.74	1.97 2.63	1.93 2.55	1.88 2.47	1.84 2.38	1.80 2.33	1.76 2.25	1.74 2.21	1.71 2.16	1.68 2.12	1.67 2.10	27
28	4.20 7.64	3.34 5.45	2.95 4.57	2.71 4.07	2.56 3.76	2.44 3.53	2.36 3.36	2.29 3.23	2.24 3.11	2.19 3.03	2.15 2.95	2.12 2.90	2.06 2.80	2.02 2.71	1.96 2.60	1.91 2.52	1.87 2.44	1.81 2.35	1.78 2.30	1.75 2.22	1.72 2.18	1.69 2.13	1.67 2.09	1.65 2.06	28
29	4.18 7.60	3.33 5.42	2.93 4.54	2.70 4.04	2.54 3.73	2.43 3.50	2.35 3.33	2.28 3.20	2.22 3.08	2.18 3.00	2.14 2.92	2.10 2.87	2.05 2.77	2.00 2.68	1.94 2.57	1.90 2.49	1.85 2.41	1.80 2.32	1.77 2.27	1.73 2.19	1.71 2.15	1.68 2.10	1.65 2.06	1.64 2.03	29
30	4.17 7.56	3.32 5.39	2.92 4.51	2.69 4.02	2.53 3.70	2.42 3.47	2.34 3.30	2.27 3.17	2.21 3.06	2.16 2.98	2.12 2.90	2.09 2.84	2.04 2.74	1.99 2.66	1.93 2.55	1.89 2.47	1.84 2.38	1.79 2.29	1.76 2.24	1.72 2.16	1.69 2.13	1.66 2.07	1.64 2.03	1.62 2.01	30
32	4.15 7.50	3.30 5.34	2.90 4.46	2.67 3.97	2.51 3.66	2.40 3.42	2.32 3.25	2.25 3.12	2.19 3.01	2.14 2.94	2.10 2.86	2.07 2.80	2.02 2.70	1.97 2.62	1.91 2.51	1.86 2.42	1.82 2.34	1.76 2.25	1.74 2.20	1.69 2.12	1.67 2.08	1.64 2.02	1.61 1.98	1.59 1.96	32
34	4.13 7.44	3.28 5.29	2.88 4.42	2.65 3.93	2.49 3.61	2.38 3.38	2.30 3.21	2.23 3.08	2.17 2.97	2.12 2.89	2.08 2.82	2.05 2.76	2.00 2.66	1.95 2.58	1.89 2.47	1.84 2.38	1.80 2.30	1.74 2.21	1.71 2.15	1.67 2.08	1.64 2.04	1.61 1.98	1.59 1.94	1.57 1.91	34
36	4.11 7.39	3.26 5.25	2.86 4.38	2.63 3.89	2.48 3.58	2.36 3.35	2.28 3.18	2.21 3.04	2.15 2.94	2.10 2.86	2.06 2.78	2.03 2.72	1.98 2.62	1.93 2.54	1.87 2.43	1.82 2.35	1.78 2.26	1.72 2.17	1.69 2.12	1.65 2.04	1.62 2.00	1.59 1.94	1.56 1.90	1.55 1.87	36
38	4.10 7.35	3.25 5.21	2.85 4.34	2.62 3.86	2.46 3.54	2.35 3.32	2.26 3.15	2.19 3.02	2.14 2.91	2.09 2.82	2.05 2.75	2.02 2.69	1.96 2.59	1.92 2.51	1.85 2.40	1.80 2.32	1.76 2.22	1.71 2.14	1.67 2.08	1.63 2.00	1.60 1.97	1.57 1.90	1.54 1.86	1.53 1.84	38
40	4.08 7.31	3.23 5.18	2.84 4.31	2.61 3.83	2.45 3.51	2.34 3.29	2.25 3.12	2.18 2.99	2.12 2.88	2.07 2.80	2.04 2.73	2.00 2.66	1.95 2.56	1.90 2.49	1.84 2.37	1.79 2.29	1.74 2.20	1.69 2.11	1.66 2.05	1.61 1.97	1.59 1.94	1.55 1.88	1.53 1.84	1.51 1.81	40
42	4.07 7.27	3.22 5.15	2.83 4.29	2.59 3.80	2.44 3.49	2.32 3.26	2.24 3.10	2.17 2.96	2.11 2.86	2.06 2.77	2.02 2.70	1.99 2.64	1.94 2.54	1.89 2.46	1.82 2.35	1.78 2.26	1.73 2.17	1.68 2.08	1.64 2.02	1.60 1.94	1.57 1.91	1.54 1.85	1.51 1.80	1.49 1.78	42

v_2	\multicolumn{19}{c}{v_1 df in Numerator}	v_2
	1	2	3	4	5	6	7	8	9	10	11	12	14	16	20	24	30	40	50	75	100	200	500	∞	
44	4.06 7.24	3.21 5.12	2.82 4.26	2.58 3.78	2.43 3.46	2.31 3.24	2.23 3.07	2.16 2.94	2.10 2.84	2.05 2.75	2.01 2.68	1.98 2.62	1.92 2.52	1.88 2.44	1.81 2.32	1.76 2.24	1.72 2.15	1.66 2.06	1.63 2.00	1.58 1.92	1.56 1.88	1.52 1.82	1.50 1.78	1.48 1.75	44
46	4.05 7.21	3.20 5.10	2.81 4.24	2.57 3.76	2.42 3.44	2.30 3.22	2.22 3.05	2.14 2.92	2.09 2.82	2.04 2.73	2.00 2.66	1.97 2.60	1.91 2.50	1.87 2.42	1.80 2.30	1.75 2.22	1.71 2.13	1.65 2.04	1.62 1.98	1.57 1.90	1.54 1.86	1.51 1.80	1.48 1.76	1.46 1.72	46
48	4.04 7.19	3.19 5.08	2.80 4.22	2.56 3.74	2.41 3.42	2.30 3.20	2.21 3.04	2.14 2.90	2.08 2.80	2.03 2.71	1.99 2.64	1.96 2.58	1.90 2.48	1.86 2.40	1.79 2.28	1.74 2.20	1.70 2.11	1.64 2.02	1.61 1.96	1.56 1.88	1.53 1.84	1.50 1.78	1.47 1.73	1.45 1.70	48
50	4.03 7.17	3.18 5.06	2.79 4.20	2.56 3.72	2.40 3.41	2.29 3.18	2.20 3.02	2.13 2.88	2.07 2.78	2.02 2.70	1.98 2.62	1.95 2.56	1.90 2.46	1.85 2.39	1.78 2.26	1.74 2.18	1.69 2.10	1.63 2.00	1.60 1.94	1.55 1.86	1.52 1.82	1.48 1.76	1.46 1.71	1.44 1.68	50
55	4.02 7.12	3.17 5.01	2.78 4.16	2.54 3.68	2.38 3.37	2.27 3.15	2.18 2.98	2.11 2.85	2.05 2.75	2.00 2.66	1.97 2.59	1.93 2.53	1.88 2.43	1.83 2.35	1.76 2.23	1.72 2.15	1.67 2.06	1.61 1.96	1.58 1.90	1.52 1.82	1.50 1.78	1.46 1.71	1.43 1.66	1.41 1.64	55
60	4.00 7.08	3.15 4.98	2.76 4.13	2.52 3.65	2.37 3.34	2.25 3.12	2.17 2.95	2.10 2.82	2.04 2.72	1.99 2.63	1.95 2.56	1.92 2.50	1.86 2.40	1.81 2.32	1.75 2.20	1.70 2.12	1.65 2.03	1.59 1.93	1.56 1.87	1.50 1.79	1.48 1.74	1.44 1.68	1.41 1.63	1.39 1.60	60
65	3.99 7.04	3.14 4.95	2.75 4.10	2.51 3.62	2.36 3.31	2.24 3.09	2.15 2.93	2.08 2.79	2.02 2.70	1.98 2.61	1.94 2.54	1.90 2.47	1.85 2.37	1.80 2.30	1.73 2.18	1.68 2.09	1.63 2.00	1.57 1.90	1.54 1.84	1.49 1.76	1.46 1.71	1.42 1.64	1.39 1.60	1.37 1.56	65
70	3.98 7.01	3.13 4.92	2.74 4.08	2.50 3.60	2.35 3.29	2.23 3.07	2.14 2.91	2.07 2.77	2.01 2.67	1.97 2.59	1.93 2.51	1.89 2.45	1.84 2.35	1.79 2.28	1.72 2.15	1.67 2.07	1.62 1.98	1.56 1.88	1.53 1.82	1.47 1.74	1.45 1.69	1.40 1.62	1.37 1.56	1.35 1.53	70
80	3.96 6.96	3.11 4.88	2.72 4.04	2.48 3.56	2.33 3.25	2.21 3.04	2.12 2.87	2.05 2.74	1.99 2.64	1.95 2.55	1.91 2.48	1.88 2.41	1.82 2.32	1.77 2.24	1.70 2.11	1.65 2.03	1.60 1.94	1.54 1.84	1.51 1.78	1.45 1.70	1.42 1.65	1.38 1.57	1.35 1.52	1.32 1.49	80
100	3.94 6.90	3.09 4.82	2.70 3.98	2.46 3.51	2.30 3.20	2.19 2.99	2.10 2.82	2.03 2.69	1.97 2.59	1.92 2.51	1.88 2.43	1.85 2.36	1.79 2.26	1.75 2.19	1.68 2.06	1.63 1.98	1.57 1.89	1.51 1.79	1.48 1.73	1.42 1.64	1.39 1.59	1.34 1.51	1.30 1.46	1.28 1.43	100
125	3.92 6.84	3.07 4.78	2.68 3.94	2.44 3.47	2.29 3.17	2.17 2.95	2.08 2.79	2.01 2.65	1.95 2.56	1.90 2.47	1.86 2.40	1.83 2.33	1.77 2.23	1.72 2.15	1.65 2.03	1.60 1.94	1.55 1.85	1.49 1.75	1.45 1.68	1.39 1.59	1.36 1.54	1.31 1.46	1.27 1.40	1.25 1.37	125
150	3.91 6.81	3.06 4.75	2.67 3.91	2.43 3.44	2.27 3.14	2.16 2.92	2.07 2.76	2.00 2.62	1.94 2.53	1.89 2.44	1.85 2.37	1.82 2.30	1.76 2.20	1.71 2.12	1.64 2.00	1.59 1.91	1.54 1.83	1.47 1.72	1.44 1.66	1.37 1.56	1.34 1.51	1.29 1.43	1.25 1.37	1.22 1.33	150
200	3.89 6.76	3.04 4.71	2.65 3.88	2.41 3.41	2.26 3.11	2.14 2.90	2.05 2.73	1.98 2.60	1.92 2.50	1.87 2.41	1.83 2.34	1.80 2.28	1.74 2.17	1.69 2.09	1.62 1.97	1.57 1.88	1.52 1.79	1.45 1.69	1.42 1.62	1.35 1.53	1.32 1.48	1.26 1.39	1.22 1.33	1.19 1.28	200
400	3.86 6.70	3.02 4.66	2.62 3.83	2.39 3.36	2.23 3.06	2.12 2.85	2.03 2.69	1.96 2.55	1.90 2.46	1.85 2.37	1.81 2.29	1.78 2.23	1.72 2.12	1.67 2.04	1.60 1.92	1.54 1.84	1.49 1.74	1.42 1.64	1.38 1.57	1.32 1.47	1.28 1.42	1.22 1.32	1.16 1.24	1.13 1.19	400
1000	3.85 6.66	3.00 4.62	2.61 3.80	2.38 3.34	2.22 3.04	2.10 2.82	2.02 2.66	1.95 2.53	1.89 2.43	1.84 2.34	1.80 2.26	1.76 2.20	1.70 2.09	1.65 2.01	1.58 1.89	1.53 1.81	1.47 1.71	1.41 1.61	1.36 1.54	1.30 1.44	1.26 1.38	1.19 1.28	1.13 1.19	1.08 1.11	1000
∞	3.84 6.63	2.99 4.60	2.60 3.78	2.37 3.32	2.21 3.02	2.09 2.80	2.01 2.64	1.94 2.51	1.88 2.41	1.83 2.32	1.79 2.24	1.75 2.18	1.69 2.07	1.64 1.99	1.57 1.87	1.52 1.79	1.46 1.69	1.40 1.59	1.35 1.52	1.28 1.41	1.24 1.36	1.17 1.25	1.11 1.15	1.00 1.00	∞

appendix table 8

Critical Values of *U* in the Mann-Whitney Test

Where the sample sizes n_1 and n_2 ($n_1 \le n_2$) are 20 or less, the computed value of U must be *less than* the following values in order to reject H_0. Unbracketed values are for one-tailed tests at $\alpha = 0.05$ or two-tailed tests at $\alpha = 0.10$, and bracketed values are for two-tailed tests at $\alpha = 0.05$ or one-tailed tests at $\alpha = 0.025$.

Where n_2 is larger than 20 a z test can be performed by computing

$$z_U = \frac{U - \dfrac{n_1 n_2}{2}}{\sqrt{\dfrac{n_1 n_2 (n_1 + n_2 + 1)}{12}}}$$

n_1 \ n_2	3	4	5	6	7	8	9	10	11	12	13	14	15	16	17	18	19	20
1	–	–	–	–	–	–	–	–	–	–	–	–	–	–	–	–	–	–
	(–)	(–)	(–)	(–)	(–)	(–)	(–)	(–)	(–)	(–)	(–)	(–)	(–)	(–)	(–)	(–)	(–)	(–)
2	–	–	0	0	0	1	1	1	1	2	2	2	3	3	3	4	4	4
	(–)	(–)	(–)	(–)	(–)	(0)	(0)	(0)	(0)	(1)	(1)	(1)	(1)	(1)	(2)	(2)	(2)	(2)
3	0	0	1	2	2	3	3	4	5	5	6	7	7	8	9	9	10	11
	(–)	(–)	(0)	(1)	(1)	(2)	(2)	(3)	(3)	(4)	(4)	(5)	(5)	(6)	(6)	(7)	(7)	(8)
4		1	2	3	4	5	6	7	8	9	10	11	12	14	15	16	17	18
		(–)	(0)	(1)	(2)	(3)	(3)	(4)	(5)	(5)	(6)	(7)	(7)	(8)	(9)	(9)	(10)	(11)
5			4	5	6	8	9	11	12	13	15	16	18	19	20	22	23	25
			(1)	(3)	(5)	(6)	(7)	(8)	(9)	(11)	(12)	(13)	(14)	(15)	(17)	(18)	(19)	(20)
6				7	8	10	12	14	16	17	19	21	23	25	26	28	30	32
				(5)	(6)	(8)	(10)	(11)	(13)	(14)	(16)	(17)	(19)	(21)	(22)	(24)	(25)	(27)
7					11	13	15	17	19	21	24	26	28	30	33	35	37	39
					(8)	(10)	(12)	(14)	(16)	(18)	(20)	(22)	(24)	(26)	(28)	(30)	(32)	(34)
8						15	18	20	23	26	28	31	33	36	39	41	44	47
						(13)	(15)	(17)	(19)	(22)	(24)	(26)	(29)	(31)	(34)	(36)	(38)	(41)
9							21	24	27	30	33	36	39	42	45	48	51	54
							(17)	(20)	(23)	(26)	(28)	(31)	(34)	(37)	(39)	(42)	(45)	(48)
10								27	31	34	37	41	44	48	51	55	58	62
								(23)	(26)	(29)	(33)	(36)	(39)	(42)	(45)	(48)	(52)	(55)
11									34	38	42	46	50	54	57	61	65	69
									(30)	(33)	(37)	(40)	(44)	(47)	(51)	(55)	(58)	(62)
12										42	47	51	55	60	64	68	72	77
										(37)	(41)	(45)	(49)	(53)	(57)	(61)	(65)	(69)
13											51	56	61	65	70	75	80	84
											(45)	(50)	(54)	(59)	(63)	(67)	(72)	(76)
14												61	66	71	77	82	87	92
												(55)	(59)	(64)	(67)	(74)	(78)	(83)
15													72	77	83	88	94	100
													(64)	(70)	(75)	(80)	(85)	(90)
16														83	89	95	101	107
														(75)	(81)	(86)	(92)	(98)
17															96	102	109	115
															(87)	(93)	(99)	(105)
18																109	116	123
																(99)	(106)	(112)
19																	123	130
																	(113)	(119)
20																		138
																		(127)

Source: Mann and Whitney (1947:52-54)

appendix table 9

Critical Values of *H* in the Kruskal-Wallis Test

Where there are three samples with sizes of up to five items each, the computed value of H must equal or exceed the following values to reject H_0. When there are more than five items per sample, H is distributed as χ^2 with df one less than the number of samples

	sizes		Significance levels	
n_1	n_2	n_3	$\alpha = 0.05$	$\alpha = 0.01$
3	2	2	4.7143	—
3	3	1	5.1429	—
3	3	2	5.3611	—
3	3	3	5.6000	7.2000
4	2	2	5.3333	—
4	3	1	5.2083	—
4	3	2	5.4444	6.4444
4	3	3	5.7273	6.7455
4	4	1	4.9667	6.6667
4	4	2	5.4545	7.0364
4	4	3	5.5985	7.1439
4	4	4	5.6923	7.6538
5	2	1	5.0000	—
5	2	2	5.1600	6.5333
5	3	1	4.9600	—
5	3	2	5.2509	6.8218
5	3	3	5.6485	7.0788
5	4	1	4.9855	6.9545
5	4	2	5.2727	7.1182
5	4	3	5.6308	7.4449
5	4	4	5.6176	7.7604
5	5	1	5.1273	7.3091
5	5	2	5.3385	7.2692
5	5	3	5.7055	7.5429
5	5	4	5.6429	7.7914
5	5	5	5.7800	7.9800

Critical Values of D in the Kolmogorov-Smirnov One-Sample Test

Sample size (n)	$\alpha =$ .20	.15	.10	.05	.01
1	.900	.925	.950	.975	.995
2	.684	.726	.776	.842	.929
3	.565	.597	.642	.708	.828
4	.494	.525	.564	.624	.733
5	.446	.474	.510	.565	.669
6	.410	.436	.470	.521	.618
7	.381	.405	.438	.486	.577
8	.358	.381	.411	.457	.543
9	.339	.360	.388	.432	.514
10	.322	.342	.368	.410	.490
11	.307	.326	.352	.391	.468
12	.295	.313	.338	.375	.450
13	.284	.302	.325	.361	.433
14	.274	.292	.314	.349	.418
15	.266	.283	.304	.338	.404
16	.258	.274	.295	.328	.392
17	.250	.266	.286	.318	.381
18	.244	.259	.278	.309	.371
19	.237	.252	.272	.301	.363
20	.231	.246	.264	.294	.356
25	.21	.22	.24	.27	.32
30	.19	.20	.22	.24	.29
35	.18	.19	.21	.23	.27
Over 35	$\dfrac{1.07}{\sqrt{n}}$	$\dfrac{1.14}{\sqrt{n}}$	$\dfrac{1.22}{\sqrt{n}}$	$\dfrac{1.36}{\sqrt{n}}$	$\dfrac{1.63}{\sqrt{n}}$

appendix table 11

Critical Values of *D* in the Kolmogorov-Smirnov Two-Sample Test

FOR SMALL SAMPLES

Where the two sample sizes are equal and less than 40, the computed value of D must exceed the following values in order to reject H_0.

	One-tailed test		Two-tailed test	
N	$\alpha = 0.05$	$\alpha = 0.01$	$\alpha = 0.05$	$\alpha = 0.01$
3	1.0000	—	—	—
4	1.0000	—	1.0000	—
5	0.8000	1.0000	1.0000	1.0000
6	0.8333	1.0000	0.8333	1.0000
7	0.7143	0.8571	0.8571	0.8571
8	0.6250	0.7500	0.7500	0.8750
9	0.6667	0.7778	0.6667	0.7778
10	0.6000	0.7000	0.7000	0.8000
11	0.5455	0.7273	0.6364	0.7273
12	0.5000	0.6667	0.5833	0.6667
13	0.5385	0.6154	0.5385	0.6923
14	0.5000	0.5714	0.5714	0.6429
15	0.4667	0.6000	0.5333	0.6000
16	0.4375	0.5625	0.5000	0.6250
17	0.4706	0.5294	0.4706	0.5882
18	0.4444	0.5556	0.5000	0.5556
19	0.4211	0.5263	0.4737	0.5263
20	0.4000	0.5000	0.4500	0.5500
21	0.3810	0.4762	0.4286	0.5238
22	0.4091	0.5000	0.4091	0.5000
23	0.3913	0.4783	0.4348	0.4783
24	0.3750	0.4583	0.4167	0.5000
25	0.3600	0.4400	0.4000	0.4800
26	0.3462	0.4231	0.3846	0.4615
27	0.3333	0.4444	0.3704	0.4444
28	0.3571	0.4286	0.3929	0.4643
29	0.3448	0.4138	0.3793	0.4483
30	0.3333	0.4000	0.3667	0.4333
35	0.3143	0.3714	0.3429	—
40	0.2750	0.3500	0.3250	—

FOR LARGE SAMPLES

Where sample sizes are larger than 40 (it is not necessary for sample sizes to be equal), the computed value of D must exceed critical values calculated as follows.

$\alpha =$	$D =$
.10	$1.22 \sqrt{\dfrac{n_1 + n_2}{n_1 n_2}}$
.05	$1.36 \sqrt{\dfrac{n_1 + n_2}{n_1 n_2}}$
.025	$1.48 \sqrt{\dfrac{n_1 + n_2}{n_1 n_2}}$
.01	$1.63 \sqrt{\dfrac{n_1 + n_2}{n_1 n_2}}$
.005	$1.73 \sqrt{\dfrac{n_1 + n_2}{n_1 n_2}}$
.001	$1.95 \sqrt{\dfrac{n_1 + n_2}{n_1 n_2}}$

appendix table 12

Critical Values of the Standard Normal Deviate z

	SIGNIFICANCE LEVEL (ONE TAILED)				
	0.1	0.05	0.01	0.005	0.001
z	1.282	1.645	2.326	2.576	3.090
$-z$	−1.282	−1.645	−2.326	−2.576	−3.090

	SIGNIFICANCE LEVEL (TWO TAILED)				
	0.1	0.05	0.01	0.005	0.001
z	1.645	1.960	2.576	2.813	3.291
$-z$	−1.645	−1.960	−2.576	−2.813	−3.291

appendix table **13**

Sample Sizes Required for 50% of Population, for 95% Certainty, and Specified Percent Accuracy

Number of Units in Study Area	Sample Size for Accuracy of					
	±1%	±2%	±3%	±4%	±5%	±10%
500	‥	‥	‥	‥	222	83
1,000	‥	‥	‥	385	286	91
1,500	‥	‥	638	441	316	94
2,000	‥	‥	714	476	333	95
2,500	‥	1,250	769	500	345	96
3,000	‥	1,364	811	517	353	97
3,500	‥	1,458	843	530	359	97
4,000	‥	1,538	870	541	364	98
4,500	‥	1,607	891	549	367	98
5,000	‥	1,667	909	556	370	98
6,000	‥	1,765	938	566	375	98
7,000	‥	1,842	959	574	378	99
8,000	‥	1,905	976	580	381	99
9,000	‥	1,957	989	584	383	99
10,000	5,000	2,000	1,000	588	385	99
15,000	6,000	2,143	1,034	600	390	99
20,000	6,667	2,222	1,053	606	392	100
25,000	7,143	2,273	1,064	610	394	100
50,000	8,333	2,381	1,087	617	397	100
100,000	9,091	2,439	1,099	621	398	100
→ ∞	10,000	2,500	1,111	625	400	100

appendix table 14

Run Statistic Critical Values for Sample Sizes n_1 and n_2 (Two-Tail Test) $\alpha = .05$

The smaller of n_1 and n_2	5	6	7	8	9	10	11	12	13	14	15	16	17	18	19	20
2								2/6	2/6	2/6	2/6	2/6	2/6	2/6	2/6	2/6
3		2/8	2/8	2/8	2/8	2/8	2/8	2/8	2/8	2/8	3/8	3/8	3/8	3/8	3/8	3/8
4	2/9	2/9	2/10	3/10	3/10	3/10	3/10	3/10	3/10	3/10	3/10	4/10	4/10	4/10	4/10	4/10
5	2/10	3/10	3/11	3/11	3/12	3/12	4/12	4/12	4/12	4/12	4/12	4/12	4/12	5/12	5/12	5/12
6		3/11	3/12	3/12	4/13	4/13	4/13	4/13	5/14	5/14	5/14	5/14	5/14	5/14	6/14	6/14
7			3/13	4/13	4/14	5/14	5/14	5/14	5/15	5/15	6/15	6/16	6/16	6/16	6/16	6/16
8				4/14	5/14	5/15	5/15	6/16	6/16	6/16	6/16	6/17	7/17	7/17	7/17	7/17
9					5/15	5/16	6/16	6/16	6/17	7/17	7/18	7/18	7/18	8/18	8/18	8/18
10						6/16	6/17	7/17	7/18	7/18	7/18	8/19	8/19	8/19	8/20	9/20
11							7/17	7/18	7/19	8/19	8/19	8/20	9/20	9/20	9/21	9/21
12								7/19	8/19	8/20	8/20	9/21	9/21	9/21	10/22	10/22
13									8/20	9/20	9/21	9/21	10/22	10/22	10/23	10/23
14										9/21	9/22	10/22	10/23	10/23	11/23	11/24
15											10/22	10/23	11/23	11/24	11/24	12/25
16												11/23	11/24	11/25	12/25	12/25
17													11/25	12/25	12/26	13/26
18														12/26	13/26	13/27
19															13/27	13/27
20																14/28

Instructions for the Use of SAS and SPSS-X in Statistical Analysis

appendix A

The examples and exercises in this text have been designed primarily for solution with a pocket or desk calculator. However, some instructors and their students probably have mainframe, minicomputers, or microcomputers available and will wish to solve these and other statistical problems with the accuracy and speed that computers provide. For those who have the machines and wish to use them, and who have available either the SAS or SPSS-X program systems, we provide the most basic commands necessary for the computer to solve the problems illustrated in Chapters 3 through 9.

Because of their wide availability and use in colleges and universities, we selected the SAS and SPSS-X program systems for computer solution of statistical problems. We recognize that there are other popular programs available, too, such as BMDP and MINITAB, but space limitations prohibit their demonstrations here. SPSS-X and SAS software can perform an impressive array of both statistical and nonstatistical functions.

If you have never used these programs before, don't expect this appendix to tell you *everything* necessary to get the system running and solving your problems. SPSS-X and SAS systems run on a wide variety of computers, and every computer has its own ways of interfacing with these programs. It will be necessary for your instructor or someone who is familiar with the available computer to demonstrate how you adapt SAS or SPSS-X *to that particular machine.* For instance, some of you have versions written for personal computers, and others have only the capability of using a large mainframe computer via a remote terminal. In the latter case, you will probably require a computer account, and will need to learn some Job Control Language (JCL) before you can proceed. Only when you get past these preliminaries will our instructions become useful.

SAS and SPSS-X programs communicate with the computer using *statements,* or lines of language that the software has been designed to understand. In SAS, for instance, each statement must end with a semicolon; SPSS-X statements end without punctuation of any kind. SAS procedures are identified with the leading word PROC, whereas SPSS-X procedural statements do not employ that key word. In any case, with the aid of primers, manuals, and more knowledgeable colleagues, you should be able to expand your knowledge quickly and use SAS and SPSS-X beyond the examples provided in this appendix.

SAS AND SPSS-X FOR THE PRODUCTION OF THE SIMPLE GRAPHS OF CHAPTER 3

We begin our instruction on the use of these programs by demonstrating their application to the problems shown in Chapter 3. As we cover the most fundamental aspects of programming, such as the rules for naming variables and when to use punctuation or when not to use blanks, the ensuing demonstrations will become less cluttered with redundant information. With few exceptions, we will make the assumption that the data values used in these programs are supplied within the program listing itself, as opposed to being in their own external data file.

How to Produce a Simple SAS Plot of One Variable Against Another (Figure 3.1a)

Figure 3.1 was not produced with either SAS or SPSS-X. In fact most graphics produced by these systems are not intended for publication, but merely to provide information to the geographer about the variables under consideration. Therefore, the quality of the computer's graphics will not often approach that of the finished charts in a journal article or the typical textbook.

Consider the following SAS program. Its purpose is to show how the proportion of consumers in a study area who visited a given shopping center over, say, the past month varied with the distance from the consumers' residential neighborhoods to the shopping center (see Figure 3.1a, for example).

```
DATA DECAY;
INPUT DRAW DISTANCE;
  LABEL DRAW = 'PROPORTION ATTRACTED';
CARDS;
31.5 2.2
33.3 2.0
.
.
69.7 6.0
;
PROC PLOT;
  PLOT DRAW*DISTANCE = '*';
  TITLE 'decay of interaction over distance';
```

SAS programs usually begin with a DATA statement. This statement simply conveys your desire to read some data and put them into a SAS data set, in this case one arbitrarily named DECAY. Data set names and variable names can contain from one to eight characters. They can contain numbers, also, but the name must begin with a letter. Note that each SAS statement (with the exception of the data lines) concludes with a semicolon. Note also that statements may be typed in either upper or lower case.

The INPUT statement tells SAS how the data values are arranged on the data lines, and what the variable names are. In this case, DRAW (the percentage of consumers drawn to the center) comes first, followed by DISTANCE. It is not necessary to specify in which columns the data appear in the data field. Merely separating the two values with a space will differentiate between the two variables. However, we could have written the INPUT statement as follows:

```
INPUT DRAW 1-4 DISTANCE 5-7;
```

in which case, it would not have been necessary to separate the data values with a space, because you have informed SAS that DRAW will always appear in columns 1–4, and DISTANCE will always appear in columns 5–7. The variable names DRAW and DISTANCE will automatically appear as the labels on the axes of the plot unless they are altered with a LABEL statement, as shown above.

The CARDS statement tells SAS that the data lines follow. There are forty-eight observations in this plot, so forty-eight pairs of values appear on the following lines. Only the first two observations and the last observation are shown above. A semicolon is entered on the line following the last data line. This informs SAS that all of the values of the data set named DECAY have now been entered.

The PROC PLOT statement tells SAS that you want to plot the two variables against each other. It is necessary to choose one variable for the vertical axis and the other for the horizontal axis. In this case, we want the consumer proportions (DRAW) on the vertical axis and distance on the horizontal axis. Thus,

```
PROC PLOT;
  PLOT DRAW*DISTANCE = '*';
  TITLE 'decay of interaction over distance';
```

will produce a plot with a title ("decay of interaction over distance") above it, and the symbol used to show the location of each pair of values will be an asterisk. If we had not specified "= '*'," the locations on the plot would have been designated by A's. In places where two observations were too close together to be represented separately, a B would have appeared. The result of this procedure is shown in Figure A.1.

Note the message below the graph stating that one observation is hidden. This means that, at the scale of this plot, one observation was too close to another observation for it to be shown separately, and since we had specified that we wanted asterisks at each location, there was no other way to show the location of both observations. There is no way to ascertain where the overlap occurred on the plot itself. It would be necessary to peruse the data to find these adjacent observations.

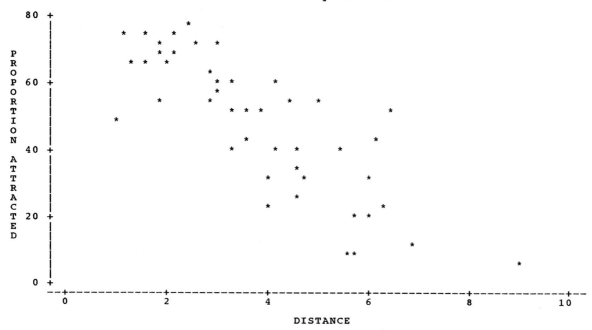

Figure A.1 A Sample Scatter Plot by SAS

How to Produce a Simple SPSS-X Plot of One Variable Against Another

We now illustrate the use of SPSS-X to accomplish the same task. The following is a set of SPSS statements:

```
DATA LIST / HS 1-4 UN 5-8
BEGIN DATA
54.6 6.7
32.219.8
.
.
27.122.7
59.4 4.6
END DATA
PLOT HSIZE = 70
  /VSIZE = 20
  /SYMBOLS = '*'
```

```
/TITLE='RELATIONSHIP OF UNEMPLOYMENT TO HIGH SCHOOL
GRADUATES'
/VERTICAL='UNEMPLOYMENT' MIN(0) MAX(30)
/HORIZONTAL='PERCENT GRADUATES' MIN (20) MAX (100)
/PLOT = UN WITH HS
FINISH
```

SPSS-X statements do not begin with a definitional statement like DATA. Instead, we may immediately specify the names and locations of the variables that are to be plotted. In this example, we wish to show the effect upon unemployment rates of the proportion of the population that has graduated from high school. The units of observation happen to be census tracts, so obviously the rates are tract averages. The term DATA LIST is separated from its specifications by a forward slash. Unlike SAS, the *default* mode for format specification is what is known as *fixed:* each variable's position on the data line must be specified. However, if the statement is specified

```
DATA LIST FREE / HS UN
```

the values for each variable do not have to fall within set columns in the field.

The SPSS-X statement that is identical to CARDS in SAS is BEGIN DATA. It is followed by the data values for HS and UN. Again, only the first and last two observations (out of twenty in this demonstration) have been reproduced. The statement following the last observation is END DATA. The PLOT command is followed by several options, each separated from the others by a forward slash. HSIZE and VSIZE control the size of the plot that will appear on the printout. In this case, the range of the data is forced into a plot that is 20 by 70 characters.

With the SYMBOLS subcommand, you can avoid the default plotting symbol and provide one of your choosing. In most installations, the default symbol for each observation will be a "1". If two observations overlap, a "2," is used, and so forth. A TITLE subcommand may be followed by the title that you wish to appear directly above the plot. The single quote marks are mandatory for this purpose. The plot's axes may also be labeled by specifying VERTICAL or HORIZONTAL, followed by the text within single quotes. MIN and MAX specify the range for each axis. If the range is not specified, the program may produce a plot that wastes space by specifying a range that is beyond that of the data provided. Finally, the PLOT subcommand specifies the vertical axis variable UN WITH HS— the variable on the horizontal axis. The last specification, FINISH, indicates the end of the SPSS-X program. The program just explained produces the plot shown in Figure A.2.

How to Produce a SAS Histogram of a Variable (Figure 3.3)

In SAS, a histogram such as that shown in Chapter 3 may be produced using a procedure called CHART. The following program statements are used to produce the histogram in Figure A.3:

```
DATA CLIMATE;
INPUT TEMP;
CARDS;
38
```

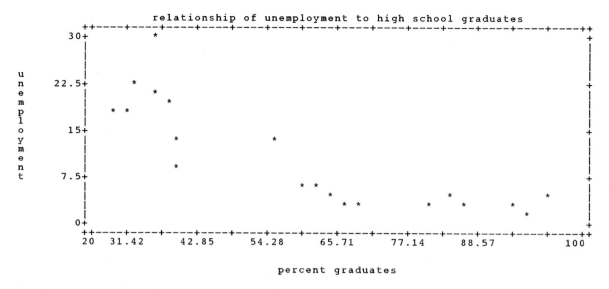

Figure A.2 A Scatter Plot Produced by SPSS-X

```
            38
             .
             .
             .
            42
            46
            ;
   PROC CHART;
     VBAR TEMP / TYPE = FREQ MIDPOINTS = 34.2 38.7 43.2 47.7;
```

The statement following CHART provides the program with your choice of either vertical or horizontal bars: VBAR indicates vertical bars; HBAR would produce a single horizontal bar for each class interval. TEMP is the variable to be charted, and TYPE indicates the kind of chart, here a frequency histogram. There are other useful types of histograms: proportions (PCT), cumulative frequencies (CFREQ), and cumulative proportions (CPCT), for example. The subcommand MIDPOINTS defines the range of values for the chart variable each bar represents by specifying the range midpoints. In this example a histogram is produced with four bars. The first bar represents the range of data values with a midpoint of 34.2; the second bar represents the range of data values with a midpoint of 38.7; and so on. These midpoints must be given in ascending order, although they need not be uniformly distributed. The statement, MIDPOINTS = 32 TO 50 BY 4.5 would have produced the same result. The resulting histogram is shown in Figure A.3.

Figure A.3 A Histogram Produced by SAS

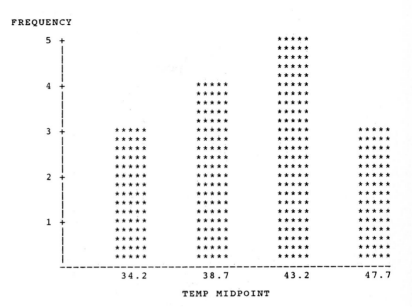

How to Produce an SPSS-X Bar Chart of a Variable

In SPSS-X, a procedure called FREQUENCIES is invoked, and the histogram is a subcommand. Consider the following set of program statements, which show the structure of the data in Appendix B-3:

```
TITLE 'a simple histogram of Maryland female unemployment
rates'
FILE HANDLE FUNEM / NAME = 'FUN.DAT'
DATA LIST FILE = FUNEM / PCT 72-80
FREQUENCIES VARIABLES = PCT
 /HISTOGRAM = INCREMENT (5)
```

    This program produced the histogram shown in Figure A.4.
    Since there were 324 observations (civil divisions) in this data set, they were not all typed into the program statement list between BEGIN DATA and END DATA statements. Instead they resided in a separate file called FUN.DAT, which was shortened to FUNEM. The percentages of female employment were in the field defined by columns 72–80 in the file. SPSS-X offers some useful features in the HISTOGRAM procedure. Minimum and maximum boundaries on the data may be specified; the horizontal axis may be scaled in percentages, and you may specify the increment (in terms of the values of the variable) at which data will be summarized. In this example, we elected to divide the range of the data into increments of 5 percent. This produced a histogram of twelve intervals. If INCREMENT had not been specified, by default, the values would have been displayed in twenty-one intervals. Another useful subcommand is NORMAL, which superimposes a normal curve over the data. These optional specifications may be entered in any order as in

```
Count    Midpoint      One symbol equals approximately    4.00 occurrences
   1      10.25    |
   0      15.25    |
   3      20.25    |*
   9      25.25    |**
  26      30.25    |*******
  98      35.25    |*************************
 133      40.25    |**********************************
  48      45.25    |************
   4      50.25    |*
   0      55.25    |
   1      60.25    |
   1      65.25    |
                   +----+----+----+----+----+----+----+----+----+----+
                   0        40       80       120      160      200
                                 Histogram frequency
```

Figure A.4   **A Histogram Produced by SPSS-X**

```
FREQUENCIES VARIABLES = PCT
  HISTOGRAM = NORMAL MIN (20) MAX (50) INCREMENT (5)
```

which would have produced a histogram of PCT with a superimposed normal curve, and interval width of 5 percent, and all values below 20 percent and above 50 percent would have been ignored.

SAS AND SPSS-X COMMANDS FOR COMPUTING THE DESCRIPTIVE STATISTICS FROM CHAPTER 4

The use of SAS and SPSS-X to produce descriptive statistics is demonstrated using the temperature data sets of Tables 4.2 and 4.3. Output from two analyses follow those lists of program statements. Then, a SAS program is displayed that demonstrates computation of centrographic statistics from the data of Table 4.8.

The SAS Procedure UNIVARIATE

```
DATA CLIMATE;
INPUT TEMP;
CARDS /* MARINE WEST COAST DATA*/;
38
38
.
.
42
46
;
PROC UNIVARIATE;
```

    This program will produce a set of descriptive statistics for the sample of marine west coast temperature data. The portion of the third statement, /* MARINE WEST COAST DATA*/ is a method for inserting comments within a program. Such a statement

Appendix A

is transparent to the computer during execution of the program, but will appear in the program listing. The appearance of /* . . . */ bounds the comment.

The SPSS-X Procedure DESCRIPTIVES

SPSS-X performs exactly the same functions as SAS with the command DESCRIPTIVES:

```
TITLE 'DESCRIPTIVE STATS ON MICROTHERMAL CLIMATE DATA'
DATA LIST / TEMP 1-2
BEGIN DATA
22
25
.
.
20
31
END DATA
DESCRIPTIVES VARIABLES = TEMP
 /STATISTICS = ALL
```

```
                                    SAS
                                 UNIVARIATE
VARIABLE=TMP

         MOMENTS                          QUANTILES(DEF=4)              EXTREMES
N             15    SUM WGTS      15    100% MAX    50    99%    50    LOWEST    HIGHEST
MEAN          41    SUM          615     75% Q3    45    95%    50       32         45
STD DEV  5.45108    VARIANCE 29.7143     50% MED   42    90%   47.6      32         45
SKEWNESS -0.357198  KURTOSIS -0.758805   25% Q1    38    10%    32       34         46
USS        25631    CSS          416      0% MIN   32     5%    32       38         46
CV       13.2953    STD MEAN 1.40746                       1%    32       38         50
                                         RANGE     18
                                         Q3-Q1     7
                                         MODE     32
```

SAS printout for marine west coast example

```
Number of valid observations (listwise) =        15.00
Variable    TMP
Mean              20.467      S.E. Mean       2.120       Std Dev        8.210
Variance          67.410      Kurtosis        -.538       S.E. Kurt      1.121
Skewness            .466      S.E. Skew        .580       Range         29.000
Minimum                8      Maximum            37       Sum          307.000
Valid observations -      15        Missing observations -         0
```

SPSS-X printout for humid microthermal example

```
DATA POPDAT;
TITLE 'Centrographic analysis, black population, Balt., 1980';
INPUT RPD $1-3 X Y WHPOP BLPOP
   /*rpd is district number, identified as alphameric variable
     in col's 1-3,
     x and y are the coordinates of district centroids,
     whpop and blpop are the weights assigned to each x-y centroid*/;
WX   = X * BLPOP;
WY   = Y * BLPOP;
WY2  = BLPOP*(y-7.00)**2;
WX2  = BLPOP*(x-4.75)**2 /* 7.0 and 4.75 are precalculated mean
     unweighted center of gravity in study area*/;
CARDS /* Input statement specifies order of variables */;
101   1.3  10.6  15831   5209
102   3.2   9.8   6670    594
103   4.7   9.9  19906    906
104   6.4  10.7  16474   7927
105   6.3   9.2   9498  27287
106   8.3   9.9  39913   4992
107   1.1   8.2   4448  47220
108   2.4   9.4   1921  40754
109   3.2   7.8    112  15325
110   4.2   8.6  15874    612
111   5.6   7.8  14908  30335
112   7.6   7.8  16307   3223
113   9.2   7.9  23400  13119
114   0.7   5.8   6533   9186
115   2.0   5.5   3780  19954
116   2.8   6.2   4386  36878
117   4.0   6.4  14170  69093
118   5.1   5.9   9045   6023
119   6.2   6.5   5847  64591
120   7.2   5.5  42690   3434
121   8.9   5.1  17804   1566
122   2.1   4.3  14584   2059
123   4.2   4.9   5357   1826
124   5.7   4.3  16123    920
125   4.2   3.1   5443  16332
126   7.0   1.0  14089   1746
PROC UNIVARIATE;
  VAR X Y      WX WY WX2 WY2;
  OUTPUT OUT = POPDAT1;
PROC PRINT DATA = POPDAT1;
PROC PRINT DATA = POPDAT /* the print commands print out the original
  data and the computed variables*/;
```

SAS statements that compute centrographic statistics

Appendix A

```
VARIABLE=X                                          VARIABLE=Y

                    MOMENTS                                             MOMENTS
N                26    SUM WGTS          26         N                26    SUM WGTS          26
MEAN        4.75385    SUM            123.6         MEAN        7.00385    SUM            182.1
STD DEV     2.44626    VARIANCE     5.98418         STD DEV     2.45642    VARIANCE     6.03398
SKEWNESS   0.119422    KURTOSIS    -0.917456        SKEWNESS  -0.416925    KURTOSIS    -0.172107
USS          737.18    CSS          149.605         USS         1426.25    CSS           150.85
CV          51.4585    STD MEAN    0.479751         CV          35.0724    STD MEAN    0.481743
T:MEAN=0    9.90899    PROB>|T|      0.0001         T:MEAN=0    14.5386    PROB>|T|      0.0001
SGN RANK      175.5    PROB>|S|      0.0001         SGN RANK      175.5    PROB>|S|      0.0001


VARIABLE=WX                                         VARIABLE=WY

                    MOMENTS                                             MOMENTS
N                26    SUM WGTS          26         N                26    SUM WGTS          26
MEAN        68742.3    SUM          1787299         MEAN         119462    SUM          3106016
STD DEV     94975.1    VARIANCE  9020279078         STD DEV      145123    VARIANCE   2.106E+10
SKEWNESS    2.27969    KURTOSIS     5.58024         SKEWNESS    1.24956    KURTOSIS    0.183017
USS       3.484E+11    CSS        2.255E+11         USS       8.976E+11    CSS        5.265E+11
CV          138.161    STD MEAN     18626.2         CV           121.48    STD MEAN     28460.9
T:MEAN=0    3.69063    PROB>|T|   0.00110279        T:MEAN=0    4.19741    PROB>|T|  .000297736
SGN RANK      175.5    PROB>|S|      0.0001         SGN RANK      175.5    PROB>|S|      0.0001


VARIABLE=WX2                                        VARIABLE=WY2

                    MOMENTS                                             MOMENTS
N                26    SUM WGTS          26         N                26    SUM WGTS          26
MEAN        81035.9    SUM          2106934         MEAN        45885.6    SUM          1193026
STD DEV      133387    VARIANCE   1.779E+10         STD DEV     66621.9    VARIANCE  4438475407
SKEWNESS    3.08566    KURTOSIS     11.4082         SKEWNESS    2.21932    KURTOSIS     4.49577
USS       6.155E+11    CSS        4.448E+11         USS       1.657E+11    CSS        1.110E+11
CV          164.603    STD MEAN     26159.4         CV          145.191    STD MEAN     13065.6
T:MEAN=0    3.09778    PROB>|T|   0.00477679        T:MEAN=0    3.51193    PROB>|T|   0.00172628
SGN RANK      175.5    PROB>|S|      0.0001         SGN RANK      175.5    PROB>|S|      0.0001
```

SAS printout for Baltimore centrographic analysis

Charts of Probability Distributions (Chapter 5)

A Poisson probability distribution like that of Figure 5.3 may be obtained using either SAS or SPSS-X. For example, consider the following list of SAS statements and hypothetical data:

```
DATA RANDOM;
INPUT FR;
CARDS;
0
0
.
.
3
4
;
```

```
PROC CHART;
  VBAR FR / TYPE=PCT MIDPOINTS = 0 1 2 3 4;
```

The last statement (TYPE=PCT) instructs SAS to produce a probability distribution of the variable FR for the occurrence of values of 0 through 4. The result is shown in Figure A.5.

The following SPSS-X program,

```
TITLE 'NORMAL CURVE HISTOGRAM OF MARYLAND FEMALE
EMPLOYMENT RATES'
FILE HANDLE FUNEM/NAME='FUN.DAT'
DATA LIST FILE=FUNEM/PCT 72-80
FREQUENCIES VARIABLES = PCT
 /HBAR = NORMAL
 /HBAR = PERCENT
```

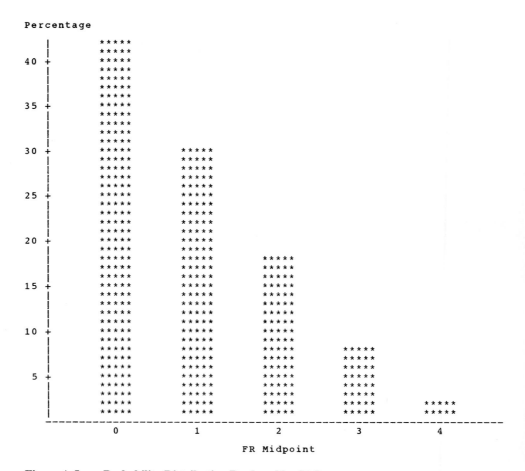

Figure A.5 **Probability Distribution Produced by SAS**

produced the histogram in Figure A.6a. The asterisks form the probability distribution. About 20 percent of the observations fall into a class interval with a midpoint of 41 (percent employed). Dots in the histogram show the location of a normal curve fitted to these data.

Finally, the following SAS program employs the same female employment data:

```
DATA EMPLOY;
INFILE FUNEM;
INPUT PCT 72-80;
PROC CHART;
    VBAR PCT / TYPE = CPCT MIDPOINTS = 28 31 34 37 40 43 46;
```

This listing produced a cumulative probability distribution, showing the cumulative proportion of the observations in intervals with the seven midpoints supplied. The results of these last two programs are shown in Figure A.6.

Computations for the *t* Test from Chapter 6

As Chapter 6 demonstrates, the Z test and t test are used to determine the probability that the difference in means is due to chance. The lower the likelihood that the difference is due to chance, the greater the likelihood that the difference is due to there being real differences between a population mean and a sample mean, or between two sample means. If the problem is to compare the mean of a random sample $(\overline{X})$ with a given population mean (μ), and known σ, then simply employ SAS UNIVARIATE or SPSS-X DESCRIPTIVES to the sample data to obtain $\overline{X}$, then carry out the Z test with a pocket calculator.

Suppose the problem involves comparing the means of two independent random samples. As a demonstration, two random samples are drawn from the data of Appendix B-2. One sample is taken only from counties that are predominantly rural and the other sample is taken from the remaining, predominantly urban, counties. Is the female employment mean of the rural sample significantly different from the mean of the urban sample?

Two Random Samples ($n_1 = n_2 = 20$) of Percent Females in Labor Force, 1980

Rural Counties		Urban Counties	
38.89	35.18	42.89	38.77
30.25	44.16	39.60	37.03
49.33	34.35	37.42	43.48
41.64	36.64	39.70	39.53
38.34	36.55	31.33	40.13
33.39	33.27	40.57	36.57
36.54	31.76	39.72	31.13
35.81	35.36	44.01	35.80
27.60	26.33	38.87	39.54
39.23	32.63	36.10	39.91

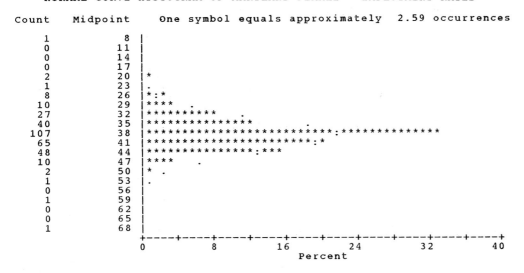

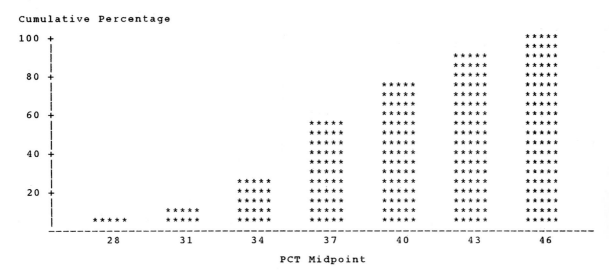

Figure A.6 Histograms produced by (a) SPSS-X, and (b) SAS

Appendix A

As has become our convention, the SAS approach to the t test will be shown first, followed by a demonstration of the SPSS-X program that enables us to test our hypothesis.

The SAS Independent Sample t Test

The SAS program begins as before, with a DATA statement, followed by the INPUT statement, and then the data:

```
DATA FEMEMPL;
INPUT SAM $ FEM;
CARDS;
R 38.89
R 35.18
.
.
U 42.89
U 38.77
.
.
U 39.91
;
PROC TTEST;
 CLASS SAM;
 VAR FEM;
```

Previously, we used only numeric variables. For the first time, we are going to show the use of an alphameric, or *character,* variable, which is a set of elements whose name contains either letters, or letters and numbers, and is interpreted by SAS literally. In this procedure, it is known as the CLASS variable. Here, we arbitrarily use the name SAM to indicate the letters R and U, which identify the sample employment rates that are in rural counties and those that are in urban counties. There will be twenty R's with their respective employment rates and twenty U's with their rates. SAS requires that all the values with the same class be together. The CLASS statement must have two, and only two, numeric or alphameric values. We could have employed "1" and "2" in place of "R" and "U." In that case, we could have eliminated the $ symbol after the name SAM, because SAS *interprets a variable name followed by $ as an alphameric variable.*

The VAR statement gives the name of the variable whose sample means are to be compared. We have again arbitrarily chosen the name FEM to represent the female employment rates of the two samples. The t statistic of Equation 6.15 is employed by SAS for testing the equality of means $\overline{X}_1$ and $\overline{X}_2$. The sample sizes, n_1 and n_2, do not have to be equal.

```
                                                    SAS
                                          TTEST PROCEDURE

VARIABLE: FEM
 SAM       N          MEAN         STD DEV        STD ERROR       MINIMUM         MAXIMUM
  r       20      35.86250000    5.35842561      1.19818039     26.33000000     49.33000000
  u       20      38.60500000    3.36749712      0.75299525     31.13000000     44.01000000
FOR H0: VARIANCES ARE EQUAL, F'=    2.53 WITH 19 AND 19 DF     PROB > F'= 0.0495

            VARIANCES          T        DF      PROB > |T|
            UNEQUAL        -1.9380    32.0       0.0615
            EQUAL          -1.9380    38.0       0.0601
```

From the printout, we see the variable name, FEM, and descriptive statistics for the two samples—n, $\overline{X}$, s, $\hat{\sigma}$, MIN, and MAX. Below this are two sets of t values, degrees of freedom, and probabilities. One is valid if the variances are assumed equal, the other if the variances are not assumed equal. You will usually find these values to be nearly identical unless the variances of the two samples vary widely. You may assess the equality of the variances by looking at the bottom, which gives us the probability that the variances are unequal due to chance. If this probability is small (say, less than .05), then we are going to reject the hypothesis that the variances are equal. In this case, $P = .0495$, which is marginal, but the values of t under either assumption are identical. The t value for the female employment experiment is -1.94, and $P = .06$. We interpret as follows: The probability that a t value of ± 1.94 could have arisen due to chance (sampling variation) is .06. The critical value of t (from Appendix Table 4) for $\nu = 38$ is ≈ 2.02. If we set our significance level at $\alpha = .05$ or lower, we could not reject the null hypothesis. We conclude that the two samples probably came from the same population with respect to female employment rates. In other words, the mean of the rural sample rates was not significantly different from the mean of the urban sample rates.

The SPSS-X Independent Sample t Test

The following is a list of SPSS-X statements that accomplish the same thing as the program just demonstrated:

```
DATA LIST FREE / SAM FEM
BEGIN DATA
1 38.89
1 35.18
  .
  .
2 42.89
2 38.77
  .
  .
2 39.91
END DATA
```

```
T-TEST GROUPS = SAM
  /VARIABLES = FEM
FINISH
```

The grouping variable in SPSS-X must be numeric, so 1's and 2's have been substituted for the R's and U's that were employed in the SAS job. In the T-TEST statement, the name GROUPS identifies the grouping variable SAM. The printout from this program provides exactly the same information as its counterpart in SAS.

```
- - - - - - - - - - - - - - - - - - - - - - - T - T E S T - - - - - - - - - - - - - - - - - - - - - - - - - - - - -
GROUP 1 - REG EQ 1
GROUP 2 - REG EQ 2
                                                          *          Pooled Variance estimate  * Separate Variance Estimate
Variable        Number              Standard   Standard   *   F    2-tail  *   t    Degrees of  2-tail *   t    Degrees of  2-tail
                of Cases   Mean     Deviation  Error      * Value   Prob.  * Value   Freedom    Prob.  * Value   Freedom    Prob.
PRECIP
  GROUP 1        20       35.8625   5.358      1.198      *                *                           *
  GROUP 2        20       38.6050   3.367       .753      *  2.53   .049   * -1.94     38       .060   * -1.94    31.98     .061
```

The Paired *t* Test in SAS

In Box 6.2, the step-by-step procedure for testing the significance of the difference between fertilizer use in two time periods is shown. The SAS listing that will perform this task follows:

```
DATA FERT;
  INPUT NATION $ EARLY LATE;
  DIFF=LATE - EARLY;
  CARDS;
ANGOL 4 4
BOTS 2 0
.
.
.
YEMEN 7 13
PAPUA 20 30
;
PROC MEANS MEAN STDERR T PRT;
  VAR DIFF;
  TITLE 'paired-comparisons t-test';
```

The regular *t* test cannot be used here since the samples are no longer independent. A variety of the *t* test referred to as a paired *t* test is used instead. PROC TTEST does not include a paired *t* test. Differences must be computed and analyzed in PROC MEANS. This statement is followed by a list of options. MEAN and STDERR cause the mean and the standard error of the mean to be printed. In this experiment, these statistics are computed for the variable DIFF. The options T and PRT give us the *t* value and its associated probability, testing if the variable DIFF is significantly different from zero.

```
                     paired-comparisons t-test
Analysis Variable : DIFF

           Mean        Std Error              T      Prob>|T|
       ----------------------------------------------------------
        19.4166667     7.7042047         2.5202688    0.0191
       ----------------------------------------------------------
```

The Paired Sample *t* Test in SPSS-X

The program listing for the SPSS-X paired sample test follows:

```
DATA LIST FREE / NATION (A) EARLY LATE
BEGIN DATA
ANGOL 4 4
BOTS 2 0
  .
  .
YEMEN 7 13
PAPUA 20 30
END DATA
T-TEST PAIRS=F70 F80
FINISH
```

SPSS-X incorporates both the independent sample and paired sample *t* tests within its T-TEST procedure. As you can see from the printout, the same essential information is forthcoming as we obtained with SAS.

```
                                        - - - t-tests for paired samples - - -

Variable    Number               Standard   Standard  |(Difference) Standard    Standard
            of Cases   Mean      Deviation  Error     |   Mean      Deviation   Error
-----------------------------------------------------+----------------------------------
F70
                       15.8333   37.803     7.717
              24                                       -19.4167      37.743     7.704
                       35.2500   73.048     14.911
F80

                  2-tail  |    t      Degrees of  2-tail
            Corr. Prob.   |  Value    Freedom     Prob.
           ---------------+----------------------------
            .967   .000   |  -2.52       23       .019
```

Analysis of Variance Tests of Chapter 7

The analysis of variance of the data on low birth weight births in Washington census tracts (Table 7.1) may be done in both SAS and SPSS-X. There are two procedures in SAS that will perform ANOVA. The lists of SAS statements needed to solve the low birth weight problem would appear as follows:

Appendix A

```
DATA BIRTH;                          DATA BIRTH;
  INPUT SAM RATE;                      INPUT SAM RATE;
  CARDS;                               CARDS;
1 139                                1 139
1 88                                 1 88
.                                    .
2 53                                 2 53
2 128                                2 128
.                                    .
3 141                                3 134
.                                    ;
3 134                                PROC ANOVA;
;                                      CLASSES SAM;
PROC NPAR1WAY;                         MODEL RATE = GROUP;
  CLASS SAM;                           MEANS GROUP;
```

As was the case in the TTEST procedure, the data must first be entered in order of subsample, with the CLASS variable taking on the values 1, 2, and 3. The birth rates are the second value on each data line. The procedures are called NPAR1WAY, or alternatively, ANOVA. NPAR1WAY also performs several nonparametric tests that involve more than two subsamples. SAS will analyze the variances of the subsamples identified by the CLASS variable.

```
              N P A R 1 W A Y   P R O C E D U R E
            Analysis of Variance for Variable RATE
                  Classified by Variable SAM
```

SAM	N	Mean	Among MS	Within MS
1	18	95.777778	3755.37063	2069.81076
2	11	114.000000		
3	6	135.166667	F Value	Prob > F
			1.814	0.1793

The means for the subsamples are printed, followed by the between-groups estimated variance (called the "among MS") and the within-groups estimated variance (called the "within MS"), with the computed F value of 1.814. When the problem is carried out with a pocket calculator, rounding off intermediate values usually produces sums of squares and estimated variances that differ from those reported in the SAS and SPSS-X printouts. You will immediately notice the discrepancies between the printouts below and the numbers shown in the text. However, the F ratios from the calculator analysis and computer analysis are identical.

The SPSS-X listing is:

```
TITLE 'LOW BIRTH WEIGHT ANOVA, WASHINGTON DATA'
DATA LIST FREE / SAM RATE
BEGIN DATA
1 139
.
2 53
```

```
   .
   .
3 134
END DATA
ONEWAY RATE BY SAM (1,3)
FINISH
```

The SPSS-X format contains more intermediate information, and the printed table appears much like Table 7.2.

```
- - - - - - - - - - - - - - - - - - - - - - - - - - - - - - - - - - - - - O N E W A Y - - -

        Variable    RATE       LOW BIRTH WEIGHT ANOVA, WASHINGTON DATA
     By Variable    SAM
                                       ANALYSIS OF VARIANCE
                                       SUM OF           MEAN              F        F
              SOURCE         D.F.      SQUARES         SQUARES          RATIO    PROB.
   BETWEEN GROUPS            2         7510.7413       3755.3706        1.8144   .1793

   WITHIN  GROUPS            32        66233.9444      2069.8108

   TOTAL                     34        73744.6857
```

The SPSS-X procedure is called ONEWAY, and the birth rates are classified BY the grouping variable SAM. The numbers (1,3) in parentheses tell SPSS-X that the subsamples are numbered 1 through 3.

Correlation and Regression in SAS and SPSS-X from Chapter 8

The speed and accuracy of computer program solutions to statistics problems is increasingly important as the complexity and size of the problem increases. Therefore, it is particularly advisable to take advantage of the computer when solving bivariate statistical problems with many observations.

SAS provides both a correlation (CORR) and a regression (GLM) procedure. We will demonstrate both procedures in one listing, using the data in Table 8.1. To obtain Pearson product-moment correlation coefficients, use the PROC CORR statement. To obtain least squares regression results for bivariate models, it is recommended that you use the PROC GLM procedure. GLM is an acronym for General Linear Model, and GLM estimates and tests hypotheses about linear models. The listing, applied to the runoff–precipitation problem, follows:

```
DATA RUNOFF;
 INPUT RUN PRECIP;
 CARDS;
20.4 60.5
23.0 60.0
  .
  .
13.1 48.0
;
```

Appendix A

```
PROC CORR;
 VAR RUN PRECIP;
 TITLE 'CORRELATION OF ANNUAL RUNOFF AND ANNUAL
 RAINFALL';
PROC GLM;
 MODEL RUN=PRECIP;
 OUTPUT OUT=RHAT P = PREDICT R = RESID;
PROC PLOT;
 PLOT RUN*PREPIC = '*' PREDICT*PRECIP = '^' / OVERLAY;
 PLOT RESID*PRECIP / VREF = 0;
```

The CORR procedure will compute correlation coefficients between all pairs of variables in the VAR list. In simple bivariate models, there is only one coefficient to report, as is shown in the printout. The GLM procedure will compute sums of squares and estimated variances for the regression component and for the error, or residual, component. An analysis of variance table is produced, and values of r^2, a, b, and $S_{Y \cdot X}$ are printed. The data needed to plot residuals is saved in a SAS file, which we have named RHAT. The predicted values of the dependent variable as well as the residuals are identified as P and R, respectively. The PLOT procedure instructs SAS to produce a scatter plot of the dependent and independent variables, and a scatter plot of standardized residuals from regression, in order that we might detect any obvious homoscedasticity in the data.

CORRELATION OF ANNUAL RUNOFF AND ANNUAL RAINFALL

Correlation Analysis

2 'VAR' Variables: RUN   PRECIP

Simple Statistics

Variable	N	Mean	Std Dev	Sum	Minimum	Maximum
RUN	16	14.4313	4.8768	230.9000	8.2000	23.0000
PRECIP	16	50.3375	6.5820	805.4000	42.0000	60.5000

Pearson Correlation Coefficients / Prob > |R| under Ho: Rho=0 / N = 16

	RUN	PRECIP
RUN	1.00000 0.0	0.97462 0.0001
PRECIP	0.97462 0.0001	1.00000 0.0

CORRELATION OF ANNUAL RUNOFF AND ANNUAL RAINFALL

General Linear Models Procedure

Dependent Variable: RUN

Source	DF	Sum of Squares	Mean Square	F Value	Pr > F
Model	1	338.87780572	338.87780572	265.39	0.0001
Error	14	17.87656928	1.27689781		
Corrected Total	15	356.75437500			

R-Square	C.V.	Root MSE	RUN Mean
0.949891	7.830223	1.1299990	14.431250

Source	DF	Type I SS	Mean Square	F Value	Pr > F
PRECIP	1	338.87780572	338.87780572	265.39	0.0001

Source	DF	Type III SS	Mean Square	F Value	Pr > F
PRECIP	1	338.87780572	338.87780572	265.39	0.0001

Parameter	Estimate	T for H0: Parameter=0	Pr > |T|	Std Error of Estimate
INTERCEPT	-21.91928617	-9.75	0.0001	2.24915995
PRECIP	0.72213630	16.29	0.0001	0.04432775

CORRELATION OF ANNUAL RUNOFF AND ANNUAL RAINFALL

Plot of RUN*PRECIP.   Symbol used is '*'.
Plot of PREDICT*PRECIP.   Symbol used is '^'.

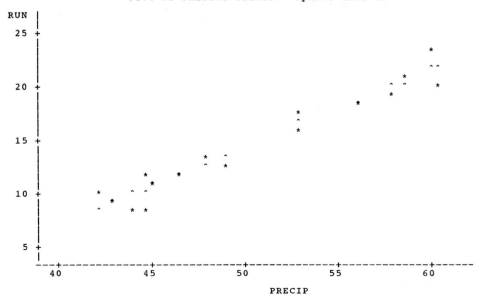

NOTE: 2 obs hidden.

Appendix A

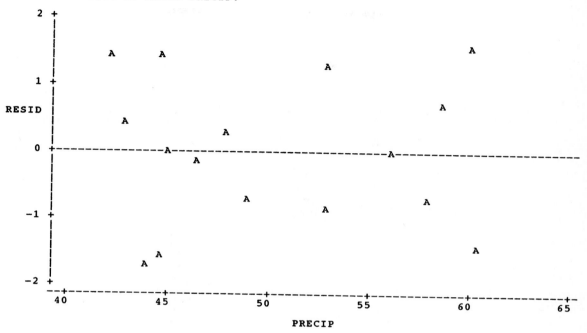

All of this material is also available from SPSS-X through the REGRESSION procedure. The listing for this program follows:

```
TITLE 'CORRELATION OF ANNUAL RUNOFF AND ANNUAL RAINFALL'
DATA LIST FREE / RUN PRECIP
BEGIN DATA
20.4 60.5
   .
   .
13.1 48.0
END DATA
REGRESSION VARIABLES RUN PRECIP
  /DEPENDENT = RUN / METHOD = FORWARD
  /RESIDUALS=DEFAULTS
  /SCATTERPLOT SIZE (SMALL) (RUN,PRECIP) (*ZRESID,PRECIP)
```

No separate correlation procedure need be executed, since the REGRESSION printout shows r, r^2, $S_{Y \cdot X}$, a, b, and the ANOVA table. It is also unnecessary to create a temporary output file to contain the residuals and predicted values. In the REGRESSION paragraph, the model is specified by establishing the DEPENDENT variable as RUN. A METHOD statement is required, even though the six options listed are aimed at analyses involving more than one independent variable. For a simple bivariate model, either FORWARD or ENTER is recommended.

The analysis of residuals requires a RESIDUALS statement, which controls the display and labeling of summary information on deviant cases as well as the display of statistics, histograms and normal probability plots. The RESIDUALS subcommand must be followed by one of several options. DEFAULTS was used here because it provides for the printout of sometimes useful plots of residuals. Specific plots are ordered with the SCATTERPLOT subcommand, in which a small size plot is specified, and two specific combinations are given—the dependent variable against the independent variable (RUN, PRECIP), standardized such that $\bar{X}$ and $\bar{Y}$ are zero, and standardized residuals against the independent variable (*ZRESID, PRECIP). If SIZE (SMALL) is not specified, these plots appear as larger diagrams, which may be useful if n is large and you wish to observe finer detail among the plots of the observations.

```
CORRELATION OF ANNUAL RUNOFF AND ANNUAL RAINFALL

                          * * * *   M U L T I P L E    R E G R E S S I O N   * * * *

Listwise Deletion of Missing Data
Equation Number 1     Dependent Variable..    RUN
Block Number   1.  Method:  Forward       Criterion    PIN   .0500

Variable(s) Entered on Step Number    1..    PRECIP

Multiple R              .97462       Analysis of Variance
R Square                .94989                          DF      Sum of Squares       Mean Square
Adjusted R Square       .94631       Regression          1            338.87781         338.87781
Standard Error         1.13000       Residual           14             17.87657           1.27690

                                     F =     265.39149       Signif F =   .0000

------------------ Variables in the Equation ------------------

Variable                B          SE B         Beta           T    Sig T
PRECIP              .722136      .044328      .974624      16.291   .0000
(Constant)       -21.919286     2.249160                    -9.746   .0000
```

Appendix A

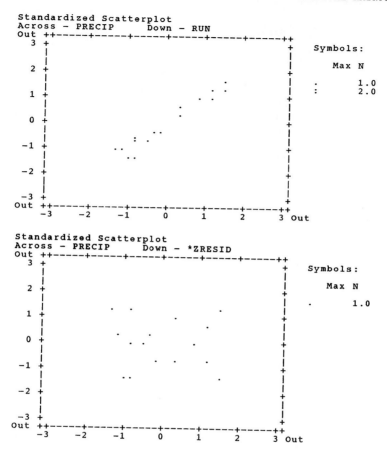

Transformations, such as the logarithmic and square root, which were discussed in Chapter 8, may be performed in the SAS and SPSS-X programs. In SAS, suppose we wished to transform both variables, Y and X in a bivariate regression analysis to base-10 logarithms. This is done immediately following the INPUT statements:

```
INPUT Y X;
LY = LOG10 (Y);
LX = LOG10 (X);
CARDS;
etc.
```

LOG10 is one of many so-called *function names* designated by SAS and SPSS to transform data. Some of the more common arithmetic and mathematical functions and their purpose are:

Function Name		
SAS	SPSS-X	Action
LOG	LN	Base *e* log
LOG10	LG10	Base 10 log
ABS	ABS	Returns absolute value
SIN	SIN	Sine (argument in radians)
COS	COS	Cosine (argument in radians)
ARSIN	ARSIN	Arcsine (inverse sine) of argument Result is in radians
INT	TRUNC	Drops off the fractional part of a number (truncation)
SQRT	SQRT	Square root

In SPSS-X, the following sequence of statements would accomplish the same task:

```
DATA LIST FREE / Y X
COMPUTE LY = LG10 (Y)
COMPUTE LX = LG10 (X)
BEGIN DATA
etc.
```

In SPSS-X data manipulation is usually performed with the COMPUTE statement; one of its most common uses is in functional transformations.

Nonparametric Tests in SAS and SPSS-X from Chapter 9

Distribution-free statistical tests are provided within SAS and SPSS-X. The following table shows the procedural subcommand in both systems for the calculation of common nonparametric tests.

Procedural Subcommand		
SAS	SPSS-X	Action
NPAR1WAY WILCOXON	NPAR WILCOXON	Wilcoxon two sample rank-sum test (SAS) Wilcoxon matched-pairs signed-ranks test (SPSS-X)
NPAR1WAY WILCOXON	NPAR MANN-WHITNEY	Mann-Whitney U test
NPAR1WAY WILCOXON	NPAR K-W	Kruskal-Wallis test
FREQ CHISQ	NPAR CHISQUARE	Chi-square one-sample test ϕ and Cramer's V
	NPAR K-S	Kolmogorov-Smirnov one and two-sample tests
CORR SPEARMAN	NONPAR CORR	Spearman's rank correlation coefficient

Some of these procedures are demonstrated employing exercises from Chapter 9. The third exercise mandates that a Wilcoxon signed-ranks test be performed on seven years of monthly stream runoff in February and March. In the SPSS-X program, seven lines of data are entered, with two variables, FEB and MAR. For this test, the procedure statement NPAR TESTS appears before the BEGIN DATA statement. The statement program appears as follows:

```
TITLE 'WILCOXON MATCHED PAIRS TEST OF RUNOFF DATA'
DATA LIST / FEB 1-3 MAR 4-6
NPAR TESTS WILCOXON = FEB MAR
BEGIN DATA
2.211.17
1.951.12
1.465.48
1.810.79
1.164.66
0.984.25
3.892.91
END DATA
```

Recall that, unless the term FREE appears in the DATA LIST statement, SPSS-X will expect to find the column specifications for each data field.

WILCOXON computes the differences between FEB and MAR, ranks the absolute differences, sums the positive and negative ranks, and computes the test statistic Z from the positive and negative rank sums. Under the null hypothesis, z is approximately normally distributed.

```
              WILCOXON MATCHED PAIRS   TEST OF RUNOFF DATA

- - - - - Wilcoxon Matched-Pairs Signed-Ranks Test

       FEB
with MAR

   Mean Rank      Cases
       2.50         4    - Ranks (MAR LT FEB)
       6.00         3    + Ranks (MAR GT FEB)
                    0      Ties  (MAR EQ FEB)
                   --
                    7      Total

        Z =   -.6761               2-Tailed P =   .4990
```

Using the procedure described in Chapter 9, a value of $W = 23.2$ is obtained, while the sum of signed ranks = 16. Thus, the null hypothesis is safely accepted. The printout from SPSS-X shows only a value of z and a two-tailed probability value. If Appendix Table 2 is consulted, it may be seen that a z value of .67, multiplied by 2, yields $P = .499$. If our level of significance, $\alpha = .05$, then we require $z = 1.96$ to reject the null hypothesis.

Thus, we may conclude that a value equaling a sum of signed ranks of 16 when $n = 7$ could arise by chance about 25 percent of the time ($.499 \div 2$). We require that chance percentage to be no more than 5 in order to reject H_0.

Exercise 4 specifies a Mann-Whitney test of the difference between peak flows for two streams. The list of statements from both SAS and SPSS-X are shown side-by-side below:

```
TITLE 'MANN-WHITNEY TEST ON        TITLE 'MANN-WHITNEY TEST ON
   STREAMFLOW DATA';                  STREAMFLOW DATA'
DATA STREAM;                        DATA LIST / GROUP 1 CFS 2-4
  INPUT GROUP CFS;                  NPAR TESTS M-W = CFS BY GROUP
                                      (1,2)
  CARDS;                            BEGIN DATA
1  98                               1 98
1  198                              1198
1  154                              .
.                                   .
.                                   2 69
2  69                               END DATA
;
PROC NPAR1WAY WILCOXON;
  CLASS GROUP;
  VAR CFS;
```

In SPSS-X, subcommand M-W performs the Mann-Whitney test. The cases are ranked in order of increasing size and the test statistic U is computed. If the samples are from the same population, the distribution of scores from the two groups in the ranked list should be random. An extreme value of U indicates a nonrandom pattern. The U value in the stream flow experiment is 43.5. Appendix Table 8 shows that for $n_1 = 10$ and $n_2 = 11$, the critical value is 26. The value of U must exceed the critical value to reject the null hypothesis. SPSS-X shows a two-tailed P of .4262, which we interpret to mean that, with these sample sizes, a value of $U = 43.5$ could arise by chance about 43 percent of the time. The probability value would need to be $\leq .05$ in order to reject the null hypothesis.

```
              MANN-WHITNEY TEST OF STREAM FLOW DATA


    - - - - - Mann-Whitney U - Wilcoxon Rank Sum W Test

        CFS
     by GROUP

       Mean Rank      Cases

          12.15         10    GROUP = 1
           9.95         11    GROUP = 2
                        --
                        21    Total

                                    Exact              Corrected for ties
            U              W       2-Tailed P           Z         2-Tailed P
           43.5          121.5       .4262           -.8101          .4179
```

The SAS printout shows the sums of the ranks ("scores") of the two independent samples. Then, a *P* value for *z* is shown. This is the probability that *S* falls outside the range defined as

$$\frac{n_1(n_1 + n_2 + 1)}{2} \pm z_{\alpha/2} \sqrt{\frac{n_1 n_2 (n_1 + n_2 + 1)}{12}}$$

Substituting for n_1 and n_2, we obtain a range for *S* of 82.2 to 137.83. The probability that *S* would fall outside that range is shown as .4384. Thus the two program systems calculate the test values somewhat differently, but the decision on H_0 is identical in either case.

```
                MANN-WHITNEY TEST ON STREAMFLOW DATA
                    N P A R 1 W A Y   P R O C E D U R E
              Wilcoxon Scores (Rank Sums) for Variable CFS
                       Classified by Variable GROUP

                          Sum of              Expected           Std Dev              Mean
   GROUP        N         Scores              Under H0           Under H0             Score
     1         10       121.500000             110.0            14.1963275         12.1500000
     2         11       109.500000             121.0            14.1963275          9.9545455
                    Average Scores were used for Ties
       Wilcoxon 2-Sample Test (Normal Approximation)
       (with Continuity Correction of .5)

         S=   121.500                   Z= 0.774848          Prob > |Z| =    0.4384

       T-Test approx. Significance =       0.4475

       Kruskal-Wallis Test (Chi-Square Approximation)
       CHISQ= 0.65621                    DF=  1             Prob > CHISQ=    0.4179
```

Exercise 5 specifies a Kruskal-Wallis test for the three independent samples of infant mortality rates. SAS again uses the WILCOXON subcommand, so it would be somewhat redundant to demonstrate that version of the test. The listing for an SPSS-X program would appear as follows:

```
TITLE 'KRUSKAL-WALLIS TEST OF INFANT MORTALITY RATES'
DATA LIST / GROUP 1 IMR 2-4
NPAR TESTS K-W = IMR BY GROUP (1,3)
BEGIN DATA
1 103
.
.
2  50
.
.
3  11
END DATA
```

The subcommand NPAR TESTS K-W is used, and the ranks (IMR) are analyzed by GROUP, which ranges from 1 to 3, as specified in parentheses. K-W tests whether all *k* samples are from the same population. The printout shows the mean ranks for the three samples. The *H* statistic has approximately a chi-square distribution, so a value of chi-square is shown along with its significance. The significance value of .0034 for the infant mortality problem may be interpreted as the probability that a value of χ^2 as large as 11.375 could have arisen by chance alone. To accept H_0, this probability must be *at least* .05. We can only conclude that the ranks of the three independent samples did not come from the same population of IMR rates.

```
    8-Feb-92      KRUSKAL-WALLIS TEST OF INFANT MORTALITY RATES
    12:51:05      SPSS-X 4.0 VAX/VMS UMBC           on UMBC2::                VMS V5.3

    - - - - - Kruskal-Wallis 1-Way Anova

            IMR
      by  GROUP

          Mean Rank        Cases

             12.50              5       GROUP = 1
              8.50              5       GROUP = 2
              3.00              5       GROUP = 3
                               --
                               15       Total

                                                             Corrected for ties
           Cases     Chi-Square    Significance       Chi-Square    Significance
             15        11.3750           .0034          11.4363           .0033
```

Chi-square goodness-of-fit tests are a bit more difficult to set up in SAS and SPSS-X. Unless you have a fairly complicated contingency table, or many such tables, to analyze, it is about as easy to compute the value of χ^2 with your pocket calculator. We will demonstrate the use of the FREQUENCY procedure in SAS to provide a chi-square statistic. The home owners distribution of Exercise 7 provides a simple example. The SAS listing follows:

```
DATA SURVEY;
TITLE 'CHI SQUARE ANALYSIS OF HOME PURCHASE DATA';
INPUT NEIGH $ CLASS $ NUM @@;
CARDS;
A L 78  A M 71  A H 100
B L 90  B M 42  B H 58
C L 41  C M 96  C H 57
D L 53  D M 41  D H 60
;
PROC FREQ;
  WEIGHT NUM;
  TABLES NEIGH*CLASS / CHISQ;
```

The @@ in the INPUT statement is called a *double trailing at-sign*. Its function is to tell SAS that each data input line contains values for several observations. You can then enter each neighborhood's data by social class identifier on a single line. The "contingencies" of this table are neighborhood (A, B, C, and D) and status (L, M, and H)—all alphabetic constants, which are so identified by the $ following NEIGH and CLASS in the INPUT statement. The frequencies for each cell in the table have arbitrarily been named NUM. This is the so called *weight* variable. The subcommand TABLES identifies the contingency variables, and a chi-square analysis is ordered following the / in that statement.

The printout shows degrees of freedom $(r - 1) \times (c - 1)$ and a χ^2 value of 52.566. Appendix Table 6 shows that when $v = 6$, the critical value of χ^2 is about 12.6. The SAS printout informs us that the probability that a value of 52.566 could arise by chance alone is 0.000. To accept H_0, this probability would have to have been at least .05. Both the Phi coefficient and Cramer's V inform us that the computed statistic is relatively weak.

```
            CHI SQUARE ANALYSIS OF HOME PURCHASE DATA
                    TABLE OF NEIGH BY CLASS

        NEIGH         CLASS

        Frequency|
        Percent  |
        Row Pct  |
        Col Pct  |H        |L        |M        |  Total
        ---------+---------+---------+---------+
        A        |    100  |     78  |     71  |    249
                 |  12.71  |   9.91  |   9.02  |  31.64
                 |  40.16  |  31.33  |  28.51  |
                 |  36.36  |  29.77  |  28.40  |
        ---------+---------+---------+---------+
        B        |     58  |     90  |     42  |    190
                 |   7.37  |  11.44  |   5.34  |  24.14
                 |  30.53  |  47.37  |  22.11  |
                 |  21.09  |  34.35  |  16.80  |
        ---------+---------+---------+---------+
        C        |     57  |     41  |     96  |    194
                 |   7.24  |   5.21  |  12.20  |  24.65
                 |  29.38  |  21.13  |  49.48  |
                 |  20.73  |  15.65  |  38.40  |
        ---------+---------+---------+---------+
        D        |     60  |     53  |     41  |    154
                 |   7.62  |   6.73  |   5.21  |  19.57
                 |  38.96  |  34.42  |  26.62  |
                 |  21.82  |  20.23  |  16.40  |
        ---------+---------+---------+---------+
        Total         275       262       250       787
                    34.94     33.29     31.77    100.00

            STATISTICS FOR TABLE OF NEIGH BY CLASS

Statistic                         DF      Value         Prob
------------------------------------------------------------
Chi-Square                         6      52.566        0.000
Likelihood Ratio Chi-Square        6      50.536        0.000
Mantel-Haenszel Chi-Square         1       2.219        0.136
Phi Coefficient                            0.258
Contingency Coefficient                    0.250
Cramer's V                                 0.183

Sample Size = 787
```

Kolmogorov-Smirnov one- and two-sample tests are performed by SPSS-X. As an example of a two-sample test, we used the tree species experiment of Table 9.16. The listing appears as follows:

```
TITLE 'KOLMOGOROV-SMIRNOV TWO-SAMPLE TEST OF TREE
SPECIES DATA'
DATA LIST / CLASS 1-2 GROUP 3 FREQ 4-5
WEIGHT BY FREQ
NPAR TESTS K-S = CLASS BY GROUP (1,2)
BEGIN DATA
 11 1
 1281
 21 3
 2234
 31 9
 3218
 .
 161 1
 162 0
END DATA
```

We arbitrarily name the diameter class CLASS and the tree type GROUP, with the frequencies of the species called FREQ. The contingency variables are CLASS and GROUP, so the cells contain the *weight* variable, FREQ. The subcommand NPAR TESTS K-S calls for an analysis of this table. The printout, below, shows $n_1 = 87$, and $n_2 = 148$. Appendix Table 11 provides for the computation of the critical value of $D_{\alpha = .05}$ as

$$D = 1.36 \sqrt{\frac{n_1 + n_2}{n_1 n_2}}$$

With our sample sizes, the critical value of $D = .18$, and since the maximum difference in the cumulative distributions is .749, the null hypothesis is rejected. The two tailed probability is well below .001 that such a difference could have been due solely to chance.

```
    kolmogorov-smirnov two-sample test of tree species data


 - - - - - Kolmogorov - Smirnov 2-Sample Test

      CLASS
   by GROUP

         Cases
           87   GROUP = 1
          148   GROUP = 2
          ---
          235   Total
```

```
       Most extreme differences
 Absolute        Positive        Negative           K-S Z       2-Tailed P
  .74922          .74922          .00000            5.546          .000
```

Single-sample Kolmogorov-Smirnov tests may also be performed in SPSS-X, where a cumulative distribution may be compared with a distribution function of a specified type, uniform, normal, or Poisson. The *SPSS-X User's Guide* gives specific guidelines on setting up the statement list.

Exercise 15 specifies a Spearman rank correlation analysis of employment data. The statement lists for SAS and SPSS-X are shown side by side below:

```
DATA SPEARMAN;                         TITLE 'SPEARMAN RANK CORRELATION
TITLE 'SPEARMAN RANK CORRELATION          ANALYSIS OF EMPLOYMENT DATA'
   ANALYSIS OF EMPLOYMENT DATA';       DATA LIST FREE / Y X
INPUT STATE $ Y X;                     BEGIN DATA
CARDS;                                 8 8
NE  8 8                                7 3
MA  7 3                                9 1
ENC 9 1                                6 6
WNC 6 6                                1 2
SA  1 2                                4 7
ESC 4 7                                5 5
WSC 5 5                                2 9
M   2 9                                3 4
P   3 4                                END DATA
;                                      NONPAR CORR VARIABLES = Y X
PROC FREQ;
   TABLES Y*X;
```

In SAS, the FREQ procedure is used, with the TABLES subcommand. Over a dozen statistics are printed out, including Spearman *r*. SPSS-X prints simply the coefficient, *n*, and the significance of *r*, which is interpreted as the probability that the test value for the coefficient could have occurred by chance. Since .33 is well above .05, we may conclude that the values of X are uncorrelated with the values of Y. As a further test, we know that, under the null hypothesis the test statistic $z = r_S\sqrt{n-1}$ can be used to evaluate $r_S = 0$. In this experiment, we compute $z = .1667\sqrt{8} = .47$. This is not significant at $\alpha = .05$, since it is far below $z = 1.96$.

```
       S P E A R M A N     C O R R E L A T I O N     C O E F F I C I E N T S

                  X                 -.1667
                                  N(     9)
                                  SIG .334
                                      Y
```

Appendix B

B-1 Des Moines, Iowa, Mean Monthly Precipitation 1900–1989

B-2 Percent Female Employment, Maryland Minor Civil Divisions, 1980

B-3 Selected Statistics from the 1980 Census Report "Provisional Estimates of Social, Economic, and Housing Characteristics," PHC 80-S1-1

B-4 AIDS Cases and Annual Rates per 100,000 Population, by State, Reported in 1989 and 1990; and Cumulative Totals, by State and Age Group, through December 1990

B-5 Selected Population Characteristics, United States, 1990

appendix B-1

Des Moines, Iowa, Mean Monthly Precipitation, 1900–1989

OBS	JA	FE	MA	AP	MY	JN	JL	AU	SP	OC	NO	DE	WIN	SPR	SUM	AUT
1	0.20	0.50	3.07	3.82	4.76	4.89	5.15	8.02	3.66	3.08	0.96	0.35	1.05	11.65	18.06	7.70
2	0.11	1.11	3.02	2.26	1.40	2.41	1.72	0.67	2.60	2.14	0.40	1.03	3.15	6.68	4.80	5.14
3	0.91	1.52	3.15	2.55	4.69	7.27	5.95	7.82	5.03	3.70	1.65	1.07	3.20	13.37	21.04	10.38
4	0.00	1.12	1.09	1.64	10.64	3.06	3.62	6.72	1.62	1.32	0.31	1.09	1.41	13.37	13.40	3.25
5	1.22	1.06	1.22	5.48	3.16	2.08	6.94	2.60	1.95	1.64	0.06	2.02	3.46	9.84	11.62	3.51
6	1.08	0.93	2.16	2.96	4.44	5.73	4.53	4.86	3.47	1.54	2.34	1.46	2.69	9.89	15.47	9.45
7	2.07	1.51	1.18	2.48	2.21	3.80	2.67	5.03	2.40	3.18	0.55	1.01	4.39	7.01	15.33	6.71
8	2.07	0.20	1.43	2.69	3.97	4.13	10.20	6.14	0.94	1.70	2.29	2.32	2.81	14.01	19.36	5.22
9	0.46	1.51	1.18	1.13	2.89	7.03	0.56	6.03	2.06	1.68	1.12	3.03	4.83	10.94	11.36	5.57
10	1.72	0.20	0.33	5.13	2.12	0.75	4.41	1.82	2.40	2.89	3.71	3.08	2.12	7.81	11.51	8.66
11	2.63	1.86	1.53	2.44	3.26	3.11	0.86	3.52	0.94	2.61	1.12	1.46	2.69	7.24	11.51	6.35
12	0.84	0.65	0.20	2.75	4.83	2.60	1.16	3.44	3.82	2.75	3.53	1.01	2.93	11.50	6.82	11.06
13	1.05	3.20	2.87	3.41	5.06	3.52	3.07	1.77	4.20	3.57	3.67	2.32	6.09	10.73	9.19	11.06
14	1.96	0.51	1.60	1.36	4.87	3.94	3.05	1.71	14.81	3.75	1.24	3.18	3.37	6.91	8.01	16.18
15	0.78	1.45	2.30	5.52	8.21	8.16	1.22	2.82	4.51	2.57	0.35	1.28	3.92	17.97	14.70	5.29
16	1.02	3.20	0.29	5.81	2.96	0.52	9.39	2.62	1.99	2.44	1.36	0.65	5.81	11.93	10.36	6.82
17	0.08	0.42	3.03	4.09	3.14	3.60	1.50	5.54	4.20	2.74	1.11	0.88	3.58	11.15	12.23	7.96
18	0.95	0.90	3.02	3.41	3.87	4.95	1.58	4.79	14.81	3.22	1.51	0.30	4.01	8.41	11.07	18.73
19	2.11	0.02	1.60	4.56	2.96	8.30	2.68	6.63	0.91	0.82	2.84	1.28	4.29	10.88	15.39	6.82
20	2.41	0.36	2.30	4.52	3.94	4.63	6.20	1.95	6.20	2.09	2.10	0.65	4.25	10.88	15.07	8.95
21	0.57	0.50	3.03	4.34	1.26	9.30	7.13	6.03	4.20	1.89	3.63	0.88	4.60	5.14	11.39	12.87
22	3.06	0.82	0.88	4.78	4.26	6.40	0.78	4.15	14.70	2.16	2.74	0.77	4.35	10.65	14.43	4.77
23	1.57	0.92	2.55	0.82	7.82	6.69	2.21	2.95	4.00	3.41	2.74	0.82	2.40	4.25	12.75	7.29
24	1.86	0.36	2.71	4.95	2.11	6.62	2.66	4.79	4.78	3.77	3.84	0.44	2.53	16.29	11.06	13.41
25	1.85	0.82	1.52	1.16	4.38	6.14	2.49	7.16	3.47	3.73	1.63	0.25	5.10	8.67	13.78	6.25
26	1.22	1.06	1.12	4.26	2.71	2.69	7.13	2.54	1.53	2.16	0.53	0.38	4.27	5.25	15.07	9.28
27	0.83	2.33	1.57	2.35	4.18	2.55	7.11	3.08	1.87	2.74	4.08	0.54	4.35	5.87	20.05	9.57
28	1.06	1.03	1.67	1.28	3.99	8.37	4.45	4.30	3.19	3.65	4.08	3.41	3.24	5.77	8.96	9.22
29	0.37	1.06	1.99	1.46	2.99	6.88	3.05	2.93	3.03	3.11	4.87	1.52	3.69	3.02	13.42	15.08
30	1.10	2.44	2.21	1.26	2.13	2.55	7.41	7.18	2.64	3.67	4.08	3.87	4.49	4.35	14.56	5.77
31	2.58	0.50	2.35	4.23	6.92	6.28	1.29	4.80	2.53	0.61	2.68	2.84	5.63	5.77	11.07	5.87
32	0.90	0.36	0.88	3.18	6.74	8.20	1.80	3.30	2.40	3.11	2.17	2.25	4.27	12.66	12.65	11.95
33	1.16	0.50	1.57	1.46	3.05	6.02	1.86	4.58	0.58	0.38	1.73	1.70	5.63	7.20	17.14	11.78
34	3.03	0.09	1.07	3.46	2.05	3.81	6.35	5.34	3.03	2.50	1.18	0.74	4.33	7.09	10.14	10.19
35	1.16	0.72	1.19	4.34	2.05	3.95	0.85	2.25	2.40	0.61	1.75	2.47	3.87	6.93	18.64	3.44
36	1.69	0.43	0.83	2.01	1.45	1.48	7.28	7.10	3.58	0.38	1.03	2.17	4.09	4.53	7.25	2.83
37	1.57	1.29	1.00	1.33	1.79	4.66	3.11	2.93	3.03	1.16	1.85	1.73	4.49	4.23	17.20	16.95
38	1.86	0.81	1.19	2.18	2.13	4.66	7.45	2.72	2.64	1.30	0.46	1.58	5.63	14.30	10.14	7.26
39	0.85	0.46	1.72	2.27	2.13	4.00	0.86	2.53	2.51	1.75	1.18	1.03	3.17	4.53	12.66	4.24
40	1.22	2.33	1.57	3.18	1.57	8.28	4.17	7.18	3.03	6.96	0.85	2.84	3.88	16.80	18.65	7.10
41	0.03	0.30	1.10	1.46	7.13	6.02	4.11	4.80	2.40	0.31	1.46	2.73	4.81	16.92	14.46	8.85
42	1.65	0.43	1.57	3.26	5.21	3.45	2.55	2.58	3.58	4.46	1.85	1.65	2.99	14.03	8.65	6.43
43	0.22	1.29	2.25	4.23	1.21	5.74	2.81	1.97	3.45	6.96	2.20	0.80	2.37	14.08	16.21	4.22
44	1.63	0.28	1.67	1.63	1.13	8.45	3.03	4.58	1.61	1.55	0.46	1.82	4.87	16.92	15.40	1.84
45	0.37	2.42	4.55	5.26	3.05	4.86	0.86	1.54	1.03	0.38	3.55	0.27	4.91	14.03	10.80	5.84
46	1.67	2.06	2.04	1.32	4.91	4.86	1.86	1.90	6.22	2.64	1.81	0.82	4.30	4.08	16.40	10.16
47	2.58	0.90	1.26	1.13	5.60	3.45	6.35	2.04	1.39	2.03	1.81	0.27	4.91	14.03	10.80	5.84
48	0.10	1.21	2.04	1.63	5.56	4.86	2.81	2.74	0.53	0.03	4.60	0.24	4.30	4.89	15.40	5.16
49	0.01	0.25	4.06	5.57	5.56	4.86	2.93	4.74	0.53	2.64	1.81	0.93	3.88	9.53	9.80	5.16
50	2.22	2.42	0.79	1.13	4.91	5.74	2.57	5.70	1.39	2.03	4.60	1.15	2.95	14.62	13.06	5.16
51	1.40	2.23	0.86	1.52	4.91	5.74	2.57	5.70	0.66	0.13	1.22	0.97	3.73	7.81	14.45	2.01
52	1.03	2.99	4.03	1.52	4.91	4.03	1.71	0.71	0.66	0.13	1.22	0.97	3.73	7.81	6.45	2.01
53	1.40	2.23	4.03	1.52	5.56	4.03	1.71	0.71	0.66	0.13	1.22	0.97	2.95	7.81	6.45	2.01
54	1.59	0.52	4.03	1.52	4.91	5.74	2.57	5.70	0.53	2.64	4.60	1.15	2.95	14.62	13.06	5.16
55	0.76	2.00	2.37	3.22	2.22	4.03	1.71	0.71	0.66	0.13	1.22	0.97	3.73	7.81	6.45	2.01

322

OBS	JA	FE	MA	AP	MY	JN	JL	AU	SP	OC	NO	DE	WIN	SPR	SUM	AUT
56	0.07	1.350	1.86	3.48	5.90	5.30	1.01	10.47	1.98	3.95	0.06	0.78	2.200	11.24	16.78	5.99
57	1.35	1.520	0.89	3.63	2.93	1.71	3.69	11.77	4.42	0.52	0.11	0.44	3.310	6.45	17.17	5.05
58	0.51	0.430	0.51	1.24	1.71	1.29	2.76	3.47	1.21	1.66	1.79	1.33	1.430	3.46	7.52	4.66
59	0.79	0.160	1.74	2.84	1.36	3.34	2.47	2.93	1.79	1.29	1.94	1.25	2.250	3.94	9.07	6.13
60	0.71	0.610	3.61	2.31	3.59	3.72	10.51	2.78	4.06	3.07	2.00	1.91	11.570	14.08	15.63	7.03
61	0.00	1.070	0.94	3.51	6.63	5.46	1.63	2.44	3.13	2.64	2.91	0.45	3.770	11.39	9.78	8.68
62	4.38	2.680	1.94	2.40	7.53	2.92	1.99	6.78	2.54	2.11	0.52	1.82	6.410	14.95	12.23	5.17
63	0.33	0.00	5.37	2.39	1.58	3.46	7.05	2.77	10.19	3.00	2.78	1.05	4.830	11.34	12.74	15.97
64	0.00	1.580	1.42	2.51	1.90	1.49	2.94	1.60	10.93	3.22	0.36	1.82	4.630	9.95	12.74	15.95
65	0.54	0.00	2.39	4.18	6.02	2.96	4.01	4.83	1.83	2.56	0.80	0.50	2.090	9.34	8.50	5.75
66	0.83	0.720	1.25	4.29	3.90	6.22	2.03	3.93	4.28	0.29	0.36	0.87	1.660	10.51	8.97	5.52
67	0.51	0.00	1.56	2.39	3.94	1.13	3.43	1.29	7.25	0.79	1.95	0.64	4.400	11.09	13.80	9.81
68	1.62	1.140	1.64	2.51	2.22	4.80	1.37	2.09	2.54	1.47	0.47	0.49	1.700	10.67	8.67	1.74
69	0.00	0.250	0.93	4.18	2.62	4.22	1.93	4.05	2.63	1.93	0.03	0.73	2.620	6.74	9.74	4.94
70	0.77	0.00	1.28	1.74	4.21	5.29	2.94	4.95	6.51	2.36	1.29	1.01	3.990	7.66	11.49	6.37
71	1.26	0.280	3.41	4.46	3.07	3.42	4.01	1.83	2.19	1.47	0.70	1.83	1.780	10.77	9.36	6.00
72	0.77	0.240	1.05	2.28	1.41	2.45	3.48	6.37	2.63	1.93	1.90	1.12	2.620	5.82	8.30	13.17
73	1.01	0.970	1.56	4.16	2.84	4.31	4.96	2.09	6.51	2.36	1.29	2.65	1.330	7.66	8.09	8.99
74	1.44	0.360	0.41	2.54	2.29	2.58	4.34	4.05	5.45	5.20	3.29	1.83	5.030	10.62	11.95	11.82
75	2.09	0.520	3.28	3.28	4.94	2.04	1.96	1.95	5.07	1.47	1.49	2.65	6.950	13.49	9.36	7.24
76	1.23	0.970	1.28	3.56	2.46	1.98	5.17	4.83	1.70	1.23	1.20	1.12	6.180	17.83	15.75	4.53
77	1.41	0.240	1.05	4.67	2.89	7.25	0.87	1.83	1.00	0.63	2.00	2.83	2.780	13.64	12.58	2.21
78	0.50	0.480	1.90	6.11	3.41	2.56	1.63	6.37	2.82	1.10	0.69	0.48	3.370	8.96	11.38	10.69
79	0.28	0.360	1.05	2.65	2.84	5.60	2.95	1.07	6.39	1.14	3.16	0.12	2.670	8.96	15.75	10.63
80	1.80	1.270	3.57	2.76	2.49	5.02	1.52	7.25	0.97	1.23	0.45	1.37	2.920	3.96	11.56	6.99
81	0.25	0.520	4.15	4.57	2.94	5.56	1.96	2.68	1.30	1.06	2.63	1.00	2.440	4.85	17.56	15.93
82	1.80	0.097	4.23	3.86	2.46	2.74	5.76	3.07	2.30	3.26	1.65	1.14	3.440	14.12	15.79	10.82
83	2.63	1.950	1.19	2.03	5.93	1.56	7.00	1.07	3.87	1.96	2.52	1.43	1.487	11.45	13.81	11.29
84	2.11	0.980	0.39	5.85	5.58	5.02	2.44	5.01	2.76	2.06	1.16	1.57	3.480	13.08	13.32	7.24
85	0.64	0.640	4.30	5.23	5.23	2.65	2.22	0.83	5.28	5.54	1.65	2.41	4.220	5.16	17.84	4.53
86	0.12	1.760	1.37	5.66	4.35	7.08	2.04	2.24	6.75	3.28	2.63	1.98	1.487	14.45	14.28	2.61
87	0.64	0.097	0.57	5.23	5.58	3.72	2.22	0.83	5.28	5.54	1.65	2.41	4.220	5.16	17.84	10.63
88	1.80	1.980	3.72	5.66	1.46	7.08	3.08	4.52	3.87	3.89	1.99	1.98	3.487	11.45	13.32	15.93
89	2.63	1.950	1.19	5.23	1.46	3.72	4.08	10.04	6.40	1.03	1.65	2.41	3.920	12.66	15.50	10.82
90	0.12	1.380	2.99	2.92	3.75	2.10	5.08	4.52	1.40	1.03	3.27	1.98	2.860	12.93	15.22	11.29
91	0.37	0.590	0.66	0.75	1.46	2.75	4.78	3.05	2.89	0.59	3.38	0.84	1.800	2.87	10.58	6.86

Percent Female Employment, Maryland Minor Civil Divisions, 1980

appendix **B-2**

OBS	CC	VD	PFEM	OBS	CC	VD	PFEM	OBS	CC	VD	PFEM	OBS	CC	VD	PFEM
1	1	1	30.7420	82	15	5	34.1286	164	27	2	38.2933	245	41	5	33.8362
2	1	2	32.9140	83	15	6	39.3848	165	27	3	39.3648	246	43	1	34.1828
3	1	3	19.2825	84	15	7	37.4923	166	27	4	38.1925	247	43	2	37.7340
4	1	4	37.1804	85	15	8	34.3521	167	27	5	43.4833	248	43	3	36.9057
5	1	5	36.9587	86	15	9	35.6303	168	27	6	42.3668	249	43	4	31.9807
6	1	6	36.7757	87	17	1	39.5335	169	29	1	33.0071	250	43	5	35.3612
7	1	7	33.2741	88	17	2	41.5644	170	29	2	31.6530	251	43	6	40.8495
8	1	8	35.7947	89	17	3	36.4220	171	29	3	37.6944	252	43	7	35.7511
9	1	9	25.5708	90	17	4	36.6438	172	29	4	47.2237	253	43	8	33.2226
10	1	10	36.5188	91	17	5	37.7526	173	29	5	38.3424	254	43	9	35.0969
11	1	11	25.7471	92	17	6	39.0096	174	29	6	43.3148	255	43	10	37.5687
12	1	12	39.1332	93	17	7	38.4064	175	29	7	41.1357	256	43	11	28.5714
13	1	13	30.3128	94	17	8	38.2275	176	31	1	38.3790	257	43	12	38.1533
14	1	14	44.8819	95	17	9	35.8779	177	31	2	37.0773	258	43	13	34.6692
15	1	16	30.0475	96	17	10	42.5802	178	31	3	38.8704	259	43	14	31.2707
16	1	17	38.3621	97	19	1	43.4673	179	31	4	42.4657	260	43	15	35.3470
17	1	18	29.5455	98	19	2	41.0828	180	31	5	42.5049	261	43	16	37.9095
18	1	19	40.0621	99	19	3	36.5422	181	31	6	39.1751	262	43	17	41.9054
19	1	20	38.1727	100	19	4	30.2469	182	31	7	42.5831	263	43	18	39.0235
20	1	21	37.6771	101	19	5	44.1624	183	31	8	39.7424	264	43	19	41.7197
21	1	22	40.5678	102	19	6	30.3630	184	31	9	43.4835	265	43	20	30.0429
22	1	23	44.0850	103	19	7	45.7113	185	31	10	39.7032	266	43	21	41.7808
23	1	24	33.3904	104	19	8	44.4186	186	31	11	42.1101	267	43	22	40.6846
24	1	26	39.4366	105	19	9	43.8066	187	31	12	37.6073	268	43	23	37.6387
25	1	28	41.1150	106	19	10	49.4444	188	31	13	45.0604	269	43	24	36.1809
26	1	29	39.8160	107	19	11	67.5676	189	33	1	42.0418	270	43	25	39.5723
27	1	30	38.4615	108	19	12	32.6683	190	33	2	44.3527	271	43	26	38.8867
28	1	31	26.3291	109	19	13	44.3452	191	33	3	39.8592	272	43	27	38.7654
29	1	34	38.2204	110	19	14	43.0127	192	33	4	42.1353	273	45	1	40.8060
30	1	35	51.5625	111	19	15	42.5000	193	33	5	41.0900	274	45	2	44.7791
31	3	1	39.5963	112	19	16	33.6449	194	33	6	47.1754	275	45	3	44.5055
32	3	2	39.0927	113	19	17	34.7305	195	33	7	41.8191	276	45	4	41.1475
33	3	3	37.9439	114	19	18	58.0000	196	33	8	34.3580	277	45	5	42.3255
34	3	4	37.0322	115	21	1	36.3366	197	33	9	38.7855	278	45	6	48.0349
35	3	5	37.7050	116	21	2	44.5105	198	33	10	43.9789	279	45	7	42.0765
36	3	6	36.5304	117	21	3	37.1067	199	33	11	38.4490	280	45	8	39.8402
37	3	8	39.8368	118	21	4	40.1681	200	33	12	47.1105	281	45	9	44.6297
38	5	1	42.5131	119	21	5	42.1073	201	33	13	47.1435	282	45	10	39.5230
39	5	2	44.4222	120	21	6	35.4520	202	33	14	41.3611	283	45	11	43.3308
40	5	3	42.4332	121	21	7	36.7005	203	33	15	40.9979	284	45	12	36.2683
41	5	4	42.5003	122	21	8	35.1225	204	33	16	44.7585	285	45	13	43.1268
42	5	5	38.1259	123	21	9	38.2329	205	33	17	47.2593	286	45	14	38.0764
43	5	6	38.2001	124	21	10	37.2978	206	33	18	45.5739	287	45	15	39.9348
44	5	7	36.5749	125	21	11	36.2916	207	33	19	42.2333	288	45	16	41.4545
45	5	8	40.0935	126	21	12	43.8326	208	33	20	44.0052	289	47	1	39.4662
46	5	9	44.2356	127	21	13	40.7407	209	33	21	44.6563	290	47	2	39.4354
47	5	10	35.3297	128	21	14	33.3023	210	35	1	32.6263	291	47	3	38.3808
48	5	11	38.9776	129	21	15	41.9675	211	35	2	36.9530	292	47	4	38.8056
49	5	12	37.9700	130	21	16	38.5022	212	35	3	42.4665	293	47	5	37.6398
50	5	13	40.6964	131	21	17	37.2215	213	35	4	37.3865	294	47	6	34.6479
51	5	14	40.1313	132	21	18	39.7541	214	35	5	37.3593	295	47	7	34.1142
52	5	15	37.3677	133	21	19	30.8394	215	35	6	37.5839	296	47	8	37.3119
53	9	1	34.2500	134	21	20	36.6298	216	35	7	36.6290	297	51	1	37.5000
54	9	2	37.1663	135	21	21	36.4283	217	37	1	37.0227	298	51	2	35.7998
55	9	3	38.4197	136	21	22	46.6523	218	37	2	36.5479	299	51	3	40.5442
56	11	1	31.4904	137	21	23	38.9201	219	37	3	37.7765	300	51	4	35.1024
57	11	2	35.1814	138	21	24	38.9960	220	37	4	36.8895	301	51	5	42.2096
58	11	3	38.6104	139	21	25	37.2439	221	37	5	33.3333	302	51	6	36.0981
59	11	4	43.8738	140	21	26	37.4187	222	37	6	33.8920	303	51	7	42.8934
60	11	5	42.5926	141	23	1	25.2269	223	37	7	36.4248	304	51	8	37.1650
61	11	6	42.8773	142	23	2	25.9202	224	37	8	31.8341	305	51	9	44.1040
62	11	7	38.2483	143	23	3	38.9474	225	37	9	44.0994	306	51	10	37.9761
63	11	8	37.9012	144	23	4	27.5964	226	39	1	41.9550	307	51	11	41.4993
64	13	1	40.3430	145	23	5	31.7597	227	39	2	49.3274	308	51	12	39.7520
65	13	2	39.2421	146	23	6	35.2132	228	39	3	31.8820	309	51	13	41.1765
66	13	3	39.5994	147	23	7	40.2795	229	39	4	36.5639	310	51	14	43.4191
67	13	4	38.5649	148	23	8	25.9380	230	39	5	43.2802	311	51	15	46.3286
68	13	5	40.1370	149	23	9	31.4079	231	39	6	34.8837	312	51	16	44.3625
69	13	6	39.2334	150	23	10	31.1334	232	39	7	41.9798	313	51	17	37.0920
70	13	7	43.0469	151	23	11	26.0870	233	39	8	38.6313	314	51	18	36.1314
71	13	8	38.8959	152	23	12	22.1875	234	39	9	41.6357	315	51	19	29.8366
72	13	9	38.3459	153	23	13	29.7398	235	39	10	19.3069	316	51	20	38.3815
73	13	10	41.8816	154	23	14	36.6412	236	39	11	30.9859	317	51	21	31.1255
74	13	11	39.7271	155	23	15	7.7465	237	39	12	30.0189	318	51	22	44.6629
75	13	12	41.3302	156	23	16	38.4286	238	39	13	38.1250	319	51	23	31.3272
76	13	13	39.9520	157	25	1	36.6372	239	39	14	26.5233	320	51	24	30.7073
77	13	14	36.5899	158	25	2	32.2078	240	39	15	47.3077	321	51	25	39.9076
78	15	1	36.6567	159	25	3	38.4907	241	41	1	44.3696	322	51	26	33.3945
79	15	2	37.8534	160	25	4	34.4396	242	41	2	41.1601	323	51	27	42.5623
80	15	3	37.1253	161	25	5	36.6284	243	41	3	42.6875	324	51	28	45.0248
81	15	4	34.1762	162	25	6	39.5398	244	41	4	38.3966				
				163	27	1	39.3175								

CC is County Code, VD is Civil Division Number. (Note: County Codes are identified at end of table.)

Appendix B

COUNTY CODE	NAME	DIVISIONS n	STATISTICS MEAN	STANDARD DEVIATION
01	ALLEGANY	30	35.73	6.59
03	ANNE ARUNDEL	7	38.25	1.29
05	BALTIMORE	15	39.97	2.79
09	CALVERT	3	36.61	2.14
11	CAROLINE	8	38.85	4.22
13	CARROLL	14	39.78	1.60
15	CECIL	9	36.31	1.86
17	CHARLES	10	38.60	2.17
19	DORCHESTER	18	42.54	9.43
21	FREDERICK	26	37.88	3.15
23	GARRETT	16	29.77	8.06
25	HARFORD	6	36.32	2.67
27	HOWARD	6	40.17	2.22
29	KENT	7	38.91	5.52
31	MONTGOMERY	13	40.67	2.49
33	PRINCE GEORGE'S	21	42.77	3.39
35	QUEEN ANNE'S	7	37.29	2.86
37	ST. MARY'S	9	36.42	3.51
39	SOMERSET	15	37.03	7.92
41	TALBOT	5	40.09	4.13
43	WASHINGTON	27	36.69	3.57
45	WICOMICO	16	42.12	2.89
47	WORCESTER	8	37.10	2.06
51	BALTIMORE CITY	28	38.79	4.74

Statistical Information on Maryland Counties: Percent Females in Labor Force

Selected Statistics from the 1980 Census Report "Provisional Estimates of Social, Economic, and Housing Characteristics," PHC80-S1-1

appendix **B-3**

SMSA	1980 Pop. (in 000s)	% Foreign Born	% Non-English Speakers	% Women in Labor Force	% College Grads	% Families Below Poverty	Median Family Income
Anaheim-Santa Ana-Garden Grove, CA	1,933	12.8	18.2	56.2	21.0	4.9	$25,482
Atlanta, GA	2,030	2.4	3.3	57.1	22.9	9.7	21,409
Baltimore, MD	2,174	3.1	4.8	52.1	16.4	9.0	21,647
Boston, MA	2,763	10.4	12.3	53.0	23.6	8.0	22,580
Buffalo, NY	1,243	5.5	8.8	46.4	13.9	7.6	20,889
Chicago, IL	7,104	10.8	17.0	52.7	15.9	9.5	22,898
Cincinnati, OH-KY-IN	1,401	1.9	2.9	48.8	16.6	8.1	21,492
Cleveland, OH	1,899	5.6	9.8	49.8	15.6	8.2	22,764
Columbus, OH	1,093	2.7	4.2	54.2	20.0	8.2	20,836
Dallas-Ft. Worth, TX	2,975	4.2	9.6	57.5	19.1	7.4	21,869
Denver-Boulder, CO	1,621	4.4	9.8	59.3	26.4	6.2	24,235
Detroit, MI	4,353	6.4	8.2	47.7	14.2	8.8	24,118
Ft. Lauderdale-Hollywood, FL	1,018	12.0	12.2	44.8	15.2	6.1	19,646
Houston, TX	2,905	7.5	17.1	56.1	20.6	8.3	23,959
Indianapolis, IN	1,167	1.6	2.8	54.3	15.8	7.6	21,842
Kansas City, MO-KS	1,327	2.3	4.2	55.8	19.1	6.0	22,715
Los Angeles-Long Beach, CA	7,478	21.6	30.9	53.3	18.9	10.1	21,334
Miami, FL	1,626	35.3	42.8	50.7	16.1	11.3	18,756
Milwaukee, WI	1,397	4.1	8.0	54.3	17.6	7.4	23,736
Minneapolis-St. Paul, MN-WI	2,114	3.4	4.9	60.1	21.8	4.4	24,793
Nassau-Suffolk, NY	2,606	8.6	11.6	49.9	20.8	4.0	26,135
New Orleans, LA	1,187	3.7	7.1	47.2	16.0	14.6	19,028
New York, NY-NJ	9,120	20.8	30.1	47.5	20.7	14.7	19,794
Newark, NJ	1,966	11.8	17.1	51.9	20.2	10.5	23,412
Philadelphia, PA-NJ	4,717	5.7	9.3	47.4	17.4	9.3	21,489
Phoenix, AZ	1,509	5.6	13.8	50.6	18.1	6.2	20,920
Pittsburgh, PA	2,264	3.4	6.5	43.1	15.9	5.7	21,950
Portland, OR-WA	1,243	4.8	6.1	54.8	19.6	6.1	22,102
Riverside-San Bernardino-Ontario, CA	1,558	8.7	17.1	45.8	13.1	8.8	19,707
Sacramento, CA	1,014	7.6	12.5	52.1	19.2	8.9	20,887
St. Louis, MO-IL	2,356	2.1	3.6	51.0	14.9	7.5	21,578
San Antonio, TX	1,072	6.9	42.5	47.9	15.7	14.5	17,473
San Diego, CA	1,862	12.3	17.7	50.7	21.2	8.2	20,259
San Francisco-Oakland, CA	3,251	15.8	20.9	55.5	26.5	6.7	25,055
San Jose, CA	1,295	12.9	20.5	59.1	25.2	5.9	26,924
Seattle, Everett, WA	1,607	7.5	7.5	55.7	24.2	4.8	25,005
Tampa-St. Petersburg, FL	1,569	7.0	8.9	42.3	13.4	8.8	16,515
Washington, DC-MD-VA	3,061	8.3	10.1	62.0	32.5	6.1	27,515

Source: American Demographics (1982) Vol. 4, No. 7.

AIDS Cases and Annual Rates per 100,000 Population, by State, Reported in 1989 and 1990; and Cumulative Totals, by State and Age Group, through December 1990

appendix **B-4**

State of residence	1989 No.	1989 Rate	1990 No.	1990 Rate	Cumulative totals Adults/adolescents	Cumulative totals Children < 13 years old	Total
Alabama	216	5.2	239	5.8	873	21	89
Alaska	17	3.1	25	4.5	97	2	9
Arizona	332	9.3	315	8.6	1,290	8	1,29
Arkansas	79	3.3	208	8.6	445	10	45
California	6,467	22.4	7,346	24.9	30,462	212	30,67
Colorado	387	11.5	364	10.7	1,585	9	1,59
Connecticut	431	13.3	425	13.0	1,805	63	1,86
Delaware	81	12.1	94	13.9	310	4	31
District of Columbia	496	80.7	741	121.1	2,672	40	2,71
Florida	3,479	27.5	4,047	31.2	13,607	386	13,99
Georgia	1,102	17.1	1,223	18.6	4,226	44	4,27
Hawaii	180	16.1	156	13.8	627	2	62
Idaho	23	2.3	28	2.8	77	2	7
Illinois	1,139	9.8	1,278	11.0	4,661	67	4,72
Indiana	397	7.1	282	5.1	1,007	10	1,01
Iowa	56	2.0	69	2.5	229	3	23
Kansas	114	4.5	137	5.4	442	3	44
Kentucky	114	3.1	189	5.0	500	5	50
Louisiana	508	11.5	703	15.8	2,251	39	2,29
Maine	66	5.4	67	5.5	219	2	22
Maryland	717	15.3	1,002	21.2	3,075	68	3,14
Massachusetts	755	12.8	844	14.2	3,270	72	3,34
Michigan	506	5.5	577	6.2	1,973	32	2,00
Minnesota	176	4.1	204	4.7	823	8	83
Mississippi	165	6.3	279	10.6	650	10	66
Missouri	442	8.6	583	11.2	1,821	12	1,83
Montana	13	1.6	17	2.1	55	1	5
Nebraska	33	2.1	58	3.6	184	2	18
Nevada	181	16.7	191	17.1	626	7	63
New Hampshire	37	3.3	66	5.9	190	5	19
New Jersey	2,230	28.7	2,464	31.5	10,091	280	10,37
New Mexico	94	6.1	109	7.0	346	3	34
New York	6,010	33.5	8,399	46.7	33,694	802	34,49
North Carolina	447	6.8	558	8.4	1,628	36	1,66
North Dakota	8	1.2	2	0.3	20	—	2
Ohio	486	4.5	660	6.1	2,259	40	2,29
Oklahoma	167	5.1	203	6.2	699	11	71
Oregon	228	8.2	335	12.0	1,014	5	1,01
Pennsylvania	1,073	8.9	1,197	9.9	4,362	75	4,43
Rhode Island	88	8.8	88	8.8	372	10	38
South Carolina	331	9.4	342	9.6	1,023	21	1,04
South Dakota	4	0.6	9	1.3	25	—	2
Tennessee	266	5.4	342	6.9	1,092	15	1,10
Texas	2,397	14.0	3,361	19.2	11,320	111	11,43
Utah	74	4.3	98	5.6	331	8	33
Vermont	20	3.6	22	3.9	77	1	7
Virginia	390	6.4	738	11.9	2,018	35	2,05
Washington	499	10.6	637	13.3	2,147	12	2,15
West Virginia	55	2.9	62	3.3	177	2	17
Wisconsin	130	2.7	209	4.3	623	4	62
Wyoming	16	3.3	3	0.6	34	—	3
U.S. total	33,722	13.6	41,595	16.6	153,404	2,620	156,02

Source: Centers for Disease Control (1991), p. 6.

Selected Population Characteristics, United States, 1990

appendix **B-5**

Region/State	X_1	X_2	X_3	X_4	X_5	X_6	X_7	X_8	X_9
New England	13,207	210	15.0	8.1	8.7	64.2	35,780	7.3	
Maine	1,228	40	13.2	7.9	1.1	73.4	27,863	10.4	211,817
New Hampshire	1,109	123	15.8	8.3	2.4	65.5	37,500	7.8	166,045
Vermont	563	61	15.3	6.8	--	72.6	31,592	--	92,755
Massachusetts	6,016	769	15.6	7.9	9.7	58.9	36,000	9.1	825,320
Rhode Island	1,003	951	14.6	8.2	8.2	59.5	30,000	--	134,061
Connecticut	3,287	675	14.5	8.9	13.5	70.1	42,362	2.9	465,465
Middle Atlantic	37,602	377	14.8	10.3	22.8	61.7	31,730	11.0	
New York	17,990	380	15.5	10.8	29.4	52.9	31,200	12.6	2,594,070
New Jersey	7,730	1,035	14.8	9.9	25.1	64.2	39,012	8.3	1,076,005
Pennsylvania	11,882	265	13.9	9.9	11.7	72.8	28,484	10.4	1,655,271
East North Central	42,009	172	15.1	10.5	16.2	67.6	29,410	12.0	
Ohio	10,847	265	15.2	9.7	12.0	70.2	29,046	10.7	1,793,411
Indiana	5,544	154	14.7	11.0	10.6	65.5	25,600	13.7	964,129
Illinois	11,431	205	15.5	11.3	26.1	62.6	31,255	12.8	1,811,446
Michigan	9,295	163	15.2	11.1	17.0	72.7	30,910	13.2	1,606,344
Wisconsin	4,892	90	14.4	8.4	6.8	67.0	29,050	8.5	772,363
West North Central	17,660	35	15.0	8.9	9.2	66.6	27,035	11.7	
Minnesota	4,375	55	15.5	7.8	6.5	67.8	30,134	11.2	34,000
Iowa	2,777	50	13.6	8.7	3.4	71.1	26,168	10.4	480,826
Missouri	5,117	74	14.8	10.1	14.4	63.0	26,155	12.7	802,060
North Dakota	639	9	17.1	10.5	--	68.6	25,075	12.3	119,004
South Dakota	696	9	15.8	10.1	10.1	65.2	24,086	13.3	126,817
Nebraska	1,578	21	15.2	9.0	7.3	67.0	26,048	12.8	268,100
Kansas	2,478	30	15.2	8.0	11.9	66.4	26,800	10.8	438,500
South Atlantic	43,566	163	15.4	11.6	28.1	67.0	27,900	12.8	
Delaware	666	345	16.5	11.8	24.5	66.2	32,254	--	95,659
Maryland	4,781	486	14.8	11.3	28.5	62.2	36,396	9.0	683,797
Dist. Columbia	607	9635	31.3	23.2	74.7	38.2	26,350	18.0	86,435
Virginia	6,187	156	15.0	10.4	24.1	70.9	34,066	10.9	979,417
West Virginia	1,793	74	12.0	9.0	4.6	70.9	21,535	15.8	344,236
North Carolina	6,629	136	15.1	12.5	24.0	70.1	26,042	12.3	1,085,976
South Carolina	3,487	115	15.4	12.3	32.2	71.2	23,820	17.1	614,921
Georgia	6,478	112	16.9	12.6	34.7	65.7	27,552	15.0	1,110,947

Appendix B

	X_1	X_2	X_3	X_4	X_5	X_6	X_7	X_8	X_9
Florida	12,938	239	15.0	10.6	29.1	65.8	26,000	12.5	1,664,774
East South Central	15,176	85	15.1	11.4	22.4	68.3	21,900	18.6	
Kentucky	3,685	93	13.7	10.7	7.9	68.4	23,512	16.1	642,696
Tennessee	4,877	119	16.2	10.8	21.5	69.1	22,706	18.4	823,783
Alabama	4,041	80	14.5	12.1	26.6	66.2	21,109	19.1	808,000
Mississippi	2,573	54	15.7	12.3	37.6	69.8	19,774	22.0	505,550
West South Central	26,704	62	17.2	9.4	33.8	64.7	24,666	18.0	
Arkansas	2,351	45	14.4	10.7	15.8	69.3	21,101	18.3	437,036
Louisiana	4,220	95	17.1	11.0	30.1	68.1	23,000	23.4	793,093
Oklahoma	3,146	46	14.5	9.0	17.2	72.4	24,132	14.8	584,212
Texas	16,987	65	18.2	9.0	40.3	61.7	25,701	17.2	3,236,787
Mountain	13,659	16	17.7	9.2	20.3	63.9	27,225	13.1	
Montana	799	5	14.1	8.7	--	68.8	23,287	15.8	152,207
Idaho	1,007	12	15.5	8.8	9.4	68.0	24,550	12.5	212,444
Wyoming	454	5	14.0	8.9	--	71.3	29,000	--	98,455
Colorado	3,294	32	16.1	9.6	20.0	56.1	26,996	12.1	560,236
New Mexico	1,515	12	18.2	10.0	43.9	67.6	22,500	19.6	287,229
Arizona	3,665	32	18.8	9.7	24.9	65.6	28,300	14.1	660,259
Utah	1,723	21	22.0	8.0	7.5	75.1	30,500	8.3	423,386
Nevada	1,202	11	17.6	8.4	17.8	54.9	29,100	10.8	168,353
Pacific	39,127	44	17.5	8.6	36.1	55.8	32,484	12.4	
Washington	4,867	73	14.7	9.0	10.1	61.2	32,030	9.6	775,755
Oregon	2,842	30	14.9	8.6	6.8	62.0	28,365	11.2	455,895
California	29,760	190	18.2	8.6	42.0	54.1	33,002	--	4,489,322
Alaska	550	1	21.1	11.6	23.5	56.8	36,029	--	105,678
Hawaii	1,108	172	17.4	7.2	73.9	57.6	35,000	11.4	169,038

Sources: Population Reference Bureau (1991); U. S. Bureau of The Census (1990), Table 83; and Centers for Disease Control (1991), p. 37.

Variables (All are as of 1990, unless otherwise indicated):

X_1: Population (thousands)
X_2: Persons per square mile of land Area
X_3: Live birth rate, per 1,000 population, 1988
X_4: Infant Mortality Rate, per 1,000 live births, 1988
X_5: Estimated percent minority population
X_6: Percent of households owning their home
X_7: Median household income
X_8: Percent of population below poverty
X_9: Number of primary and secondary students

Answers to Selected Exercises

appendix **C**

Chapter 2

5. Fahrenheit temperatures 50, 60, 70, 80, 90, and 100 become Celsius temperatures 10, 15.6, 21.1, 26.7, 32.2, and 37.8.

6. Primary and secondary student rates per 1,000 population for Michigan and Virginia are 172.8 and 192.0, respectively.

8. The distance between control points 1 and 4 is computed as

$$c = \sqrt{(X_1 - X_2)^2 + (Y_1 - Y_2)^2}$$
$$= \sqrt{(1 - .8)^2 + (7 - 3.5)^2} = 3.51$$

10. The area of Subarea 15 is computed as demonstrated in Box 2.2. Nine line segments were fitted to the shape of the subarea as shown at the right.

Sum of contributions to area of polygon ≈ 7.8.
$A = 3.9$ square miles.

11. The length of the diagonal is traced from outline point 4 to point 8.

$$d = 2\sqrt{\frac{A}{\pi}} = 2\sqrt{\frac{3.9}{3.14}} = 2.2$$

then, $S = 2.2 \div 2.8 = .8$.

Chapter 3

6. A contour map of the hypothetical study area should appear approximately as shown here.

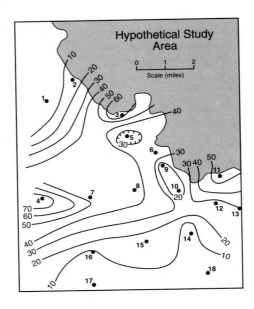

Chapter 4

2. Following the *empirical rule,* about 238 students.

10. For Boston:

$$CV = \frac{s}{\bar{x}} \times 100 = \frac{7.88}{41.6} \cdot 100 = 18.9$$

and for San Diego:

$$CV = \frac{s}{\bar{x}} \times 100 = \frac{4.12}{7.88} \cdot 100 = 52.3$$

11. For suspended particulate, $w\bar{X} = 5.5$, $w\bar{Y} = 6.7$, and $SD_w = 2.22$.
For sulfur dioxide, $w\bar{X} = 5.7$, $w\bar{Y} = 6.8$, and $SD_w = 2.12$.

Chapter 5

1. B.   **2.** B.

4. $(4 \div 52) + (4 \div 52) = .15$; the *addition rule.*

5. $(64{,}167 \div 69{,}524) = .92$.

6. $(5{,}357 \div 69{,}524) = .08$.

Appendix C   **333**

9. $(209 \div 5{,}357) \times (2{,}834 \div 5{,}357) = .57$; the *multiplication rule*.

12. $q^5 + 5q^4p + 10q^3p^2 + 10p^3q^2 + 5p^4q + p^5$.

15. First, compute z for 52 feet:

$$= 52 - 67.56 \div 25.7 = -.61;$$

corresponding proportion (from Appendix Table 2) is .2291; the proportion of buildings above the mean is .50; therefore, the proportion of buildings under 52 feet $= 1.0 - (.5 + .2291) = .2709$.

18. Compute z for 25 percent:

$$= 25 - 58.25 \div 12.55 = -2.65;$$

corresponding proportion (from Appendix Table 2) is .496; 25 percent is close to the left tail of the distribution; $.5 + .496$ percent of the area of the distribution lie above this point, so the probability of a 25 percent change is $1.0 - (.5 + .496) = .004$.

21. Maryland's death rate in 1986 was 29.2. The national mean was 30.9. First compute z score for the difference:

$$z = 29.2 - 30.9 \div 6.74 = -.25$$

Find the probability in Appendix Table 2: .0987. Since Maryland is close to the national mean, this rate is relatively likely to be selected by a random draw. The highest probability in the table (half the curve) would be .5; a rate equal to that of Maryland would then yield a probability of $.5 - .0987 = .4013$.

23. $Sk = 3(\mu - Md) \div \sigma = 3(30.9 - 29.5) \div 6.74 = .62$. The distribution is slightly skewed to the right, since the median value is lower than that of the mean.

Chapter 6

4. A one-tailed test is employed when an alternative hypothesis is specified with a given direction for the outcome. For instance

$$H_1 = X_1 > X_2$$

specifies that one mean is greater than the other.

7. Use Equation 6.6: **a.** 11.   **c.** 2.

9. Use Equation 6.15, solve for μ_1 and μ_2:

$$\overline{X}_1 - \overline{X}_2 = -3.19; \quad t_{\alpha=.01, v=33} = 2.75; \quad \hat{\sigma} = 1.73;$$
$$\mu_1 = -7.94, \quad \mu_2 = 1.57$$

11. Step 1: There was no significant difference between percent black and percent white female labor force in Maryland's urban districts in 1980; $\overline{X}_W = \overline{X}_B$.

Step 2: For a comparison of means from a normally distributed population, use a t test.

Step 3: The conventional level of significance is $\alpha = .05$.

Step 4: Sample sizes are $n_B = 24$, $n_W = 26$, pooled with two degrees of freedom loss, $\nu = 48$; critical value of $t_{\alpha=.05, \nu=48} = |2.01|$.

Step 5: Since s_W and s_B are known, begin with Equation 6.10:

$$SS_W = \nu_W(s_W^2) = 25 \cdot 5.99^2 = 897.00$$
$$SS_B = \nu_B(s_B^2) = 23 \cdot 6.37^2 = 933.27$$

Next, compute pooled variance, using Equation 6.11:

$$s_{1+2}^2 = \frac{SS_W + SS_B}{(n_1 + n_2 - 2)} = \frac{897.0 + 933.27}{48} = 38.13$$

Next, compute the standard error of the difference, using Equation 6.12:

$$\hat{\sigma}_{\overline{X}_W - \overline{X}_B} = \sqrt{s_{1+2}^2 \left(\frac{1}{n_W} + \frac{1}{n_B}\right)} = \sqrt{38.13 \cdot .08} = 1.75$$

Finally, use Equation 6.14 to compute value of t:

$$t = \frac{\overline{X}_W - \overline{X}_B}{\hat{\sigma}_{\overline{X}_W - \overline{X}_B}} = \frac{39.11 - 36.09}{1.75} = 1.73$$

Step 6: Decision: Since the computed absolute value of t (1.73) is less than the critical value (2.01), conclude that we cannot reject the null hypothesis. From Appendix Table 4, we can see that for $\nu \approx 48$, a value of 1.73 has a probability of nearly .10 of occurring by sampling variation alone. For this experiment, we established .05 as our critical probability for chance variation.

18. The null hypothesis is that the AIDS rates did not change significantly between the earlier and the later reported periods. H_0: $\overline{d}(X_1 - X_2) = 0$ (two-tailed test).

Use the paired comparison t test of Equation 6.20. The level of significance is again $\alpha = .05$. The decision will be to accept H_0 if $|t| \leq 2.06$ ($\nu = 24$, $\alpha = .05$).

First compute $\overline{d} = -267.5 \div 25 = -10.7$;

next, compute the variance with Equation 6.17:

$$S_d = \{[25 \cdot 8368.43] - (-267.5)^2\} \div 600 = 229.42;$$

next, compute the standard error with Equation 6.18:

$$\hat{\sigma}_{\overline{d}} = \sqrt{229.42} \div \sqrt{25} = 3.03$$

Then, compute the value of t, using Equation 6.19:

$$t = -10.7 \div 3.03 = -3.53$$

Decision: Since the computed absolute value of t (3.53) is greater than the critical value (2.06), conclude that we cannot accept the null hypothesis. From Appendix Table 4, we can see that for $v \approx 24$, a value of 3.53 has a probability of between .005 and .001 of occurring by sampling variation alone. For this experiment, we established .05 as our critical probability for chance variation.

Chapter 7

1. The null hypothesis states that there is no significant difference between shape ratios of unglaciated, Illinoisan and Wisconsin glaciated basins. $\mu_1 = \mu_2 = \mu_3$.

    Since there are more than two means and the population of basin shapes is normally distributed, a one-way ANOVA is called for. The significance level will again be set to $\alpha = .05$.

    The test statistic is F. The critical value for (numerator =) 2 and (denominator =) 49 degrees of freedom is 3.185.

    First obtain the sums of each subsample:

	Unglaciated	Illinoisan	Wisconsin
$\Sigma X =$	17.53	45.46	5.29

    Next, obtain the sum of the squared observations: $\Sigma X^2 = 90.44$.

    Next, compute the grand sum $= 17.53 + 45.46 + 5.29 = 68.28$, and the grand mean $= 68.28 \div 52 = 1.31$.

    Next, compute the sum of the squared group totals, each divided by its sample size:

    $$(17.53^2 \div 13) + (45.46^2 \div 35) + (5.29^2 \div 4) = 89.69$$

    Next, obtain the grand total squared and divided by total sample size = correction term (CT):

    $$CT = (68.28)^2 \div 52 = 89.66$$

    Then, compute $SS_T = \Sigma X^2 - N\bar{X}^2 = 90.44 - (52 \cdot 1.31^2) = 1.0$.
    Then, use Equation 7.4 to compute SS_B: $(23.64 + 59.05 + 7.0) - CT = 89.69 - 89.66 = .03$.
    Then, $SS_W = SS_T - SS_B = 1.0 - .03 = .97$.

    ANOVA Table

Source of Variation	v	Sum of Squares	Estimated Variance
Between groups	2	.03	.015
Within groups	49	.97	.0197
Total	51	1.0	

    Compute F ratio (Between-Group Variance to Within-Group Variance) = .76.

    Decision: Since the computed F ratio is well below the critical value (3.18), the null hypothesis cannot be rejected. Conclude that there is no statistically significant difference between basin shapes in these geologic categories.

Chapter 8

1. The statistics (based on all thirty-eight observations) for all three relationships were:

	(a)	(b)	(c)
a	19.08	3.09	3730.33
b	−.21	1.21	354.27
r	.39	.82	.69
r^2	.15	.67	.47
$S_{Y \cdot X}$	2.40	5.73	1813.40
4. t_r	−2.55	8.55	5.69

All t statistics are significant at the $\alpha = .01$ level.

8. Substitute .85 in the linear equation for X:

$$Y = 19.08 - (.21 \cdot 85) = 1.23 \text{ percent}$$

See Appendix A, pp. 312–319, for solutions to most of the exercises in that chapter.

List of Equations

appendix **D**

The important equations presented in the text are listed here for reference. Explanation of the symbols is given in the appropriate chapter as indicated by the number in parentheses to the right of each equation.

A. Descriptive Statistics

1. Computation of the sample mean

$$\bar{X} = \frac{\sum_{i=1}^{n} X_i}{n} \tag{4.2}$$

2. Computation of the sample variance (the standard deviation is the positive square root of the variance)

$$s^2 = \frac{n \sum_{i=1}^{n} X_i^2 - \left(\sum_{i=1}^{n} X_i\right)^2}{n(n-1)} \tag{4.6}$$

3. Coefficient of variation

$$CV = \frac{s}{\bar{X}} \times 100 \tag{4.8}$$

Appendix D

4. Pearson's coefficient of skewness

$$Sk = \frac{3(\overline{X} - Md)}{s} \tag{4.9}$$

5. Kurtosis

$$Km = \frac{\Sigma(X_i - \overline{X})^4/n}{(s^2)^2} - 3 \tag{4.10}$$

6. Weighted mean center $(w\overline{Y}, w\overline{X})$

$$w\overline{X} = \frac{\sum_{i=1}^{n} w_i \cdot X_i}{\sum_{i=1}^{n} w_i} \tag{4.11}$$

$$w\overline{Y} = \frac{\sum_{i=1}^{n} w_i \cdot Y_i}{\sum_{i=1}^{n} w_i} \tag{4.12}$$

7. Weighted standard distance

$$SD_w = \sqrt{\frac{\sum_{i=1}^{n} w_i \cdot (X_i - \overline{X})^2}{\sum_{i=1}^{n} w_i} + \frac{\sum_{i=1}^{n} w_i \cdot (Y_i - \overline{Y})^2}{\sum_{i=1}^{n} w_i}} \tag{4.16}$$

B. z Score (Computed from the Sample)

$$z = \frac{(X_i - \overline{X})}{s} \tag{5.3}$$

C. Testing Hypotheses about Sample Mean Using Normal Curve Areas

1. The z test

$$z = \frac{(\overline{X} - \mu)}{\hat{\sigma}_{\overline{X}}} \tag{6.3}$$

Appendix D

2. Confidence interval for the mean, using z

$$\overline{X} \pm z_{(1-\alpha)} \, \sigma_{\overline{X}} \tag{6.4}$$

D. The *t* Distribution

1. Testing hypotheses about sample mean using t

$$t = \frac{\overline{X} - \mu}{\hat{\sigma}_{\overline{X}}} \tag{6.6}$$

2. Confidence limits of μ based on $\overline{X}$

$$\mu = \overline{X} \pm t \, \hat{\sigma}_{\overline{X}} \tag{6.7}$$

3. Estimate of standard error of difference between sample means

$$\hat{\sigma}_{\overline{X}_1 - \overline{X}_2} = \sqrt{s_{1+2}^2 \left(\frac{1}{n_1} + \frac{1}{n_2} \right)} \tag{6.12}$$

4. t test of hypothesis about difference between sample means

$$t = \frac{\left(\overline{X}_1 - \overline{X}_2 \right)}{\hat{\sigma}_{\overline{X}_1 - \overline{X}_2}} \tag{6.14}$$

5. Confidence limits of $\mu_1 - \mu_2$

$$\overline{X}_1 - \overline{X}_2 = (\mu_1 - \mu_2) \pm t \, \hat{\sigma}_{\overline{X}_1 - \overline{X}_2} \tag{6.15}$$

6. t test of hypothesis about mean difference between matched pair samples

$$t = \frac{\overline{d} - 0}{\hat{\sigma}_{\overline{d}}} \tag{6.19}$$

7. Standard error of the mean of $\overline{d}$

$$\hat{\sigma}_{\overline{d}} = \frac{s_d}{\sqrt{n}} \tag{6.18}$$

E. One-Way Analysis of Variance

1. Between-groups sum of squares

$$SS_B = \frac{\left(\sum_{i=1}^{n_1} X_i\right)^2}{n_1} + \frac{\left(\sum_{i=1}^{n_2} X_i\right)^2}{n_2} + \ldots + \frac{\left(\sum_{i=1}^{n_k} X_i\right)^2}{n_k} - \frac{\left(\sum_{i=1}^{N} X_i\right)^2}{N} \quad (7.4)$$

$$= \sum_{j=1}^{k} \left[\frac{\left(\sum_{i=1}^{n_j} X\right)^2}{n_j}\right] - \frac{\left(\sum_{i=1}^{N} X_i\right)^2}{N}$$

2. Within-group sum of squares

$$SS_W = \sum X^2 - \frac{\left(\sum_{i=1}^{n_1} X_i\right)^2}{n_1} + \frac{\left(\sum_{i=1}^{n_2} X_i\right)^2}{n_2} + \ldots + \frac{\left(\sum_{i=1}^{n_k} X_i\right)^2}{n_k} \quad (7.5)$$

$$= \sum X^2 - \sum_{j=1}^{k} \left[\frac{\left(\sum_{i=1}^{n_j} X_i\right)^2}{n_j}\right]$$

F. Bivariate Correlation and Regression

1. Correlation coefficient

(8.3)

$$r = \frac{n\sum_{i=1}^{n} X_{1i}X_{2i} - \left(\sum_{i=1}^{n} X_{1i}\right)\left(\sum_{i=1}^{n} X_{2i}\right)}{\sqrt{n\sum_{i=1}^{n} X_{1i}^2 - \left(\sum_{i=1}^{n} X_{1i}\right)^2} \sqrt{n\sum_{i=1}^{n} X_{2i}^2 - \left(\sum_{i=1}^{n} X_{2i}\right)^2}}$$

2. t test for significance of Spearman's product moment correlation coefficient

$$t = r\sqrt{\frac{n-2}{1-r^2}} \quad (8.4)$$

3. Estimating equation for regression of X on Y

$$\hat{Y}_i = a + bX_i \quad (8.5)$$

Appendix D

4. Slope of regression line

$$b = \frac{\Sigma XY - \bar{X}(\Sigma Y)}{\Sigma X^2 - \bar{X}(\Sigma X)} \quad (8.9)$$

5. Value of intercept a

$$a = \bar{Y} - b\bar{X} \quad (8.10)$$

6. Standard error of estimate

$$S_{Y \cdot X} = \sqrt{\frac{\sum_{i=1}^{N}\left(Y_i - \hat{Y}\right)^2}{n-2}} \quad (8.13)$$

G. Wilcoxon Signed Ranks Test

1. Variance of the distribution of ΣR

$$\sigma_W^2 = \frac{n(n+1)(2n+1)}{6} \quad (9.3)$$

2. Statistic for range of W (two-tailed test)

$$W = \pm Z_{\alpha/2} \sqrt{\frac{n(n+1)(2n+1)}{6}} \quad (9.4)$$

H. Mann-Whitney Test

1. U statistic

$$U_1 = n_1 n_2 + \frac{n_1(n_1+1)}{2} - \Sigma R_1 \quad (9.6)$$

$$U_2 = n_1 n_2 + \frac{n_2(n_2+1)}{2} - \Sigma R_2 \quad (9.7)$$

I. Kruskal-Wallis Test

$$H = \left[\frac{12}{N(N+1)}\left(\frac{\sum_{i=1}^{n_1} R_1^2}{n_1} + \frac{\sum_{i=1}^{n_2} R_2^2}{n_2} + \ldots + \frac{\sum_{i=1}^{n_k} R_k^2}{n_k}\right)\right] - 3(N+1) \quad (9.11)$$

J. Chi-Square Test of Observed versus Expected Frequencies

$$\chi^2 = \sum_{1}^{k} \frac{(f_O - f_E)^2}{f_E} \qquad (9.12)$$

K. *c* Test for Difference of Means in Nearest Neighbor Analysis

$$c = \frac{\bar{d}_o - \bar{d}_E}{SE_{\bar{d}}} \qquad (9.20)$$

L. Standard Error of Difference of Means in *c* Test

$$SE_{\bar{d}} = \frac{0.26136}{\sqrt{n^2/A}} \qquad (9.21)$$

M. Spearman Rank Correlation Coefficient

$$r_S = 1 - \frac{6\Sigma D^2}{n(n^2 - 1)} \qquad (9.25)$$

Index

a, Y-intercept, 208
α, type I error, 154, 157–58, 229
Absolute value, 31
Acceptance region, 157, 229
Accuracy, 14
Addition, rule of, 121
AIDS cases and rates, U.S., 327
Alternative hypothesis, 153, 170, 191, 204
Analysis of variance, 183, 304–6
    assumptions, 184–85
    inferential test in regression analysis, 217
    one-way (single classification), 183
    partitioning of total sum of squares and degrees of freedom, 186, 189, 217
    table, 186, 189, 217
ANOVA. *See* Analysis of variance.
Area, 25
    computation of, 32–36
Areal association. *See* Correlation.
Area sample, 41
Area under curve, 140
Arithmetic mean. *See* Mean.
Array, 87, 142
Association, areal. *See* Correlation.
    nonparametric tests for, 219

Autocorrelation, 207
Average. *See* Mean.
Azimuth, 39

b, regression coefficient, 208–12
β, type II error, 154
Bell-shaped distribution, 135
Between-group sum of squares, 187
Between-group variance, 185
Between-subsample variation, 184
Bimodal distribution, 87
Binary, 17
Binomial
    distribution, 132
    expansion, 133
Bivariate
    correlation analysis, 198
    mean center, 104
    normal probability, 199
    regression analysis, 207
    relationship, 197
    scattergram, 200, 209
χ^2, chi-square, 238–45
$\chi^2_{\alpha[\nu]}$, critical chi-square at probability level α, and degrees of freedom ν, 239–40

CV, coefficient of variation, 101, 204, 215
Cartesian, 21, 28, 44, 104
Causation, 205
Cell frequencies in chi-square, 240, 243
Census data for U.S. metropolitan areas, 326
Central limit theorem, 136, 151, 156
Central tendency, 134
   measure of, 83
Centrographic statistics, 104
Centroid, 105
Chance. *See* Significance or hypothesis.
Chebyshev theorem, 98
Chi-square distribution, 239
Chi-square test
   for goodness of fit, 238–45, 248, 316–17
   strength of phi statistic, 244
Choropleth map, 71
Classification, 58–60
Class interval, 26, 60–63
Climate change, 2
Clustered pattern, 250
Cluster sampling, 46
Coefficient of
   correlation. *See* Correlation coefficient.
   determination r^2, 215
   rank correlation r_S, 260–63
   variation, 101, 204
Combinations and probability, 123
Compactness, 35
Comparative dispersion, 112
Components of variance. *See* Variance.
Composite outcome, 121
Computer graphics, 56
   histogram, 293, 298, 300
   residual plot, 309, 311
   scatter plot, 290–92, 308
Concept, 16
Conditional probability, 124
Confidence
   bands, 143
   interval, 160
   limits, 143, 157
      difference between means, 160, 165–66, 169
   regression slope, 208
Confounded explanations, 17
Confounded source of error, 184
Constant, 15
Contingency table, 240–44
Continuous probability distribution, 134

Continuous variables, 15
Contour map, 27, 69
   fitting procedure, 71, 75
Control points, 104
Correlation
   areal association, 199, 205
   chance association, 206
   coefficient, 199–200
   computer programs, 306–12
   Pearson's *r*, 199
   and regression, contrasted, 198, 207
   relationship with causality, 205
   significance test in, 204
   Spearman's r_S, 260–63
Covariance, 199
Cramer's *V*, 245
Critical
   values of Mann-Whitney statistic, 234
   value of rank sum in Wilcoxon's signed ranks test, 230
   value in test distribution, 157, 204
Cumulative
   frequency distribution, 139
   frequency histogram, 64
   probability distribution, 136

Data
   primary, 14
   secondary, 14
   sets, appendices 322–29
Deciles. *See* Range.
Degrees of freedom, 163, 169f, 240
Density, 38
Density function, 134, 190, 192
Dependent variable, 57, 208
Descriptive statistics, 83, 294–97
Des Moines, Iowa, precipitation data, 322
Determination, coefficient of, 215
Deviate
   standard normal. *See z* scores.
Deviation
   from regression, mean square for, 213, 217
   standard, 98
Dichotomous, 17
Difference
   between two means, 158, 162
      confidence limits of, 160
Digitizing tablet, 32
Direction, 39

Direction of relationship, 198–99
Discontinuous variable, 15
Discrete location, 15
Discrete probability distribution, 125
Discrete variable, 15
Dispersion, measures of, 89, 92, 96, 112
Distance, 27
   decay, 56–58
   economic, 31, 69
   Euclidean, 27, 29
   Manhattan, 31
   standard, 108
   statistical. *See* Statistical distance.
Distribution
   binomial, 132
   bivariate, 200
   chi-square. *See* Chi-square distribution.
   continuous, 15
   discrete. *See* Discrete probability distribution.
   frequency. *See* Frequency distribution.
   leptokurtic, 104, 219
   of means, 151
   mesokurtic, 104
   normal, 88, 100, 134
   platykurtic, 104
   Poisson, 127
   probability, 125, 297–99
   rectangular. *See* Rectangular distribution.
   skewed, 88, 102, 220
   Student's t, 162–63, 204
Distribution-free methods. *See* Nonparametric statistics.
Dot density map, 38
Dot map, 66

e, base of Napierian logarithms, 129
Ecological fallacy, 10, 204
Economic distance, 31
Empirical evidence and inference, 149
Empirical rule, 100, 136
Error
   regression, 213
   sampling. *See* Sampling error.
   standard. *See* Standard error.
   sum of squares, 189, 213
   types I and II, 154, 157
   variance in ANOVA, 186–87, 189, 217
Estimating equation, 208
Euclidean distance, 29
Events, independence of, 120, 241

Expected
   distribution, 200
   frequencies, 238
   Poisson frequencies, 131, 248
   value for Y given X, 210
Experiment, 120
Experimental error in ANOVA, 184
Extreme values on scatter plot, 206

f_E, expected frequency, 239
f_O, observed frequency, 239
F, variance ratio, 191, 217
F distribution, 190
F test, 191
Fisher, R. A., 162, 190
Fractile diagram, 140
Freedom, degrees of. *See* Degrees of freedom.
Frequencies, 15, 131, 238
   small, 240
Frequency distribution, 60
   of continuous variable, 134
   histogram, 60
   normal, 89
   ogive, 65, 140
Frequency polygon, 62
Function
   power, 218
   probability density, 129, 136, 190, 192
Functional relationship, 197

Geographic sampling, 41
Gerrymandering, 36
Global warming, 2
Goodness-of-fit tests, 237
   chi-square, 238–45, 248
   Kolmogorov-Smirnov, 254–60
   regression, 212
Gossett, W. S., 162
Grand mean, 187
Graphical methods
   frequency polygons, 62
   histograms, 60
   maps, 66
   ogive, 65
   residual plot, 222
   scatter plot, 197
Groups
   variance between, 187
   variance within, 188

H, statistic for Kruskal-Wallis test, 234–37
H_0, null hypothesis, 151
H_1, alternative hypothesis, 153
Heteroscedasticity, 221, 254
Hexagonal lattice, 250
Hierarchy, 18
Histogram, 60–62
   probability, 126
Homogeneity of variances, 155
   assumption in analysis of variance, 184
Homoscedasticity, 200
Hypothesis
   alternative, 153
   chance association, 206
   null, 151, 200
   testing, 151

Independence
   in classification, 241
   in probability, 120
   of samples, 42, 207
Independent sample, 161
Independent sample means test. *See t* test.
Independent samples tests (nonparametric), 232, 234
Independent stochastic process, 246
Independent variable, 57, 205, 208
Indicator, 17
Inferential statistics, 149, 158
Intercept in regression, 208–9
Interdependent relationships, 206
Interquartile range. *See* Range.
Interval
   class. *See* Class interval.
   confidence, 161
Interval scale of measurement, 21
Inverse power function, 218
Inverse relationship, 198
Isolines, 69

Joint probability, 122

k, subscript indicating class number, 186, 188, 239
Kolmogorov-Smirnov test for goodness of fit, 254–60, 318
Kolmogorov-Smirnov two-sample test, 258–60
Kruskal-Wallis Test, 234–37, 315–16
Kurtosis, 104, 156, 220

λ, lamda, 129, 239

Large-number methods, 175
Least squares, 210
Length, 35
Leptokurtic distribution, 104, 219
Level of significance, 156–57
Line, 25
Linear regression. *See* Regression.
Logarithmic distance, 30
Logarithmic transformation, 219
Logarithms, 129
Log-normal (log-linear) model, 222

MS, mean square, 186, 217
μ, parametric mean, 86, 155
   confidence limits for, 169
Manhattan distance, 31
Mann-Whitney U test, 232–34, 314–15
Maps
   dot, 66
   choropleth, 71
   contour, 69
Mean
   arithmetic, 84–87
   used to compute skewness, 103
   methods for computing, 86
   sample, 86
Mean center, bivariate, 104
   weighted, 105
Mean square variation, 189
Means
   more than two independent means, test of difference, 234–37
   two independent means, t test of difference, 166
Measurement
   definition, 13, 16
Measurement scales, summary, 24
Measures of central tendency, 83–89
   mean, 83, 104
   median, 87
   mode, 87
Measures of variability, 91
   interquartile range, 91
   quartile, 90
   range, 90
   standard deviation, 98
   variance, 96
Median, 87
   used to compute skewness, 103
   runs test above and below, 253

Median, continued
  sign test for, 227–29
Mesokurtic distribution, 104
Metric scale in parametric statistics, 155
Metric variable, 15, 21
Mode, 87
Model, nonlinear, 218
Model, regression, 211
Multivariate statistics, 197, 206, 212
Mutually exclusive events, 122

n, sample size, 46, 161
v, degrees of freedom, 163
Napierian logarithm, 129
Nearest neighbor analysis, 248–53
Negatively skewed distribution. *See* Skew.
Negative relationship, 198–99
Network map, 68
Nominal scale of measurement, 17
Nonlinear relationship, 207, 218
Nonparametric statistics, 155, 225
  advantages and disadvantages, 226
  computer programs, 312–19
  See Specific test.
Normal curve
  areas of, 100, 139
  cumulative frequencies, 65, 141
  deviates, 139
  distribution, 88, 134
    bivariate, 104
    conditions for, 88, 104
    standard deviation units, 99–102
    variance, 96
    z (standard) score, 138
  probability
    bivariate, 199
    calculation, 137, 139
    density function, 136
    graph, 140, 142
    table, 139
Normality, 155
  assumption of, in analysis of variance, 184
  graphic test on probability paper, 140
Null hypothesis, 151, 199
Number-of-runs test, 253–54

Observation, 59
Observed frequencies, 131, 238
Ogive, 65, 140

One-tailed test, 157–58, 169, 204
One-way analysis of variance, 183
Order of events, 123
Ordinal scale of measurement, 17
Outlier on scatter plot, 206

P, probability, 120
ϕ, phi statistic, 244–45
Paired comparisons
  nonparametric test, t test for, 172–75
Parameter 83, 155
Parametric
  mean, 86
  product-moment correlation coefficient, 199
  regression coefficient, 208
  statistics, 134
  test, 155, 169, 226
  variance, 96
Partitioning
  between-group sum of squares, 187
  total sum of squares and degrees of freedom, in ANOVA, 186
  variation in regression, 214–17
  within-group sum of squares, 188
Pearson's measure of skewness, 103
Pearson's product-moment correlation coefficient, 199
Percent female employment data, 324
Permutation, 129
Phi statistic, 244–45
Platykurtic distribution, 104
Point data, 24
Point pattern analysis, 245–53
Point sample, 41
Poisson distribution, 126
  calculation of expected frequencies, 131, 248
  density function, 129
  randomness, 129, 245
  tables, 131
Polar coordinate system, 39
Polygon, 32
Pooled variance, 168, 174
Population
  characteristics, 1990, U.S., 328
  correction, 161f
  definition, 59
  and inference, 149
  mean, 86
  and nonparametric tests, 226

Population, continued
    standard deviation, 98
    variance, 96
Positively skewed distribution. *See* Skew.
Positive relationship, 198–99
Power function, 218
Power of a test, 225
Precipitation data, Des Moines, Iowa, 322
Precision, 14
Primary data, 15
Probability, 119
    addition rule, 121
    bivariate normal, 199
    conditional, 124
    density function, 129, 136, 190, 192
    distribution, 126, 134, 297–99
    graph. *See* Normal curve probability graph.
    histogram, 126
    multiplication rule, 123
    mutually exclusive events, 122
    normal density function, 100, 136
    Poisson. *See* Poisson distribution.
    of significant difference, 150
    unconditional, 120
Product-moment correlation coefficient, 199
Proportionate sample, 45
Pythagorean theorem, 28

Quadrat, 39, 44, 246, 258
Quartile range. *See* Range.
Quintile. *See* Range.

r, product-moment correlation coefficient, 199
r^2, coefficient of determination, 215
r_S, Spearman's coefficient of rank correlation, 260–63
Random
    error, 151, 184
    events, discrete, 127, 245
    experiment, 120, 258
    numbers, 43
    pattern, 250, 258
    sampling, 42, 120
    variable, 120
Randomizing observations, 184
Randomness, tests for, 249, 253
Range, 90–91
Rank correlation
    Spearman's coefficient of, 260–63, 319

Ranked data, 18
Rate of change constant, 208
Rates, 23
Ratio scale of measurement, 22
Raw data, 56
Rectangular distribution, 62
Relative variability, 101
Region
    acceptance. *See* Acceptance region.
    rejection. *See* Rejection region.
Regression
    coefficient, 208
    computer programs, 306–12
    equation, 208
    estimation of Y from X, 208
    introduction to, 207
    least squares, 210
    linear, 208–12
    model, 211
    nonlinear, 218
    significance test in, 217
    standard error, 213
Regular pattern, 250
Rejection region, 157, 229
Relative dispersion, 112
Reliability, 14
Replacement rule, 43, 125
Replicates, 183
Research hypothesis, 153
Residuals, 216
    normality of, 222
    plot, 222
    randomness of, 253
Residual sum of squares, 189, 213
Residual variance, 215
Robustness, 225, 229, 232
Rule of
    addition, 121
    multiplication, 123
Runs test, 253–54

s, sample standard deviation, 98, 155
s^2, sample variance, 96, 155
$S_{Y.X}$, standard error of estimate, 215
SS, sum of squares, 94, 186
σ, parametric standard deviation, 98, 155
σ^2, parametric variance, 96, 155
$\hat{\sigma}$, standard error, 153
$\hat{\sigma}_{\bar{X}}$, standard error of mean, 162

Index

Sample, 41, 59, 149
  error, 142
  mean, 86, 151, 156
  representative, 41
  size, 46
    n required for a test, 161
  space, 120
  statistic, 154–56
  variance, 96
Sampling error, 142
Sampling, geographic
  cluster, 46
  point and area, 41
  random, 42
  stratified, 45, 46, 169
  systematic, 44
  unit, 41
Sampling variation, 151, 165, 189
SAS computer programs, 56, 287. *See also* Computer graphics.
Scalar, 26
Scales of measurement, 17
  interval, 21
  metric, 21
  ordinal, 17
  ratio, 22
Scatter. *See* Variability.
Scatter plot (or scattergram), 197–98, 209, 216
Scientific law, 7
Secondary data, 15
Shape index, 35–38
Sigma hat, 153
Sigma notation
  summation, 84
  standard deviation, 98
  standard error, 153
Signed-ranks test, Wilcoxon's, 229
Significance
  of correlation coefficient, computation, 204
  of the difference between two means, 169, 175
  levels, 156–58
  statistical, 6, 150
  tests
    in correlation, 204, 262
    of regression statistics, 217
Sign test, 227–29
Simple regression analysis, 207
Size, sample. *See* Sample size.

Skew, 88, 102, 219
  computation of, 103
Slope of regression line, 208
Small-number methods, 175
Spatial autocorrelation, 207
Spatial distributions, 25
  statistical measures
    bivariate mean, 104
    standard distance, 108–12
Spatial structure, 66
Spearman's cofficient of rank correlation, 260–63, 319
SPSS-X computer programs, 56, 287. *See also* Computer graphics.
Square root transformation, 219
Standard deviation
  use in Chebyshev's theorem, 98
  population, 98
  sample, 98
Standard distance, 108–12
Standard error, 151, 153
  of difference between two means, 162, 167–68
  of estimated value Y, along regression line, 208, 213–14
Standard normal deviate, 139
Standard score, 138
Statistical distance, 94
Statistical hypotheses, 149
Statistics, 6–11
  definition, 83, 155
  descriptive, 83, 293–97
  inferential, 149, 158
  population, 59
  sample, 41, 59, 149
  use of, 3
Stochastic process, 246
Stratified sampling, 46, 169
Stream order, 18
Strength of relationship, 198
Structure, 6
Student's t distribution, 162–63, 204
Subarea, 42
Subsamples in ANOVA, 184–85
Summation notation, 84–85
Sum of squares, 94, 96, 167
  between groups in ANOVA, 187
  explained, 213
  total, 186, 215
  unexplained, 213
  within groups in ANOVA, 188

Surfaces, 26
Surrogate, 16
Surveys, 8
System, 2–5
Systematic sampling, 45
Systematic variation, 184, 187

$t_{\alpha[\nu]}$, critical values of Student's distribution for ν degrees of freedom, 165
t distribution, Student's, 161–64
t tables, 163, 165
t test for
    difference between means, 156, 162, 169, 299–303
    paired comparisons, 172–75, 303–4
    significance of correlation coefficient, 204, 262
Table, ANOVA, 189
Table, contingency, 241
Tail of curve, 139
Test
    for goodness of fit, 212, 248, 254
    of hypothesis, 151
    one-tailed, 157, 169
    of significance. *See* Specific test name.
        for correlation coefficients, 204, 262
    statistic, 158
    two-tailed, 157
Thematic map, 55, 66
Theoretical distribution, 126
Theory, 6, 175
Thunderstorm, 5
Ties in ranks, 229, 233, 235
Time distance, 31
Topographic map, 26
Topological map, 69
Total sum of squares, 186, 215
Total variation, 187, 214
Transect, 45
Transformation, 207
    logarithmic, 219
    reciprocal, 220
    square root, 220
Traverse, 45
Two-tailed test, 157, 174
Two-way analysis of variance
    definition, 184
Type I and II errors, 154, 157, 174, 229

Unconditional probability, 120
Unit area, 25

V statistic, 245
Validity, 16
Variability, 91
Variable
    continuous, 15
    definition, 15
    dependent, 208
    discrete, 15
    independent, 205, 208
    random, 121, 246
Variance
    components in ANOVA, 184–86, 217
    definition, 96
    equality, 200
    error, 184, 187
    experimental error, 184
    pooled, 168
    population, 96
    regression, 217
    residual, 217
    sample, 96
    systematic, 184, 187
Variation
    chance, 189, 206
    coefficient of. *See CV*.
    mean square, 189
    partitioning in regression, 214–17
Vector arithmetic, 40

Weighted mean center, *See* Bivariate mean center.
Wilcoxon signed-ranks test, 229, 313
Within-group sum of squares, 188
Within-subsample variation, 184

$\bar{X}$, sample mean, 84, 155
    mean of independent variable, 203

Y intercept, 209
Y, mean of dependent variable, 203

z, scalar, 26
z scores, 138
z test, 159–60, 262

ISBN 0-675-21338-X